Endothelin

CLINICAL PHYSIOLOGY SERIES

Endothelin
Edited by Gabor M. Rubanyi

Hypoxia, Metabolic Acidosis, and the Circulation
Edited by Allen I. Arieff

Response and Adaptation to Hypoxia: Organ to Organelle
Edited by Sukhamay Lahiri, Neil S. Cherniack, and Robert S. Fitzgerald

Clinical Physiology of Sleep
Edited by Ralph Lydic and Julien F. Biebuyck

Atrial Hormones and Other Natriuretic Factors
Edited by Patrick J. Mulrow and Robert Schrier

Physiology of Oxygen Radicals
Edited by Aubrey E. Taylor, Sadis Matalon, and Peter A. Ward

Effects of Anesthesia
Edited by Benjamin G. Covino, Harry A. Fozzard, Kai Rehder, and Gary Strichartz

Interaction of Platelets With the Vessel Wall
Edited by John A. Oates, Jacek Hawiger, and Russell Ross

High Altitude and Man
Edited by John B. West and Sukhamay Lahiri

Disturbances in Neurogenic Control of the Circulation
Edited by Francois M. Abboud, Harry A. Fozzard, Joseph P. Gilmore, and Donald J. Reis

New Perspectives on Calcium Antagonists
Edited by George B. Weiss

Secretory Diarrhea
Edited by Michael Field, John S. Fordtran, and Stanley G. Schultz

Pulmonary Edema
Edited by Alfred P. Fishman and Eugene M. Renkin

Disturbances in Lipid and Lipoprotein Metabolism
Edited by John M. Dietschy, Antonio M. Gotto, Jr., and Joseph A. Ontko

Disturbances in Body Fluid Osmolality
Edited by Thomas E. Andreoli, Jared J. Grantham, and Floyd C. Rector, Jr.

Endothelin

EDITED BY

Gabor M. Rubanyi

Institute of Pharmacology
Schering AG, Research Center
Berlin, Germany

New York   Oxford
Published for the American Physiological Society
by OXFORD UNIVERSITY PRESS
1992

Oxford University Press

Oxford New York Toronto
Delhi Bombay Calcutta Madras Karachi
Petaling Jaya Singapore Hong Kong Tokyo
Nairobi Dar es Salaam Cape Town
Melbourne Auckland

and associated companies in
Berlin Ibadan

Copyright © 1992 by the American Physiological Society

Published for the American Physiological Society by
Oxford University Press, Inc., 200 Madison Avenue, New York, New York 10016

Oxford is a registered trademark of Oxford University Press

All rights reserved. No part of this publication may be reproduced,
stored in a retrieval system, or transmitted, in any form or by any means,
electronic, mechanical, photocopying, recording, or otherwise,
without the prior permission of the American Physiological Society.

Library of Congress Cataloging-in-Publication Data
Endothelin / edited by Gabor M. Rubanyi.
p. cm. (Clinical physiology series)
Includes bibliographical references and index.
ISBN 0-19-506641-3
1. Endothelins. I. Rubanyi, Gabor M., 1947–
II. American Physiological Society (1887–   ) III. Series.
[DNLM: 1. Endothelins. QU 68 E565]
QP552.E54E53 1992
612.1—dc20 DNLM/DLC   91-2763

9 8 7 6 5 4 3 2 1

Printed in the United States of America
on acid-free paper

Preface

Vascular endothelial cells play an essential role in a variety of physiological and pathophysiological processes. The ability of these cells to synthesize and release biologically active substances contributes in a significant way to their functions. The discovery of endothelium-derived relaxing factors (EDRFs) and endothelium-derived contracting factors (EDCFs) in the past decade revolutionized our understanding of the role of endothelial cells in the local control of vascular function in health and disease.

Under physiological conditions a delicate balance exists between EDRFs and EDCFs. Due to their vasodilator, antiplatelet, and cytoprotective properties, prostacyclin and EDRF/NO contribute to the maintenance of the fluidity of the blood and to adequate tissue perfusion. In various pathological situations, activation and injury of endothelial cells can lead to their dysfunction, characterized by an imbalance between EDRFs and EDCFs, and causes pathophysiological consequences such as thrombosis, vasospasm, hypertension, and vascular wall proliferation/remodeling.

In contrast to the worldwide research interest and activity with EDRF during the 1980s, only a few laboratories studied the nature and biological activity of EDCFs. This situation changed dramatically with the discovery of a peptidergic EDCF produced by cultured endothelial cells and by the subsequent isolation and identification of this EDCF as a unique 21 amino acid peptide, named endothelin. The availability in a pure, synthetic form of this endothelium-derived potent vasoconstrictor peptide met a scientific community already enthusiastic about the progress in this field. It was therefore not unexpected that research with endothelin exploded. In a relatively short time it was discovered that three different endothelin genes exist in the human genome, encoding three different isoforms of the peptide, each with a host of biological activities. Although far from complete, a large amount of information has accumulated since 1988 on the molecular biology, bioprocessing, localization, receptor subtypes, biological activity, mechanism of action, and potential physiological and pathophysiological significance of this unique peptide family.

This book summarizes the progress made in endothelin research and is based on a symposium sponsored by the American Physiological Society at the 74th Annual Meeting of the Federation of American Societies for Experimental Biology, Washington, D. C., 1990. The book starts with an overview of endothelium-dependent vasoconstriction and an authentic report on the discovery of a peptidergic endothelium-derived contracting factor, and continues with in-depth reviews of the molecular biology, structure and structure–activity relationship, receptor subtypes, tissue specific expression and action, histochemical localization, biological actions, interactions with other endothelium-derived vasoactive substances (EDRF/NO, PGI_2), and potential clinical significance of endothelins. The book ends with an attempt to integrate the known properties of endothelins into our present knowledge of the homeostatic function of the endothelial cell and into the existing concepts of local and neurohumoral control of the cardiovascular system.

The editor and the contributors are aware that a book on endothelin research can be neither complete nor in all aspects up-to-date because of the rapid pace of research. Nonetheless, the book is the first comprehensive review and systematic recollection of the progress achieved in the first three years of endothelin research. It should be a valuable reference for both basic scientists and clinicians, who are either actively involved in endothelin research or interested in this peptide family.

The editor expresses his gratitude to the authors for their excellent contributions, to the American Physiological Society for including the book in its Clinical Physiology Series, and to the staff of Oxford University Press for their efficient publication of the book.

July 1991   *G.M.R.*

Contents

Contributors ix

Introduction: The Discovery of Endothelin xi
Tomoh Masaki and Gabor M. Rubanyi

1. Endothelium-Derived Contracting Factors 3
   Paul M. Vanhoutte, Thomas Gräser, and Thomas F. Lüscher

2. From Endotensin to Endothelin: The Discovery and Characterization of an Endothelial Cell–Derived Constricting Factor 17
   Robert F. Highsmith

3. Molecular Biology of Endothelins 31
   Penny E. Phillips, Christina Cade, Lynne H. Parker Botelho, and Gabor M. Rubanyi

4. Endothelin Structure and Structure–Activity Relationships 41
   Paul W. Erhardt

5. Endothelin Receptors and Receptor Subtypes 58
   Hitoshi Miyazaki, Motohiro Kondoh, Yasushi Masuda, Hirotoshi Watanabe, and Kazuo Murakami

6. Tissue Specificity of Endothelin Synthesis and Binding 72
   Lynne H. Parker Botelho, Christina Cade, Penny E. Phillips, and Gabor M. Rubanyi

7. Cellular Actions of Endothelin in Vascular Smooth Muscle 103
   Tommy A. Brock and N. Raju Danthuluri

8. Interaction between Endothelin and Endothelium-Derived Relaxing Factor(s) 125
   Thomas F. Lüscher, Chantal Boulanger, Zhihong Yang, and Yasuaki Dohi

9. Endothelin as a Growth Factor in Vascular Remodeling and Vascular Disease 137
   Victor J. Dzau, Richard E. Pratt, and John P. Cooke

10. Endothelin and the Heart 144
    Winifred G. Nayler and Xin Hua Gu

11. Renal and Systemic Hemodynamic Actions of Endothelin 158
    Andrew J. King and Barry M. Brenner

12. Endothelin, A Ubiquitous Peptide: Morphological Demonstration of Immunoreactive and Synthetic Sites and Receptors in the Respiratory Tract and Central Nervous System   179
    S.J. Gibson, D.R. Springall, and Julia M. Polak

13. Endothelin in Human Disease. I: Essential Hypertension, Vasospastic Angina, Acute Myocardial Infarction, and Chronic Renal Failure   209
    Yoshihiko Saito, Kazuwa Nakao, Hiroo Imura

14. Endothelin in Human Disease. II: Shock, Pulmonary Hypertension, and Congestive Heart Failure   221
    Duncan J. Stewart

15. Endothelin and the Homeostatic Function of the Endothelial Cell   238
    Regina M. Botting and John R. Vane

16. Hypothetical Role of Endothelin in the Control of the Cardiovascular System   258
    Gabor M. Rubanyi and John T. Shepherd

Index   273

Contributors

Lynne H. Parker Botelho, Ph.D.
Department of Pharmacology
Berlex Laboratories, Inc.
Cedar Knolls, New Jersey

Regina M. Botting, Ph.D.
The William Harvey Research Institute
St. Bartholomew's Hospital Medical College
London, England

Chantal Boulanger, Ph.D.
Laboratory of Vascular Research
Department of Medicine
University Hospital
Basel, Switzerland

Barry M. Brenner, M.D., Ph.D.
Renal Division and Department of Medicine
Brigham and Women's Hospital
Harvard Medical School
Boston, Massachusetts

Tommy A. Brock, Ph.D.
Department of Medicine
University of Alabama
College of Medicine
Birmingham, Alabama

Christina Cade, Ph.D.
Department of Pharmacology
Berlex Laboratories, Inc.
Cedar Knolls, New Jersey

John P. Cooke, M.D., Ph.D.
Division of Vascular Medicine
Brigham and Women's Hospital
Harvard Medical School
Boston, Massachusetts

N. Raju Danthuluri, Ph.D.
Department of Cell Biology
University of Alabama
School of Medicine
Birmingham, Alabama

Yasuaki Dohi, M.D.
Department of Internal Medicine
Nagoya City University Medical School
Nagoya, Japan

Victor J. Dzau, M.D.
Falk Cardiovascular Research Center
Stanford University School of Medicine
Stanford, California

Paul W. Erhardt, Ph.D.
Department of Medicinal Chemistry
Berlex Laboratories, Inc
Cedar Knolls, New Jersey

Sarah J. Gibson, Ph.D.
Department of Histochemistry
Royal Postgraduate Medical School
London, England

Thomas Gräser, M.D., Ph.D.
Institute of Pharmacology
Schering AG Research Center
Berlin, Germany

Xin Hua Gu, M.B.B.S., M.Sc.
Department of Medicine
University of Melbourne
Heidelberg, Victoria
Australia

Robert F. Highsmith, Ph.D.
Department of Physiology and Biophysics
University of Cincinnati
College of Medicine
Cincinnati, Ohio

Hiroo Imura, M.D.
Department of Medicine
Kyoto University
School of Medicine
Kyoto, Japan

Andrew J. King, M.D.
Division of Nephrology
New England Medical Center
Boston, Massachusetts

Motohiro Kondoh, M.Sc.
Institute of Applied Biochemistry
University of Tsukuba
Tsukuba, Ibaraki
Japan

Thomas F. Lüscher, M.D.
Division of Cardiology
University Hospital
Basel, Switzerland

Tomoh Masaki, M.D., Ph.D.
Institute of Basic Medical Sciences
University of Tsukuba
Tsukuba-shi, Ibaraki
Japan

Yasushi Masuda, M.Sc.
Institute of Applied Biochemistry
University of Tsukuba
Tsukuba-shi, Ibaraki
Japan

Hitoshi Miyazaki, Ph.D.
Institute of Applied Biochemistry
University of Tsukuba
Tsukuba-shi, Ibaraki
Japan

Kazuo Murakami, Ph.D.
Institute of Applied Biochemistry
University of Tsukuba
Tsukuba-shi, Ibaraki
Japan

Kazuwa Nakao, M.D., Ph.D.
Department of Medicine
Kyoto University
School of Medicine
Kyoto, Japan

Winifred G. Nayler, Ph.D.
Department of Medicine
University of Melbourne
Heidelberg, Victoria
Australia

Penny E. Phillips, Ph.D.
Department of Pharmacology
Berlex Laboratories, Inc.
Cedar Knolls, New Jersey

Julia M. Polak, D.Sc., M.D.
Department of Histochemistry
Royal Postgraduate Medical School
Hammersmith Hospital
London, England

Richard E. Pratt, Ph.D.
Falk Cardiovascular Research Center
Stanford University School of Medicine
Stanford, California

Gabor M. Rubanyi, M.D., Ph.D.
Institute of Pharmacology
Schering AG, Research Center
Berlin, Germany

Yoshihiko Saito, M.D.
Department of Medicine
Kyoto University
School of Medicine
Kyoto, Japan

John T. Shepherd, M.D., D.Sc.
Department of Physiology and Biophysics
Mayo Clinic and Foundation
Rochester, Minnesota

David R. Springall, Ph.D.
Department of Histochemistry
Royal Postgraduate Medical School
Hammersmith Hospital
London, England

Duncan J. Stewart, M.D., F.R.C.P.(C)
Department of Medicine
McGill University
Faculty of Medicine
Montreal, Canada

Sir John R. Vane, F.R.S.
The William Harvey Research Institute
St. Bartholomew's Hospital Medical
* College*
London, England

Paul M. Vanhoutte, M.D., Ph.D.
Center for Experimental Therapeutics
Baylor College of Medicine
Houston, Texas

Hirotoshi Watanabe, M.Sc.
Institute of Applied Biochemistry
University of Tsukuba
Tsukuba-shi, Ibaraki
Japan

Zhihong Yang, M.D.
Laboratory of Vascular Research
Department of Medicine
University Hospital
Basel, Switzerland

Introduction: The Discovery of Endothelin

TOMOH MASAKI AND GABOR M. RUBANYI

The discovery that endothelial cells synthesize and release potent vasorelaxant (prostaglandin I_2 and endothelium-derived relaxing factor [EDRF])[2,6] and vasoconstrictor (endothelium-derived contracting factor [EDCF])[1,8] substances introduced a novel regulatory mechanism of vascular smooth muscle tone and opened new perspectives for vascular biology, pharmacology, pathology, and therapy. In the early 1980s, Furchgott and Zawadzki's observation[2] of endothelium-dependent vasorelaxation generated worldwide interest, and several laboratories entered into a race to identify the elusive, nonprostanoid relaxing substance EDRF. In 1982, combining their expertise with cultured endothelial cells (R. Highsmith and his student, K. Hickey) and vascular smooth muscle (G.M. Rubanyi and R.J. Paul), this group at the University of Cincinnati decided to *bioassay* EDRF. The idea was to test the biological activity of the culture medium of bovine aortic endothelial cells on isolated pig coronary arteries. It was expected that cultured endothelial cells release EDRF into the culture medium, which will then relax the smooth muscle preparation. This experiment was designed before the extreme lability of EDRF was discovered in later, more properly performed bioassay studies.[7] Therefore, no relaxing activity was observed when the culture medium was added to the smooth muscle preparation, because the donor (endothelial cells) and acceptor (smooth muscle) were in two different laboratories. Instead, the culture medium triggered slowly developing and long-lasting contraction of vascular smooth muscle, which could not be attributed to any known vasoconstrictor mediators and was shown to be of peptidergic nature.[3] Thus, in 1983, an *improperly* designed bioassay study led to the discovery of a peptidergic EDCF (for more details, see Ch. 2).

Masaki and his group at the University of Tsukuba, whose main interest at this time was the biochemistry and molecular biology of muscle proteins, were first exposed to the rapid development of endothelial research when they participated in a symposium on vascular endothelium held at the annual meeting of the Japanese Pharmacological Society in 1986. They were looking for a project where their expertise in molecular biology could be utilized. In May 1987, after reading the paper of Hickey et al.,[3] they recognized the opportunity to identify the peptidergic EDCF. Yanagisawa, Kurihara, Goto, and Kimura, the excellent peptide chemist who discovered neurokinin, were the initial participants in the project. They finished the purification of this peptide by the end of July. Both Dr. Yazaki's and Dr. Mitsui's groups helped in supplying the conditioned medium of the cultured endothelial cells. The initial sequence analysis of the peptide was carried out in collaboration with the Applied Biosystem Co., Ltd., because the university had no peptide sequencer at that time. The whole sequence, except five residues, was determined by the middle of August. Kimura through that these undetermined residues were cysteines. The sequence was analyzed again on the new apparatus in the National Institute of Basic Biology at Okazaki. The whole sequence of the peptide was determined by the next day. Meanwhile, Yanagisawa started the preparation of mRNA of endothelial cells and cDNA cloning and determined the whole sequence of the preproform of the peptide within 3 weeks. At the end of August, Dr. Fujino of the Takeda Pharmaceutical Company was asked to make a 20 amino acid residue peptide. They synthesized it within 1 week, the biological activity was tested, but the results were negative. Kimura made a new amino acid analysis and looked for Trp according to the result of the sequence of the preproform of the peptide. These analyses confirmed the existence of Trp in the peptide. The 21 residue peptide, including Trp at the 21 position, was synthesized and matched the native peptide both chemically and pharmacologically. It was the end of October 1987 when the peptidergic EDCF was isolated, purified, sequenced, and synthesized. It was named "endothelin," and a paper describing it was sent to and later published in *Nature*.[9]

In contrast to the earlier description of the peptidergic EDCF, this paper stimulated immediate interest in the scientific community because of the unique structural and pharmacological properties of endothelin. Endothelin has no similarity in its sequence to the known peptides of mammalian origin. However, half a year later, the sequence of a rare snake venom, sarafotoxin, was reported that was very similar to that of endothelin.[5] One of the most remarkable steps in the progress of endothelin research, following the initial publication, was the discovery of isotypes of endothelin. Analysis of human genomic sequence revealed the existence of three distinct genes of endothelin, which encode three distinct endothelin peptides.[4]

Because of the high interest and very active research in this field in the last 3 years, a large amount of information has been generated about the molecular biology, chemistry, pharmacology, and potential physiological and pathological importance of this peptide family. In this first book on endothelin, leading experts summarize the state of the art and future direction of research in this exciting new field.

REFERENCES

1. DeMey, J. G., and P. M. Vanhoutte. Heterogenous behavior of the canine arterial and venous wall: importance of the endothelium. *Circ. Res.* 51: 439–447, 1982.
2. Furchgott, R. F., and J. V. Zawadzki. The obligatory role of endothelial cells in the relaxation of arterial smooth muscle by acetylcholine. *Nature* 288: 373–376, 1980.
3. Hickey, K. A., G. M. Rubanyi, R. Paul, and R. F. Highsmith. Characterization of a coronary vasoconstrictor produced by cultured endothelial cells. *Am. J. Physiol.* 248 (*Cell Physiol.* 17): C550–C555, 1985.
4. Inou, A., M. Yanagisawa, S. Kimura, Y. Kasuya, T. Miyauchi, K. Goto, and T. Masaki. The human endothelin family: three structurally and pharmacologically distinct isopeptides predicted by three separate genes. *Proc. Natl. Acad. Sci. USA* 86: 2863–2867, 1989.
5. Kloog, Y., and M. Sokolovsky. Similarities in mode and sites of action of sarafotoxins and endothelins. *Trends Pharmacol. Sci.* 10: 212–214, 1989.

6. MONCADA, S., R. GRYGLEWSKI, S. BUNTING, and J. R. VANE. An enzyme isolated from arteries transforms prostaglandin endoperoxidase to an unstable substance that inhibits platelet aggregation. *Nature* 263: 663–665, 1976.
7. RUBANYI, G. M., R. R. LORENZ, and P. M. VANHOUTTE. Bioassay of endothelium-derived relaxing factor(s): inactivation by catecholamines. *Am. J. Physiol.* 249: (*Heart Circ. Physiol.* 18): H95–H101, 1985.
8. RUBANYI, G. M., and P. M. VANHOUTTE. Hypoxia releases a vasoconstrictor substance from the canine vascular endothelium. *J. Physiol. (Lond.)* 364: 45–56, 1985.
9. YANAGISAWA, M., H. KURIHARA, S. KIMURA, Y. TOMOBE, M. KOBAYASHI, Y. MITSUI, Y. YAZAKI, K. GOTO, and T. MASAKI. A novel potent vasoconstrictor peptide produced by vascular endothelial cells. *Nature* 332: 411–415, 1988.

Endothelin

1
Endothelium-Derived Contracting Factors

PAUL M. VANHOUTTE, THOMAS GRÄSER, AND THOMAS F. LÜSCHER

In an early exploration of endothelium-dependent relaxations in the canine femoral artery, we noted that the removal of the endothelium reduced the contractions evoked by increasing concentrations of potassium ions.[5] This lead to the suggestion that the endothelial cells may release vasoconstrictor substances. Later work in isolated arteries and veins demonstrated that indeed rapid endothelium-dependent increases in tension can be obtained (Fig. 1.1), which are explained best by the production of mediators that activate the underlying vascular smooth muscle. By analogy with endothelium-derived relaxing factor,[9] the endothelial substances leading to rapid increases in tension have been termed "endothelium-derived contracting factor(s)" (EDCF).[24,41,42]

CYCLOOXYGENASE-DEPENDENT,
ENDOTHELIUM-DEPENDENT CONTRACTIONS

Arachidonic Acid and Its Metabolites

Exogenous arachidonic acid induces endothelium-dependent contractions in canine veins.[6,27] The contractions are prevented by inhibitors of cyclooxygenase (Fig. 1.2), but not those of prostacyclin or thromboxane synthetase. In the unstimulated rabbit aorta, arachidonic acid evokes more pronounced increases in tension in rings with endothelium[36,37]; the endothelium-dependent component of the contractions induced by arachidonic acid is inhibited by indomethacin. In the canine basilar artery, the endothelium-dependent contractions evoked by arachidonic acid are reduced or abolished by inhibitors of cyclooxygenase, thromboxane synthetase, or prostaglandin receptors.[15,17]

Other Stimuli

In certain blood vessels, endothelium-dependent contractions, which are prevented by inhibitors of cyclooxygenase, can be evoked with agonists that usually evoke endothelium-dependent relaxations. Thus, in the aorta of the spontaneously hypertensive rat, the renal artery of the normal rat, and the basilar artery of dog and rabbit, acetylcholine induces endothelium-dependent contractions that can be blocked by indomethacin (Fig. 1.3).[1,17,21,23,33,34] Likewise, in the basilar artery of the dog, norepinephrine, the Ca^{2+} ionophore A23187

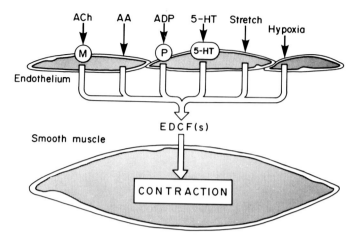

FIGURE 1.1. A number of physiochemical stimuli, neurohumoral mediators, and the calcium ionophore A23187 can evoke endothelium-dependent contractions in certain blood vessels, presumably because they evoke the release of endothelium-derived contracting factor(s) (EDCF). ACh, acetylcholine; AA, arachidonic acid; ADP, adenosine diphosphate; 5-HT, 5-hydroxytryptamine; M, muscarinic receptor; P, purinergic receptor. (From Lüscher and Vanhoutte.[24])

(Fig. 1.4), and nicotine cause endothelium-dependent contractions that are prevented by inhibitors of cyclooxygenase.[4,15,39,40]

Little is known about the cellular mechanisms underlying the release of the cyclooxygenase-dependent EDCF. The observation that the Ca^{2+} ionophore A23187 evokes its release suggests a Ca^{2+}-dependent process. Endothelium-

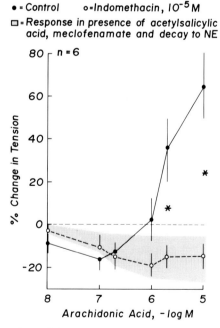

FIGURE 1.2. Demonstration that inhibitors of cyclooxygenase abolish the contractions caused by increasing concentrations of arachidonic acid (which are endothelium dependent) in canine femoral veins with endothelium contracted by norepinephrine (NE). Asterisks indicate that the effect of indomethacin is statistically significant. (From Miller and Vanhoutte.[27])

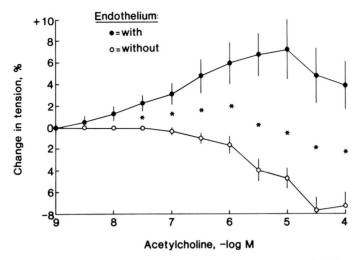

FIGURE 1.3. In the canine basilar artery, increasing concentrations of acetylcholine cause contractions in preparations with, but not in those without, endothelium. Asterisks indicate that the difference between rings with and without endothelium is statistically significant. (From Katusic et al.[17])

dependent, tetrodotoxin-sensitive contractions have been observed with electrical stimulation in cat cerebral arteries, which may imply that fast Na^+ channels are involved in the release of the cyclooxygenase-dependent EDCF.[11]

Mediator(s)

Thromboxane A_2 is the major vasoconstrictor product of cyclooxygenase; its formation from endoperoxides is catalyzed by the enzyme thromboxane syn-

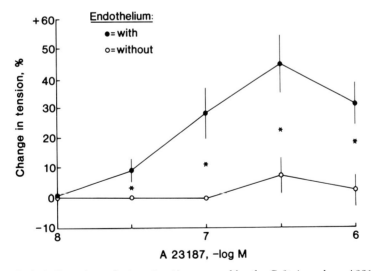

FIGURE 1.4. Endothelium-dependent contractions caused by the Ca^{2+} ionophore A23187 in canine basilar arteries. These experiments suggest that an increase in cytosolic Ca^{2+} concentration triggers the release of endothelium-derived contracting factors. Asterisks indicate that the difference between rings with and without endothelium is statistically significant. (From Katusic et al.[17])

thetase,[8,28] and can occur in endothelial cells. Thus the prostanoid is a likely candidate to be a cyclooxygenase-dependent EDCF. This actually may be the case in the isolated pulmonary artery of the rabbit, where thromboxane synthetase inhibitors reduce endothelium-dependent contractions evoked by acetylcholine, norepinephrine, and nicotine.[1,35,38,40] Likewise, in the basilar artery of the dog, endothelium-dependent contractions to norepinephrine, nicotine, and ionophore A23187 are blocked by dazoxiben, an inhibitor of thromboxane synthetase, suggesting that endothelium-derived thromboxane A_2 mediates the contractions (Fig. 1.5).[17,33–35] However, the production of thromboxane A_2 cannot explain the endothelium-dependent contractions evoked by arachidonic acid in canine veins or by acetylcholine and endoperoxides in canine cerebral arteries.[17,27,38]

Although the endothelium-dependent contractions evoked by acetylcholine, arachidonic acid, and ionophore A23187 depend on cyclooxygenase, the attempts to bioassay a cyclooxygenase-dependent EDCF have failed. In the aorta of the rat, the amounts of vasoconstrictor prostanoids detected in the lumen during stimulation with acetylcholine are not sufficient to explain the contractions.[22] Thus other nonprostanoid substances, in particular, oxygen-derived free radicals formed during activation of the cyclooxygenase pathway, must be considered.[28,44] Indeed, activation of endothelial cyclooxygenase generates superoxide anions.[20] The oxygen-derived free radicals contract the canine basilar artery; superoxide dismutase, on the other hand, prevents endothelium-dependent contractions evoked by ionophore A23187 in that blood vessel (Fig. 1.6). Because superoxide dismutase is a large molecule that is unlikely to enter endothelial cells, its mechanism of action must involve inactivation of superoxide released from these cells. Thus, in the canine basilar artery, superoxide anion appears to be the cyclooxygenase-dependent EDCF (Fig.

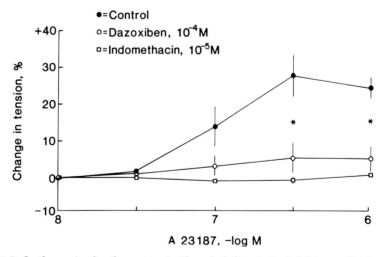

FIGURE 1.5. In the canine basilar artery (with endothelium), the inhibitors of both cyclooxygenase (indomethacin) and thromboxane synthetase (dazoxiben) significantly (*) reduce the contractions caused by the Ca^{2+} ionophore A23187. These experiments demonstrate that the metabolism of arachidonic acid through cyclooxygenase plays a key role in the endothelium-dependent response to the ionophore. (From Katusic et al.[17])

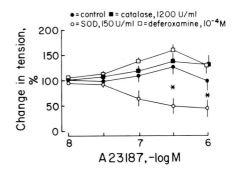

FIGURE 1.6. In the canine basilar artery with endothelium, the contractions evoked by the Ca^{2+} ionophore A23187 are prevented by the scavenger of superoxide anions (superoxide dismutase [SOD]), but not by those of hydrogen peroxide (catalase) or of hydroxyl radicals (deferoxamine). These experiments indicate that superoxide anions mediate the endothelium-dependent contractions evoked by the ionophore. (From Katusic and Vanhoutte.[19])

1.7).[18,44] The sensitivity for inhibitors of thromboxane synthetase implies that is the formation of thromboxane A_2 that generates the superoxide anions that diffuse from the endothelial cells to the underlying smooth muscle. However, this is not necessarily the case in all blood vessels. Thus, in the aorta of the spontaneously hypertensive rats, the endothelium-dependent contractions caused by acetylcholine are not prevented by inhibitors of thromboxane synthetase or superoxide dismutase, but are abolished by antagonists of thromboxane–endoperoxide receptors.[2] In this preparation, it is likely that the endothelial cells release hydroxyl radicals rather than superoxide anion and that these oxygen-derived free radicals switch on the production of endoperoxides in vascular smooth muscle that then activate the contractile process.

HYPOXIC ENDOTHELIUM-DEPENDENT CONTRACTIONS

An abrupt decrease in partial pressure of oxygen (anoxia) augments the contractions of isolated canine peripheral, coronary, and cerebral arteries (Fig. 1.8).[6,7,18,31] this anoxic facilitation of contraction is abolished or reduced by the removal of the endothelium. In "sandwich" preparations of the canine coronary artery contracted with prostaglandin $F_{2\alpha}$, the hypoxic facilitation can be transferred from a donor tissue with endothelium to a coronary artery strip without endothelium, demonstrating that the hypoxic response involves a diffusible substance (Fig. 1.9).[14,31] The hypoxic endothelium-dependent contractions are

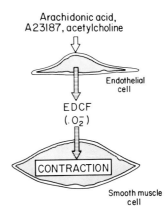

FIGURE 1.7. Proposal that superoxide anion ($.O_2^-$) is the endothelium-derived contracting factor (EDCF) released by arachidonic acid, the Ca^{2+} ionophore A23187, and acetylcholine from the endothelial cells of the canine basilar artery. (Modified from Vanhoutte and Katusic.[44])

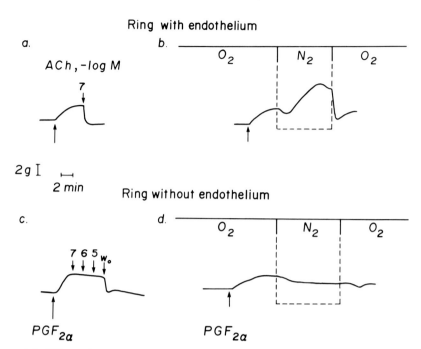

FIGURE 1.8. Effects of sudden anoxia (achieved by switching the bubbling of the organ chamber from a gas mixture containing 95% O_2 and 5% CO_2 to 95% N_2 and 5% CO_2) on rings of canine coronary arteries with (a, b) and without (c, d) endothelium. The preparations were contracted first with prostaglandin F_{2a} (PGF_{2a}). The presence and absence of endothelium was checked by demonstrating the occurrence and the lack, respectively, of relaxation to acetylcholine (ACh; left). Note that hypoxia caused a transient relaxation followed by a strong contraction in the ring with endothelium; in the ring without endothelial cells, only the relaxation was noted. This experiment demonstrates that anoxia causes endothelium-dependent contractions. w_o, Washout. (From Rubanyi and Vanhoutte.[31])

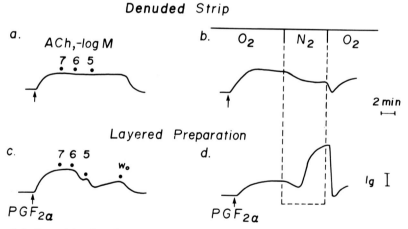

FIGURE 1.9. In a strip of canine coronary artery without endothelium, acetylcholine (ACh; a) causes no relaxation during a contraction evoked by prostaglandin F_{2a} (PGF_{2a}); sudden anoxia (N_2) causes a decrease in tension (b). If the strip without endothelium is "sandwiched" (layered) with a strip of artery with endothelium (intimal side against intimal side), the sandwich relaxes to acetylcholine (c) and contracts when made anoxic (d). This experiment demonstrates that a diffusible substance plays a key role in the endothelium-dependent contractions evoked by anoxia. w_o Washout. (From Rubanyi and Vanhoutte.[31])

augmented by indomethacin presumably because it prevents the hypoxia-induced release of vasodilator prostanoids (Fig. 1.10).[31] The anoxic endothelium-dependent contractions are not blocked by inhibitors of phospholipase A_2 or lipoxygenase, which rules out a product of the metabolism of arachidonic acid as the mediator. Methylene blue (to block the production of cyclic guanosine monophosphate) and hydroquinone (an inhibitor of endothelium-derived relaxing factor) also do not reverse the contractions; thus these cannot be attributed to a lesser release, or to a blunted effect, or endothelium-derived relaxing factor.[31] The mediator of the hypoxic response remains elusive; it is not likely to be endothelin (Fig. 1.11).[43] This conclusion is based on the following observations: *(1)* anoxic endothelium-dependent facilitation is not observed in peripheral veins of the dog, although their endothelium releases EDCF (Fig. 1.12), while their smooth muscle is exquisitely sensitive to endothelin (Fig. 1.13);[6,26,30] *(2)* Ca^{2+} antagonists prevent endothelium-dependent contractions

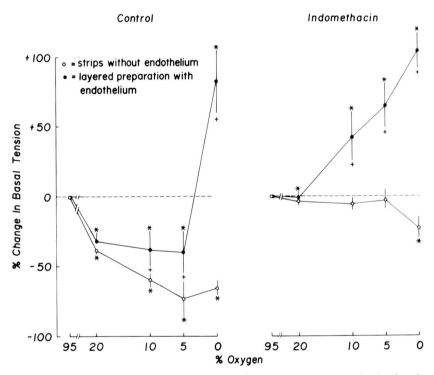

FIGURE 1.10. Experiments performed with strips of canine coronary arteries in the absence (*left*) and presence (*right*) of indomethacin (10^{-5} M, to inhibit cyclooxygenase). The effects of progressive decreases in the concentration of oxygen (indicated as the percentage of oxygen contained in the gas mixture bubbling the organ chambers) on instimulated strips (without endothelium) and on strips without endothelium "sandwiched" (layered) with strips containing endothelial cells. Note that moderate decreases in oxygen content cause relaxations in preparations both with and without endothelium (*left*) that are prevented by indomethacin (*right*); these can be attributed to the production of vasodilator prostanoids mainly in the vascular smooth muscle. Inhibition of cyclooxygenase unmasks endothelium-dependent contractions induced by stepwise increases in oxygen content (*right*); because these are seen only in the sandwiches with endothelium, they can be attributed to the progressive release of endothelium-derived contracting factor. +, The difference between strips without and sandwiches with endothelium is statistically significant; *, the difference from the control response (dotted line) is statistically significant. (From Rubanyi and Vanhoutte.[31])

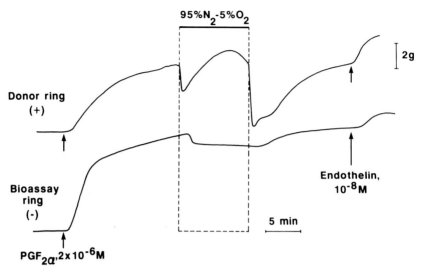

FIGURE 1.11. Experiment in which a ring of canine coronary artery (donor) with endothelium (+) is continuously superfused and the superfusate bioassayed with a ring without endothelium (−) of the same artery; isometric tension is measured in both simultaneously. Prostaglandin $F_{2\alpha}$ ($PGF_{2\alpha}$) is infused upstream of the donor ring (first arrows); both preparations contract. The superfusate is then made anoxic (by switching from 95% O_2 and 5% CO_2 to 95% N_2 and 5% CO_2 in the aerating gas mixture), the donor tissue, after an initial relaxation, contracts, and the bioassay tissue only relaxes. After return to control conditions, a concentration of endothelin is injected upstream of the donor ring, which causes it to contract to the same level as during anoxia; the bioassay ring now also contracts. This experiment is hard to reconcile with the hypothesis that endothelin is the endothelium-derived contracting factor released during anoxic endothelium-dependent contractions. (From Vanhoutte et al.[43])

evoked by anoxia in canine blood vessels (Fig. 1.14), but not contractions evoked by endothelin (Fig. 1.15);[14,16,25] and *(3)* the anoxic EDCF cannot be bioassayed under conditions in which the bioassay tissue contracts because of exogenous endothelin (Fig. 1.11).[43]

In the canine coronary artery, the endothelium-dependent facilitation by anoxia is inhibited by Ca^{2+} antagonists. Under bioassay conditions, the hypoxic contractions are prevented in preparations in which only the smooth muscle has been treated with the long-acting Ca^{2+} antagonist flunarizine (Fig. 1.16). This indicates that Ca^{2+} antagonists inhibit endothelium-dependent hypoxic contractions at the level of the vascular smooth muscle rather than by preventing the release of EDCF.[14]

ENDOTHELIN

Cultured bovine aortic endothelial cells release potent vasoconstrictor peptide(s).[10,12,13,28] The substance(s) contract a variety of vascular smooth muscle from different species. It is a small peptide (or peptides) that has been characterized as endothelin.[25,45,46] To date there is no evidence that endothelin contributes to the cyclooxygenase or anoxic acute changes in tension mediated by endothelial cells in freshly isolated blood vessels. Thus, stricto sensu, it cannot

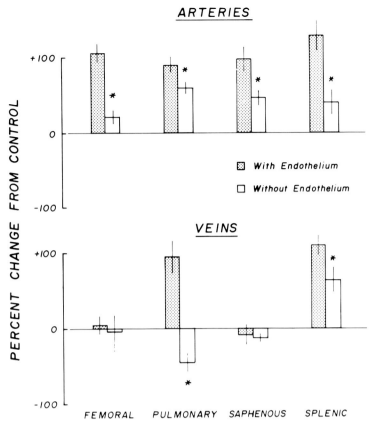

FIGURE 1.12. Effect of anoxia (10 min) on the contractile responses (ED_{30}) to norepinephrine in canine arteries (*top*) and veins (*bottom*). Rings with and without endothelium were studied in parallel. Changes in tension are expressed as percentage of the contractile response to norepinephrine and are shown as means ± SEM. Asterisks indicate that the difference between rings with and without endothelium is statistically significant. (From DeMey and Vanhoutte.[6])

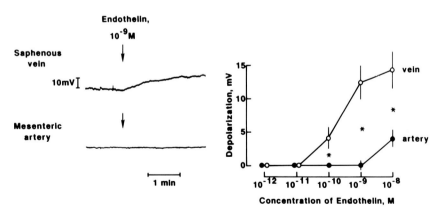

FIGURE 1.13. *Left:* Changes in membrane potential in canine saphenous veins and mesenteric arteries in response to 10^{-9} M endothelin. *Right:* Concentration-dependent effects of endothelin on amplitude of depolarization in canine saphenous veins and mesenteric arteries with endothelium. Data are shown as means ± SE of four to ten impalements obtained in four to seven different tissues. Asterisks indicate significant difference between veins and arteries (areas under curves were compared by a one-way analysis of variance, $P < 0.05$). (From Miller et al.[26])

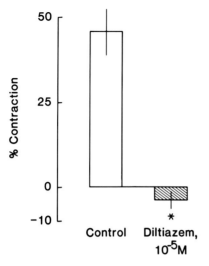

FIGURE 1.14. Effect of diltiazem (10^{-5} M) on anoxic contractions of canine basilar arteries with endothelium. Data are shown as means ± SEM (N = 6) and expressed as percentages. Asterisk indicates that the effect of diltiazem was statistically significant ($P < 0.05$). (From Katusic and Vanhoutte.[18])

be regarded as an EDCF. However, it could contribute to the long-term regulation of the basal tone of the blood vessels in response to stimuli such as thrombin, vasopressin, or angiotensin II, which cause delayed but long-lasting stimulation of the production of the peptide.[3,32,45] It could even be more the case that endothelium-derived nitric oxide blunts the release of the peptide[3] and is a potent antagonist of the activation of vascular smooth muscle that it causes (Fig. 1.17).[26,43] In view of the very low levels of circulating endothelin reported

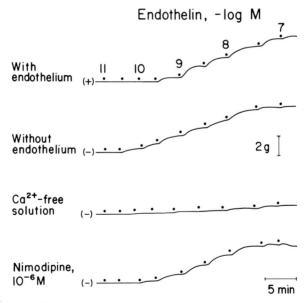

FIGURE 1.15. Tracings of contractions to endothelin in canine femoral veins with and without endothelium. Contractions were enhanced by removal of the endothelium and were inhibited by incubation in calcium-free solution containing 1 mM EGTA. The calcium antagonist nimodipine (10^{-6} M) did not inhibit the contractions. All experiments were conducted in the presence of indomethacin (10^{-5} M), phentolamine (10^{-5} M), and propranolol (5×10^{-6} M). Similar results were obtained in five other experiments. (From Vanhoutte et al.[43])

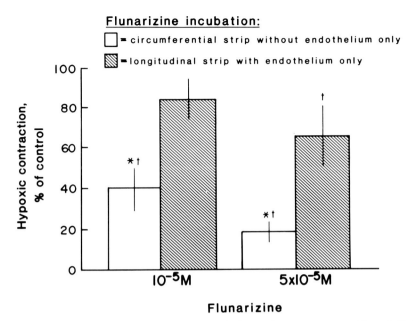

FIGURE 1.16. Anoxic contractions in "sandwiches" (layered preparations with endothelium) whereby either the denuded strip or the strip with only endothelium have been incubated with the long-acting Ca^{2+} antagonist flunarizine prior to layering. At 10^{-5} M, incubation of the smooth muscle–part of the sandwich with flunarizine markedly reduces the anoxic contractions; if given only to the endothelial cells, this concentration of the Ca^{2+} antagonist does not significantly affect the response to lowering the oxygen concentration. The difference persists at the higher concentration of flunarizine tested. These studies demonstrate that Ca^{2+} antagonists inhibit endothelium-dependent contractions caused by anoxia at the level of the vascular smooth muscle, but not by affecting the release of endothelium-derived contracting factor. *, The difference between open and hatched bars is statistically significant; †, the effect of flunarizine as compared with control (data not shown) is statistically significant. (From Iqbal and Vanhoutte.[14])

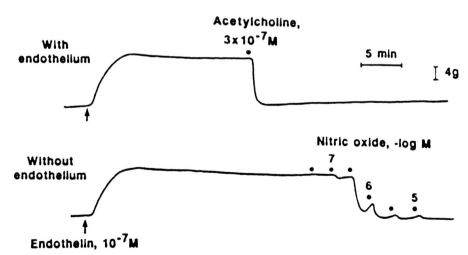

FIGURE 1.17. Demonstration in the canine coronary artery that the release of endothelium-derived relaxing factor by acetylcholine (top, ring with endothelium) or exogenous nitric oxide (bottom, ring without endothelium) causes potent and sustained relaxations during contractions evoked by endothelin. These studies imply that endothelium-derived relaxing factor may be a physiological antagonist of endothelin-induced contractions. (From Vanhoutte et al.[43])

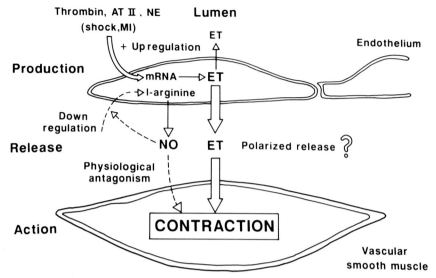

FIGURE 1.18. Interaction of endothelium-derived vasorelaxing and vasoconstricting factors. AT II, angiotensin II; MI, myocardial infarction; NE, norepinephrine; NO, nitric oxide; ET, endothelin.

thus far in different species, any proposal for a role of the peptide in vasomotor regulation implies that it is released preferentially in abluminal direction (Fig. 1.18).[24,43]

REFERENCES

1. ALTIERE, R. J., J. A. KIRITSY-ROY, and J. D. CATRAVAS. Acetylcholine-induced contractions in isolated rabbit pulmonary arteries: role of thromboxane A_2. *J. Pharmacol. Exp. Ther.* 236: 535–541, 1986.
2. AUCH-SCHWELK, W., Z. S. KATUSIC, and P. M. VANHOUTTE. Thromboxane A_2 receptor antagonists inhibit endothelium-dependent contractions. *Hypertension* 15: 699–703, 1990.
3. BOULANGER, C., and T. F. LÜSCHER. Release of ET from the porcine aorta: inhibition by endothelium-derived NO. *J. Clin. Invest.* 85: 587–590, 1990.
4. BRASHERS, V. L., M. J. PEACH, and C. E. ROSE. Augmentation of hypoxic pulmonary vasoconstriction in the isolated perfused rat lung by in vitro antagonists of endothelium-dependent relaxation. *J. Clin. Invest.* 82: 1495–1502, 1988.
5. DEMEY, J. G., and P. M. VANHOUTTE. Role of the intima in cholinergic and purinergic relaxation of isolated canine femoral arteries. *J. Physiol.* 376: 347–355, 1981.
6. DEMEY, J.G., and P. M. VANHOUTTE. Heterogeneous behavior of the canine arterial and venous wall: importance of the endothelium. *Circ. Res.* 51: 439–447, 1982.
7. DEMEY, J. G. and P. M. VANHOUTTE. Anoxia and endothelium-dependent reactivity in canine femoral artery. *J. Physiol. (Lond.)* 335: 65–74, 1983.
8. ELLIS, E. F., O. OELZ, L. J. ROBERTS II, N. A. PAYNE, B. J. SWEETMAN, A. S. NIES, and J. A. OATES. Coronary arterial smooth muscle contraction by a substance released from platelets: evidence that it is thromboxane A_2. *Science* 93: 1135–1137, 1976.
9. FURCHGOTT, R. F., and J. V. ZAWADZKI. The obligatory role of endothelial cells in the relaxation of arterial smooth muscle by acetylcholine. *Nature* 288: 373–376, 1980.
10. GILLESPIE, M. N., J. O. OWASOYO, I. F. MCMURTRY, and R. F. O'BRIEN. Sustained coronary vasoconstriction provoked by a peptidergic substance released from endothelial cells in culture. *J. Pharmacol. Exp. Ther.* 236: 339–343, 1986.
11. HARDER, D. R., and J. A. MADDEN. Electrical stimulation of the endothelial surface of pressurized cat middle cerebral artery results in TTX-sensitive vasoconstriction. *Circ. Res.* 60: 831–836, 1987.
12. HICKEY, K. A., G. M. RUBANYI, R. J. PAUL, and R. F. HIGHSMITH. Characterization of a cor-

onary vasoconstrictor produced by cultured endothelial cells. *Am. J. Physiol.* 248 *(Cell Physiol.* 17): C550–C556.
13. HIGHSMITH, R. F., D. AICHHOLZ, O. FITZGERALD, R. PAUL, G. M. RUBANYI, and K. HICKEY. Endothelial cells in culture and production of endothelium-derived constricting factors. In: *Relaxing and Contracting Factors: Biological and Clinical Research,* edited by P. M. Vanhoutte. Clifton, NJ: Humana, 1988, p. 137–158.
14. IQBAL, A., and P. M. VANHOUTTE. Flunarizine inhibits endothelium-dependent hypoxic facilitation in canine coronary arteries through an action on vascular smooth muscle. *Br. J. Pharmacol.* 95: 789–794, 1988.
15. KATUSIC, Z. S., and J. T. SHEPHERD. Endothelium-dependent responses of cerebral arteries. In: *Relaxing and Contracting Factors: Biological and Clinical Research,* edited by P. M. Vanhoutte. Clifton, NJ: Humana, 1988, p. 333–345.
16. KATUSIC, Z. S., J. T. SHEPHERD, and P. M. VANHOUTTE. Anoxic contractions in isolated cerebral arteries: Contribution of endothelium-derived factors, metabolites of arachidonic acid and calcium entry. *J. Cardiovasc. Pharmacol.* 8(Suppl. 8): 97–101, 1986.
17. KATUSIC, Z. S., J. T. SHEPHERD, and P. M. VANHOUTTE. Endothelium-dependent contractions to calcium ionophore A23187, arachidonic acid and acetylcholine in canine basilar arteries. *Stroke* 19: 476–479, 1988.
18. KATUSIC, Z. S., and P. M. VANHOUTTE. Anoxic contractions in isolated canine cerebral arteries: contribution of endothelium-derived factors, metabolites of arachidonic acid, and calcium entry. *J. Cardiovasc. Pharmacol.* 8(Suppl 8): 97–101, 1986.
19. KATUSIC, Z. S., and P. M. VANHOUTTE. Superoxide anion is an endothelium-derived contracting factor. *Am. J. Physiol.* 257 *(Heart Circ. Physiol.* 26): H33–H37, 1989.
20. KONTOS, H. A. Oxygen radicals in cerebral vascular injury. *Circ. Res.* 57: 508–516, 1985.
21. LÜSCHER, T. F., D. DIEDERICH, F. R. BÜHLER, and P. M. VANHOUTTE. Interactions between platelets and the vessel wall: role of endothelium-derived vasoactive substances. In: *Hypertension: Pathophysiology, Diagnosis and Management,* edited by J. H. Laragh and B. Brenner. New York: Raven, 1989.
22. LÜSCHER, T. F., J. C. ROMERO, and P. M. VANHOUTTE. Bioassay of endothelium-derived vasoactive substances in the aorta of normotensive and spontaneously hypertensive rats. *J. Hypertens.* 4:(Suppl. 6): 81–83, 1986.
23. LÜSCHER, T. F., and P. M. VANHOUTTE. Endothelium-dependent contractions to acetylcholine in the aorta of the spontaneously hypertensive rat. *Hypertension* 8: 344–348, 1986.
24. LÜSCHER, T. F., and P. M. VANHOUTTE. *The Endothelium: Modulator of Cardiovascular Function.* Boca Raton, FL: CRC, 1990, p. 1–228.
25. MASAKI, T. The discovery, the present state, and the future prospects of endothelin. *J. Cardiovasc. Pharmacol.* 13(Suppl. 5): 1–4, 1989.
26. MILLER, V. M., K. KOMORI, J. C. BURNETT, and P. M. VANHOUTTE. Differential sensitivity to endothelin in canine arteries and veins. *Am. J. Physiol.* 257: H1127–H1131, 1989.
27. MILLER, V. M., and P. M. VANHOUTTE. Endothelium-dependent contractions to arachidonic acid are mediated by products of cyclooxygenase in canine veins. *Am. J. Physiol.* 248 *(Heart Circ. Physiol.)* H432–H437, 1985.
28. MONCADA, S., and J. R. VANE. Pharmacology and endogenous roles of prostaglandins endoperoxides, thromboxane A_2 and prostacyclin. *Pharmacol. Rev.* 30: 293–331, 1979.
29. O'BRIEN, R. F., and I. F. MCMURTY. Endothelial cell supernates contract bovine pulmonary artery rings, abstracted. *Am. Rev. Respir. Dis.* 129: 337, 1984.
30. ROSEN, G. M., and B. A. FREEMAN. Detection of superoxide generated by endothelial cells. *Proc. Natl. Acad. Sci. USA* 81: 7269–7273, 1984.
31. RUBANYI, G. M., and P. M. VANHOUTTE. Hypoxia releases a vasoconstrictor substance from the canine vascular endothelium. *J. Physiol. (Lond.)* 364: 45–56, 1985.
32. SCHINI, V. B., H. HENDRICKSON, D. M. HEUBLEIN, J. C. BURNETT, JR., and P. M. VANHOUTTE. Thrombin enhances the release of endothelin from cultured porcine aortic endothelial cells. *Eur. J. Pharmacol.* 165: 333–334, 1989.
33. SHIRAHASE, H., M. FUJIWARA, H. USUI, and K. KURAHASHI. A possible role of thromboxane A_2 in endothelium in maintaining resting tone and producing contractile response to acetylcholine and arachidonic acid in canine cerebral arteries. *Blood Vessels* 24: 117–119, 1987.
34. SHIRAHASE, H., H. USUI, K. KURAHASHI, M. FUJIWARA, and K. FUKUI. Possible role of endothelial thromboxane A_2 in the resting tone and contractile responses to acetylcholine and arachidonic acid in canine cerebral arteries. *J. Cardiovasc. Pharmacol.* 10: 517–522, 1987.
35. SHIRAHASE H., H. USUI, K. KURAHASHI, M. FUJIWARA, and K. FUKUI. Endothelium-dependent contraction induced by nicotine in isolated canine basilar artery—possible involvement of a thromboxane A_2 (TXA_2) like substance. *Life Sci.* 42: 437–445, 1988.

36. SINGER, H. A., and M. J. PEACH. Endothelium-dependent relaxation of rabbit aorta. I. Relaxation stimulated by arachidonic acid. *J. Pharmacol. Exp. Ther.* 226: 790–795, 1983.
37. SINGER, H. A., and M. J. PEACH. Endothelium-dependent relaxation of rabbit aorta. II. Inhibition of relaxation stimulated by methacholine and A23187 with antagonists of arachidonic acid metabolism. *J. Pharmacol. Exp. Ther.* 227: 796–801, 1983.
38. TODA, N., S. INOUE, K. BIAN, and T. OKAMURA. Endothelium-dependent and independent responses to prostaglandin H_2 and arachidonic acid in isolated dog cerebral arteries. *J. Pharmacol. Exp. Ther.* 244: 297–302, 1988.
39. USUI, H., K. KURAHASI, K. ASHIDA, and M. FUJIWARA. Acetylcholine-induced contractile response in canine basilar artery with activation of thromboxane A_2 synthesis sequence. *IRCS Med. Sci. Physiol.* II: 418–419, 1983.
40. USUI, H., K. KURAHASI, H. SHIRAHASE, K. FUKUI, and M. FUJIWARA. Endothelium-dependent vasocontraction in response to noradrenaline in the canine cerebral artery. *Jpn. J. Pharmacol.* 44: 228–231, 1987.
41. VANHOUTTE, P. M. Endothelium-dependent contractions in arteries and veins. *Blood Vessels* 24: 141–144, 1987.
42. VANHOUTTE, P. M. Endothelium-dependent contractions in veins and arteries. In: *Relaxing and Contracting Factors: Biological and Clinical Research*, edited by P. M. Vanhoutte. Clifton, NJ: Humana, 1988, p. 27–39.
43. VANHOUTTE, P. M., W. AUCH-SCHWELK, C. BOULANGER, P. A. JANSSEN, Z. S. KATUSIC, K. KOMORI, V. M. MILLER, V. B. SCHINI, and M. VIDAL. Does endothelin-1 mediate endothelium-dependent contractions during anoxia? *J. Cardiovasc. Pharmacol.* 13(Suppl. 5): 124–128, 1989.
44. VANHOUTTE, P. M., and Z. S. KATUSIC. Endothelium-derived contracting factor: endothelium and/or superoxide anion? *Trends Pharmacol. Sci.* 9: 229–230, 1988.
45. YANAGISAWA, M., A. INOUE, T. ISHIKAWA, Y. KASUQA, S. KIMURA, S. KUMAGAYE, K, NAKAJIMA, T. X. WATANABE, S. SAKAKIBARA, K. GOTO, and T. MASAKI. Primary structure, synthesis, and biological activity of rat endothelin, an endothelium-derived vasoconstrictor peptide. *Proc. Natl. Acad. Sci. USA* 85: 6964–6967, 1988.
46. YANAGISAWA, M., H. KURIHARA, S. KIMURA, Y. MITSUI, M. KOBAYASHI, T. X. WATANABE, and T. MASAKI. A novel potent vasoconstrictor peptide produced by vascular endothelial cells. *Nature* 332:411–415, 1988.

2

From Endotensin to Endothelin: The Discovery and Characterization of an Endothelial Cell–Derived Constricting Factor

ROBERT F. HIGHSMITH

A quick review of the short but frenzied history of the endothelial cell–derived vasoconstrictor endothelin reflects science at its best. Pioneering observations were followed by the application of the modern tools of molecular biology and by rapid commercial availability. In turn, these events quickly led to the bandwagon description of the actions and mechanisms of action of this endothelial cell–derived factor. As the flow of basic information begins to yield unified ideas, and with specific quantitative approaches, we can start to evaluate the physiological and pathological importance of this unique bioactive peptide.

What is needed now is a historical perspective in reviewing the early experimentation that directly led to the isolation and sequencing of the endothelin molecule. Accordingly, most of the discussion here emphasizes historical developments rather than experimental details, because the latter have been published in original papers or review articles. Within this framework, the early characterization and properties of this unique peptide as well as the important confirmation of these findings by other investigators will be reviewed. Some original thoughts on the mechanisms of action of the constrictor and the later events culminating in the isolation and purification of endothelin will be discussed.

DISCOVERY OF AN ENDOTHELIAL CELL–DERIVED CONSTRICTING FACTOR

Approximately 10 years ago, our studies were focused on the role of the endothelial cell (EC) in the regulation of the fibrinolytic or clot-dissolving enzyme system. Kristine Hickey (née Agricola), a graduate student working under my guidance, had completed a portion of her Ph.D. thesis dealing with the effects of changes in pH and PO_2 on plasminogen activator synthesis by cultured ECs. In the fall of 1982, as a part of our weekly literature review sessions, we discussed the pioneering finding by Furchgott and Zawadzki[4] concerning the obligatory role of the endothelium in modulating the relaxation of arterial smooth muscle by acetylcholine. We were particularly intrigued by the uncertainty surrounding the chemical nature of the diffusible second messenger (endothelium-derived relaxing factor [EDRF]), which was proposed to mediate the relaxation response. At that time, EDRF was thought to be a humoral agent, a lipoxygenase derivative, or a free radical. A few weeks later, Hickey proposed a series of experiments to define more rigorously the chemical nature of this endothelium-derived factor. These studies were to serve as the second portion of her Ph.D. thesis. We agreed to devote a few weeks to some pilot studies because of their potential importance and because we had already established and characterized an EC culture system. My colleagues and eventual collaborators, Drs. Paul and Rubanyi, had also kindly agreed to lend their expertise in smooth muscle physiology and pharmacology as well as some equipment for measuring blood vessel contractility.

The first of the pilot experiments, begun in February 1983, was to design a bioassay system for quantifying EDRF in the conditioned medium from cultured ECs. (We were unaware at that time that the very short half-life of EDRF would preclude its measurement in static cultures; ironically, however, this property of EDRF and our naiveté would subsequently allow for the detection of the stable vasoconstrictive products of cultured ECs.) The objective was accomplished by measuring isometric force in denuded and precontracted rings of porcine or bovine coronary arteries following addition of EC-conditioned media. After only a few experiments, it became quite clear that no detectable vasodilatory signals were evident in EC-conditioned media. In fact, contrary to our expectations, we noted a small yet consistent vasoconstriction response in vessels that were submaximally contracted. Upon testing the EC-conditioned media in vessels with minimum resting tension, it was quite obvious from the dose–response relationships that the media contained a very potent vasoconstrictor substance (Fig. 2.1). Our initial experiments also indicated very clearly that the vasoconstriction response did not require an intact endothelium on the bioassay vessel and that it was rather unique in being slow in onset and having a substantial tonic component. (The further characterization of the vasoconstrictor substance, done entirely with bioassay contractility measurements, would be delayed considerably by the mere difficulty in washing out the contractile responses.) Since I have learned many times over that true discoveries in science are rare, I thought our finding might have already been reported in the rapidly growing literature on EC biology. However, a tentative search for such a publication proved negative.

Our excitement with the apparent discovery of a novel vasoconstrictor substance produced by ECs was temporarily dampened after control experiments were performed. Much to our dismay, the first such experiment seemed to indicate that conditioned media from control pituitary cells and fibroblasts cultured identically in parallel with ECs, as well as nonconditioned control media, also elicited contractions of bioassay vessels. After several anxious days it was determined that the control contractions, which were quite different from those observed with EC-conditioned media, were solely attributable to the presence of vasoactive material in a new lot of serum that had been purchased as a culture supplement. Fortunately, use of a different lot of serum and serum-free cultures abolished the control contractions but did not affect those obtained in response to EC-conditioned media.

After this brief setback, we proceeded with the preliminary biochemical and pharmacological characterization of the vasoconstrictor substance. The

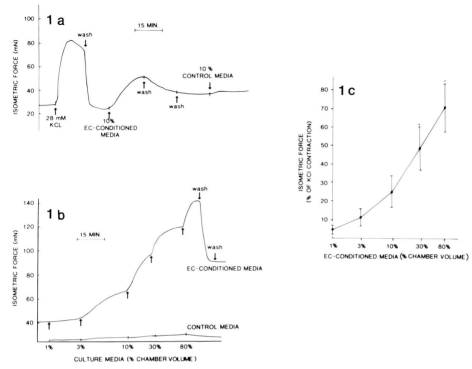

FIGURE 2.1. Isometric force response of de-endothelialized porcine coronary arteries to endothelial cell (EC)–conditioned media. Dosages represent the final percentage (v/v) concentration of media in the vessel chamber. a: Representative tracing of a typical experiment. b: Cumulative dose–response curve obtained in one coronary artery. c: average (mean ± SEM) dose–response relationship obtained in ten different coronary artery preparations. Asterisks indicate significantly different ($P < 0.05$) response as compared with the preceding lower dose.[6] (Reproduced with permission from the American Physiological Society.)

findings of these studies were even more exciting than our initial observations in that they uniformly suggested that a *novel* vasoactive *peptide* of EC origin had been discovered. The substance elicited contractions in every vessel tested, was not inhibited by any antagonists directed against the well-characterized agonist–receptor systems, yet was quite sensitive to agents and conditions known to affect protein structure and and/or function. Thus we had growing confidence that the constricting factor was a unique peptide apparently produced only by ECs and that the findings were not an artifact of cell culture.

Before submitting our results for publication, an intense review of the literature was undertaken to determine if there was any prior evidence supporting the phenomenon of endothelium-dependent vasoconstriction or the involvement of endothelium in modulating vasoconstriction in response to other agonists. We felt that it would be important to show (or at least to suggest evidence) that vasoactive secretory products of cultured ECs indeed bear a functional relationship to the normal influence of intact endothelium in contact with vascular smooth muscle (VSM). We found that the intact endothelium had been shown to facilitate the contraction of VSM under several conditions. In isolated canine arteries and veins, anoxia had been shown to augment the contractile response to several different vasoconstrictor agents.[17,18] DeMey and Vanhoutte[2] compared the responses of isolated arteries and veins and reported that mechanical removal of the endothelium reduced the maximal contractile response to norepinephrine and decreased the augmentation of the response to norepinephrine caused by anoxia. DeMey and Vanhoutte[3] also noted that cyclooxygenase inhibition did not prevent the potentiating response of anoxia on norepinephrine-induced tone in the canine femoral artery. Holden and McCall[8] had also reported in a preliminary communication at the 1983 American Thoracic Society meeting that hypoxic or anoxic vasoconstriction of porcine pulmonary arteries was endothelium dependent. Thus the generation of a substance by the intact endothelium that facilitates contraction of VSM would be a likely explanation of the results. These findings, which were unknown to us at the onset of our experiments, provided suggestive evidence that the intact endothelium may somehow activate contraction of VSM in addition to expressing vasodilatory signals. We thought that the peptidergic factor we had discovered might play a key role as a potential mediator of endothelium-dependent vasoconstriction. These early studies and those of Rubanyi and Vanhoutte[16] also provided confidence that EC culture would be a useful and appropriate model of the intact endothelium.

In January 1984 we submitted our characterization of the novel EC-derived peptidergic factor for publication in *Science*. In that manuscript we coined the term "endotensin" to designate the source and action of the peptide. In April, we were notified that the paper would not be published, and an opportunity for resubmission was not apparent. Despite an overall positive review, two major criticisms were raised. First, both reviewers felt that there was a lack of data to support specificity of the cell type involved in the secretion and/or release of the factor. Similarly, the reviewers questioned whether the results might be merely an artifact of cell culture and thought that we needed to present more evidence that such a substance might be released in vivo. No objection was raised to our use of the term "endotensin." After much discussion with my colleagues and with the editor, we decided to resubmit a revised paper to *Science* because the necessary control experiments to demonstrate specificity using non-EC cultures had been completed and because several preliminary reports were emerging that had a very positive and supportive impact on our findings.

We published the first abstract of our findings in the proceedings of the annual FASEB meeting held in St. Louis in April 1984.[1] Kristine Hickey also gave the first public presentation of our results at that meeting. Shortly after her talk, Dr. Richard O'Brien of the University of Colorado informed us that he and Dr. Ivan McMurtry had just obtained nearly identical results using conditioned media from bovine pulmonary arterial or aortic ECs and that the results would soon appear as an abstract[11] for the American Thoracic Society meetings to be held the following month (May 1984). Dr. William Holden at the Oregon Health Sciences University had also just published a paper[9] early in 1984 confirming their earlier findings[8] that hypoxia-induced contraction of pulmonary arteries was endothelium dependent. Dr. Rubanyi, then working with Dr. Vanhoutte at the Mayo Medical School, had also completed a study, building on the earlier findings of Holden and McCall, that demonstrated that hypoxic vasoconstriction of isolated canine coronary arteries was also endothelium dependent. Importantly, Rubanyi and Vanhoutte had further noted that the condition of oxygen deprivation released a diffusible constrictor substance (rather than inhibiting the release of a vasodilator) from the intact endothelium. Although not stated per se, the pharmacological properties of the vasoconstrictor were consistent with the substance being a polypeptide identical to "endotensin." Rubanyi and Vanhoutte[15] would present these findings at the annual American Heart Association meeting at Miami in the fall of 1984. Their results had been submitted for publication,[16] and, because Dr. Rubanyi was also a collaborator in our initial studies, he allowed us to enclose a copy of the manuscript with our resubmission to *Science* as evidence for the release of a similar if not identical substance from the intact endothelium. We were convinced that the results of several independent studies utilizing quite different approaches were focused on the same vasoconstrictor substance, and, with our additional control studies, we could adequately address the reviewers' concerns.

With renewed confidence, yet with awareness that the race was on, the revised paper was resubmitted to *Science* in June 1984. At the end of the summer, we were notified that the referees had recommended that it be published but that the editors had decided to reject it once again "because of an exceptionally large backlog of accepted manuscripts." After nearly 1 year of frustration, our dealings with *Science* came to an abrupt and painful end. In October 1984, our results, along with copies of the prior review from *Science,* were sent to the *American Journal of Physiology*. After minor revisions, the manuscript was accepted in February 1985 and published as a rapid communication in May.[6] The only criticism raised in the final review was of our use of the term *endotensin* to refer to the constricting factor. A reviewer and the editor felt that it was premature to ascribe a specific term to a "factor" that had not been purified, and we had not yet demonstrated that the factor resulted in an elevation in systemic vascular resistance. By the time the reviews were finished we had indeed partially purified the factor and had demonstrated its hypertensive effect in vivo. However, weary from the prolonged reviews yet thankful that the paper would finally be published, we reluctantly surrendered and settled for "endothelial cell–derived constricting factor" or "EDCF."

CHARACTERIZATION AND PROPERTIES OF EDCF

One of our early efforts confirmed that EDCF was expressed by cultured ECs derived from each of the following sources: bovine aorta and pulmonary artery, porcine aorta and left anterior descending coronary artery, human temporal and umbilical artery and vein, rat microvasculature (epididymal and retroperitoneal fat), and rat atrium. Based on bioassays of conditioned media, four sources of cultured ECs resulted in an apparent lack of EDCF production: bovine corneal ECs, bovine aortic ECs cultured on collagen-coated microcarrier beads, human ECs isolated from omental fat, and ECs derived from rat ventricle. Specificity of cell type involved in producing EDCF was evaluated by testing the vasoactivity of media obtained from control cultures of rat pituitary cells, fibroblasts (mouse 3T3 cells and human dermis), and VSM (rat aorta and the cell lines A7r5 and A-10). Conditioned media obtained from equal numbers of these control cells had no effect on vessel tone. Interestingly, of the media tested from cells of nonendothelial origin, those from canine tracheal epithelial cells also had significant EDCF-like activity.

After 24 h of culture, conditioned media (with or without serum) from bovine aortic ECs elicited a significant vasoconstriction in all bovine, canine, and porcine vessel preparations tested. The potency of culture media in inducing coronary vasoconstriction increased significantly with time in culture, reaching a maximum on days 3–4 of incubation. Comparing the vasoactive responses of aliquots of media taken with or without volume replacement indicated that the increase in vasoconstrictor activity as a function of time in culture was due to the progressive accumulation of the constrictor material in the medium. The production rate of EDCF paralleled the EC growth curve: maximum production occurring during log-phase growth and contact inhibition of production as the cells approached confluency. This production cycle was repeatable for at least 17–20 serial passages; longer term cultures were not investigated because of the questionable functional changes that occur upon continuous long-term propagation of this cell type.

In most vessels tested, the contractile response to EC-conditioned media was typically of slow onset, with a latency of 0.5–5 min. Maximal tension was usually achieved within 15 min but occasionally required 30 min or more. In contrast to K^+-induced contractions or to those elicited by most other agonists, the contractile response to EC-conditioned media was characteristically difficult to wash out (Fig. 2.2). Nearly all vessels tested failed to relax completely

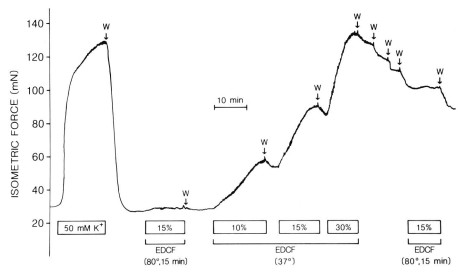

FIGURE 2.2. Changes in isometric tension in a de-endothelialized porcine coronary artery following treatment with K^+ or endothelium-derived constricting factor (EDCF). The time of exposure and concentrations of the agents added are indicated in bars below the tracing. Washing (W) of the vessel chamber is indicated by arrows. Isometric force is expressed in millinewtons (mN), where 10 mN = 1 g of tension.[7]

following extensive washing, thus illustrating a substantial tonic component to the contractile response. Significant tachyphylaxis to EDCF was observed in most vessels. The vasoconstriction did not require an intact endothelium, because complete denudation of the bioassay vessels as confirmed by electron microscopy did not affect the response. The constriction required extracellular Ca^{2+} and was attenuated by the Ca^{2+} antagonist verapamil. Although we initially reported that verapamil abolished EDCF-induced contractions, subsequent experiments revealed only a marked attenuation. Similar results were obtained with other organic Ca^{2+} channel blockers.

Dose–response relationships were obtained by recording individual steady-state tensions of coronary artery segments in response to fixed concentrations of the EC-conditioned media (Figs. 2.1, 2.2). Isometric tension increased in a dose-dependent fashion as the percentage of EC-conditioned media was increased in the muscle chamber. Saturation of the response did not occur at higher concentrations of the vasoconstrictor, suggesting that a maximum response was not yet obtained. Repeated challenges to different arterial segments with the same media gave reproducible responses. Thus the variability in the response of the vessels to a fixed concentration of media was most likely due to differences in the concentrations of EDCF in each particular pool of media. On several occasions, with higher concentrations of EDCF, a transient vasodilation preceded the constriction response, but only in bioassay vessels containing an intact endothelium. The same result was obtained in the presence of cyclooxygenase inhibitors and suggested that EDCF, at least at higher concentrations, may stimulate the release of EDRF.

One of the first objectives in biochemically defining EDCF was the estimation of molecular weight. EC-conditioned medium was subjected to calibrated gel filtration at neutral pH, and the resulting fractions were bioassayed for vasoconstricting activity using de-endothelialized vessels. In most cases, a prominent peak of vasoconstrictor activity eluted midway between pancreatic trypsin inhibitor and cytochrome c. A selectivity curve based on the calculated partition coefficients of standard marker proteins indicated an average molecular weight of 8500 ± 1500. Although these results were very reproducible, we felt a bit uneasy about the accuracy of the determination, because preliminary experiments using ultrafiltration and dialysis membranes with defined exclusion limits had suggested that the EDCF molecule was <5000 daltons. Also, upon gel filtration, a smaller peak of activity occasionally eluted in the range corresponding to a molecular weight of 2500. The lower molecular weight peak was particularly evident when the separations were performed at acidic pH. (Subsequent to these experiments we determined that EDCF-like immunoreactivity was associated with both peaks, suggesting anomalous behavior of the peptide in gel filtration and/or pH-dependent multimer formation.) In addition to estimating molecular weight, we found that treatment of ECs with cycloheximide blocked EDCF production and exposure of EDCF to 80°–100°C (Fig. 2.2), Na-dodecyl-SO_4, trypsin, alkali, and acid hydrolysis abolished the vasoconstrictive effect on vessel segments (Table 2.1).

The protein nature of the vasoconstrictor was further supported by pharmacological experiments in which we evaluated the effects of various antagonists on the activity of EDCF (Table 2.2). The vasoconstriction induced by

TABLE 2.1. Agents and treatments that block the production or action of EDCF[a]

Time of treatment or addition of agent	Agent (dosage) or treatment
During endothelial cell culture	Cycloheximide (2.5 mg/ml)
Post-culture, to endothelial cell–conditioned media	Na-dodecyl-SO$_4$ (0.1%, 25°C, 60 min) Trypsin (10^{-5} M, 37°C, 45 min) Alkali (pH 11.5, 25°C, 60 min) Acid (pH 2.5, 100°C, 30 min)
Directly into muscle chamber	EGTA (10^{-3} M) Verapamil (3×10^{-6} M)

[a]Agents were added or treatments were performed at the times indicated, and contractile responses were recorded as described in the footnote to Table 2-2. Responses were compared with those from untreated controls, and in each case the coronary artery contractile responses were markedly attenuated by the above agents or treatments.

EC-conditioned media or by partially purified EDCF was unaffected by the presence of inhibitors directed against various nonprotein agonist–receptor systems, including the α-adrenergic (phentolamine), β-adrenergic (propranolol), serotonergic (methysergide), H$_1$-histaminergic (pyribenzamine), and cholinergic (atropine) receptors. Also, the activity of EDCF applied to vessels with or without endothelium was unaffected by the EDRF antagonists methylene blue and hemoglobin. We also determined that EDCF was not a prostanoid produced in the endothelium or generated in VSM, because the mechanical response to the vasoconstrictor was unaffected by the inclusion of phenidone, indomethacin, or meclofenamate during endothelial cell culture or in the muscle chamber.

In summary, our initial characterization indicated that the vasoconstriction induced by EDCF was not related to arachidonic acid metabolism and was unaffected by inhibitors to the major agonist–receptor systems, including the serotonergic, histaminergic, cholinergic, and the α- and β-adrenergic pathways. The constrictive action of EDCF was completely blocked by several treatments known to affect protein structure and/or conformation. These data, cou-

TABLE 2.2. Antagonists that do not affect the production or action of EDCF[a]

Time agent added	Agent and dosage
During endothelial cell culture	Phenidone (10^{-5} M) Indomethacin (5×10^{-6} M) Meclofenamate (6×10^{-6} M)
Directly into muscle chamber	Phentolamine (3×10^{-6} M) Propranolol (4×10^{-7} M) Methysergide (3×10^{-7} M) Pyribenzamine (10^{-5} M) Atropine (8×10^{-9} M) Phenidone (10^{-5} M) Indomethacin (5×10^{-6} M) Meclofenamate (6×10^{-6} M)

[a]Agents were added at various times and dosages as indicated, and contractile responses were recorded isometrically from de-endothelialized rings of porcine left anterior descending coronary artery. Responses were compared with those from untreated controls, and no significant differences were apparent. Effectiveness of antagonists was confirmed by noting that the responses to their respective agonists were abolished.

pled with the results of the calibrated gel filtration experiments, strongly suggested that EDCF was a polypeptide with a molecular weight of approximately 2.5 kd.

CONFIRMATORY EVIDENCE

Independent confirmation of most of the above properties of EDCF was obtained in 1986 and 1987 by Drs. Gillespie and Owasoyo at the University of Kentucky and by Drs. O'Brien, Robbins, and McMurtry of the University of Colorado. Using concentrated media from bovine aortic ECs and the retrogradely perfused rabbit heart, Gillespie et al.[5] reported in 1986 that inhibition of cyclooxygenase or lipoxygenase had no effect on EDCF-induced increases in coronary perfusion pressure, thereby excluding a role for metabolites of arachidonate. The action of EDCF in this preparation was also unaffected by blockade of α-adrenergic, serotonergic, or histaminergic receptors. The protein nature of EDCF was also suggested by their finding that EDCF-induced increases in coronary flow resistance were markedly attenuated by prior treatment of the constrictor with trypsin and that the trypsin inhibitor aprotinin protected EDCF from inactivation. In addition to confirming the peptidergic nature of EDCF in an isolated heart preparation, their studies ruled out angiotensin II as a potential candidate for EDCF because saralasin, at a dose fully blocking the action of angiotensin II, failed to antagonize the increases in perfusion pressure elicited by EDCF. Gillespie et al.[5] were also the first group to refer formally to the endothelial cell-derived vasoconstrictor factor as EDCF.

A more direct confirmation of our initial findings[6] was reported in 1987 by O'Brien et al.[12] Conditioned media from cultured bovine aortic and pulmonary artery ECs elicited sustained dose-dependent contractions of bovine coronary and pulmonary arteries as well as rat and guinea pig pulmonary arteries and aortae. The endothelial cell-derived substance also caused pronounced constriction when infused into isolated rabbit hearts and rat kidneys. The contractile responses to EDCF were both qualitatively and quantitatively similar to what we had reported and were unaffected by inhibitors of arachidonate metabolism or by antagonists of serotonergic, histaminergic, α-adrenergic, opioid, leukotriene, angiotensin II, or substance P receptors. Cycloheximide abolished the production of EDCF, and trypsin, neutral protease, and verapamil blocked the activity of the vasoconstrictor. They further reported a molecular weight of 3,000, and, in contrast to our findings, they noted that EDCF was heat stable. These investigators were also the first to report that EDCF production and/or release was stimulated by thrombin but unaffected by hypoxia. They also postulated an abluminal release pathway for EDCF that might act as a local autocoid or paracrine mechanism in mediating the tone of the underlying smooth muscle cell.

In summary, the studies of Gillespie et al.[5] and O'Brien et al.[12] fully supported our initial characterization of EDCF. Remarkably good agreement was evident in the data obtained from the three laboratories despite considerable differences in experimental design (different species and vessels, origins of ECs, subcultured vs. primary cultures of ECs, bioassay preparations, and so forth). Collectively, the results established the existence of EDCF as a unique and potent vasoactive peptide that could elicit profound increases in isometric tension in isolated vessels as well as marked reductions in blood flow in the intact circulation.

EARLY CLUES ON THE MECHANISM OF ACTION OF EDCF

Experiments in my laboratory in 1986–1987 focused on the actions and mechanisms of action of partially purified EDCF. These studies were carried out by graduate students, fellows, and collaborators whom I had enlisted. Like our first paper, however, the results of these studies were very difficult to publish as well, but for different reasons. By the time that most of the papers were in review, EDCF (as endothelin) had become commercially available. Our use of whole media conditioned by ECs or partially purified EDCF, rather than the purified peptide, made our results somewhat outdated and less appealing to the reviewers. With great delays, we were often faced with no other alternatives than to resubmit the data elsewhere[10,13] or to repeat the entire set of experiments with purified endothelin before the papers were acceptable for publication.[14] Therefore, early attempts at elucidating the mechanism(s) of action of EDCF on VSM were thwarted by the potential presence of contaminants in the crude EDCF preparations derived from EC-conditioned medium. We had argued, albeit unsuccessfully, that studies using this material were quite relevant because they assessed the action of the peptide in the presence of other "naturally occurring" substances or cofactors expressed by ECs. Despite these constraints, several observations were made during the characterization of EDCF that provided insight into the actions of this peptide. Most of the findings in this regard would subsequently remain valid even when assessed with purified material.

In our initial paper[6] we noted that the vasoconstrictor response required extracellular Ca^{2+} in that vessels were unresponsive to EDCF in Ca^{2+}-free solution containing EGTA. (We also found, but did not publish, that modest yet detectable contractions in Ca^{2+}-free medium could be elicited but only at very high concentrations of EDCF.) The inhibition was fully reversible upon removal of the Ca^{2+} chelator and restoration of normal Ca^{2+} levels. In addition, we noted that verapamil markedly attenuated the VSM response to EDCF. These results led us to conclude that the influx of extracellular Ca^{2+} through voltage-sensitive Ca^{2+} channels was probably an important event underlying the contractile response to EDCF.

In collaboration with Dr. David Pang, we began to evaluate more closely the role of Ca^{2+} in VSM contractions elicited by EDCF.[13] Using the A-10 smooth muscle cell line, EDCF enhanced Ca^{2+} uptake in a time-dependent fashion to a value about sevenfold greater than nonconditioned control medium within 5 min of incubation and was maintained for at least an additional 15 min period. EDCF increased Ca^{2+} uptake into the cultured cells in a dose-dependent fashion. Preheating of EDCF at 80°C for 5 min abolished the enhanced Ca^{2+} uptake. The effects of three Ca^{2+} antagonists—bepridil, verapamil, and nitrendipine—on EDCF-induced changes in contractile force and

Ca^{2+} uptake were then studied. Preincubation with each of the antagonists significantly and proportionally attenuated both force development and Ca^{2+} uptake in response to EDCF. The inhibition of EDCF-induced contractions by verapamil was partially reversible. The ability of the Ca^{2+} antagonists to inhibit contractions in response to EDCF was not altered by the presence of an intact endothelium on the bioassay vessel. These results supported and extended our initial findings that simply suggested that normal contractions in response to EDCF required the influx of extracellular Ca^{2+}.

In collaboration with Drs. Ousterhout and Sperelakis, we investigated the electrophysiological effects of EDCF on freshly isolated rat aortic VSM.[7] Although control nonconditioned medium caused slight membrane depolarization, EDCF significantly depolarized and increased the excitability of VSM, with nearly all of the EDCF-treated cells generating action potentials upon electrical stimulation. The evoked action potentials were attenuated by verapamil. These results, coupled with the Ca^{2+} uptake data, further suggested that a locus of action of the ECDF was depolarization and gating of the voltage-sensitive Ca^{2+} channel.

In light of the key role of Ca^{2+} in the contractile response to EDCF, we also examined the changes in cytosolic Ca^{2+} levels in Fura 2–loaded rat aortic VSM cells before and after treatment with EDCF.[7] Two different patterns were observed. In fresh primary cells with lower resting Ca^{2+} levels, EDCF resulted in a rapid and transient increase in intracellular Ca^{2+} ($[Ca^{2+}]_i$,) followed by quick expulsion or sequestration so that a new higher steady-state level was obtained. We also observed that some VSM cells, particularly those maintained in culture for several days and having lower resting $[Ca^{2+}]_i$ levels, lacked the rapid Ca^{2+} transient and displayed only the sustained increase in $[Ca^{2+}]_i$ in response to EDCF. In these cells, the increment in $[Ca^{2+}]_i$ typically required several minutes to achieve a new steady-state value, which was well-sustained. We hypothesized that the signal transduction mechanisms responsible for the rapid Ca^{2+} transient were absent or deficient in long-term cultures of VSM, or, alternatively, that these cells possessed a more efficient Ca^{2+} extrusion mechanism. On the basis of these preliminary experiments, we concluded that (1) heterogeneity exists in the resting value of $[Ca^{2+}]_i$ in cultured VSM and that cells with lower resting $[Ca^{2+}]_i$ require a higher concentration of EDCF to achieve the same percentage increment in $[Ca^{2+}]_i$, and (2) the increase in VSM steady-state $[Ca^{2+}]_i$ was dependent on the dose of EDCF and temporally preceded both the development of significant force and the influx of extracellular Ca^{2+} observed in other experiments.

Based on the results of the above studies, we felt that the rapid elevation of $[Ca^{2+}]_i$ was most likely attributable to EDCF-induced formation of Ca^{2+}-mobilizing second messengers such as the inositol phosphates. In collaboration with Drs. Rapoport and Stauderman, we investigated the relationship between contraction and phosphatidylinositol hydrolysis induced by EDCF in rat aorta.[14] EDCF elicited both a time- and dose-dependent increase in the incorporation of ^3H-inositol into inositol mono- and polyphosphates that correlated with the development of tension in rat aortic tissue. Using high-performance liquid chromatography, we also found a significant increase in the formation of inositol bis- and trisphosphate within a minute of stimulation with EDCF.

These results indicated that the phospholipase C–mediated hydrolysis of phosphatidylinositol was an early cellular event in the VSM response to EDCF and that inositol trisphosphate formation most likely accounted for the early mobilization of internal stores of Ca^{2+}. In an additional study, the dihydropyridine Ca^{2+} channel agonist [+]-S202-791 similarly increased phosphatidylinositol hydrolysis and elicited contractions of rat aorta. Contractile responses to [+]-S202-791 and EDCF were unaffected by indomethacin. Interestingly, however, the increase in inositol monophosphate labeling caused by [+]-S202-791 was completely blocked by indomethacin, while that obtained in response to EDCF was unaffected by the cyclooxygenase inhibitor. These results suggested that EDCF most likely did not act as a Ca^{2+} agonist and possessed a different mode of action on VSM than that of a dihydropyridine-sensitive Ca^{2+} channel agonist.

In summary, our initial results eventually lead us to propose[7] that EDCF resulted in a prolonged elevation of intracellular Ca^{2+} in VSM and subsequent tonic contractions via a biphasic mechanism: *(1)* a rapid, phospholipase C–mediated increase in phosphatidylinositol turnover with subsequent mobilization of internal Ca^{2+} stores and *(2)* a slower and sustained influx of extracellular Ca^{2+} through voltage-dependent Ca^{2+} channels. We also felt that the slower second phase was initially driven by EDCF-induced activation of a cation channel that was permeable to Ca^{2+} yet was not blocked by the L-type Ca^{2+} antagonists. Activation of this channel would allow for the influx of Ca^{2+} in amounts sufficient to initiate contraction and to depolarize the VSM membrane modestly to a value that would then trigger activation of the voltage-dependent, L-type Ca^{2+} channel. After gating of the L-type channel, a rapid influx of extracellular Ca^{2+} would then initiate the more tonic, sustained phase of the contraction. Finally, the elevated intracellular Ca^{2+} and diacylglycerol formed from the increase in phosphatidylinositol hydrolysis might signify an important role for the activation of protein kinase C in the mechanism of vasoconstriction. The potential involvement of protein kinase C was suggested early on by the similarity in the contractile responses to EDCF and phorbol esters, both being tonic in nature and difficult to wash out. Also, we had observed that H-7, a putative inhibitor of protein kinase C, attenuated the contractile responses to EDCF.

PURIFICATION OF EDCF AND THE DISCOVERY OF ENDOTHELIN

Having received background training in protein biochemistry, I was particularly enthusiastic about pursuing the isolation and purification of EDCF. With the purified molecule, immunochemical techniques could be developed that would then enable studies dealing with the physiological and pathological roles of the peptide. Using conditioned medium from ECs cultured on microcarrier beads and conventional biochemical techniques, we were successful in purifying to homogeneity small quantities of EDCF that were used for amino acid analysis and as an immunogen for the production of antibodies to the peptide. The immediate goal was to develop an immunoaffinity procedure for the preparative scale purification of EDCF. These experiments were very time-consuming because of limitations in equipment and personnel trained in these methodologies. Despite taking about 2 years to complete, we had made good progress in determining the single-chain structure, the terminal sequences, and that the molecule had 21–23 amino acids. We thought that the peptide had four cysteine residues with one located C-terminally, but were uncertain of the intrachain disulfide bond topology. We did not attempt to publish the partial sequence, however, because we thought that only a few more months would be necessary to complete the study.

In about mid-April 1988, a colleague informed me that there was an article[19] in a recent issue of *Nature* (March 31, 1988) by a large group of investigators from Japan that I *might* be interested in. Later that evening I read the paper: "A novel potent vasoconstrictor peptide produced by vascular endothelial cells." Although the title was simple enough, the content of the article was overwhelming. Drs. Yanagisawa, Masaki, and colleagues had done it all—the EC-derived vasoconstrictor had been isolated and the cDNA cloned and sequenced. Endotensin had become endothelin!

The feeling of frustration soon gave way to a sincere appreciation for the quality and quantity of work that had been completed by the Japanese group in such a short time. Indeed, in the few short years since our initial discovery and their seminal work, continuing remarkable progress in endothelin research has been achieved. Despite the competitive nature of research surrounding the discovery of any "new" molecule, the accomplishments of researchers in this area have been fueled by the open sharing of both data and ideas on the part of scientists from around the world. Hopefully this spirit will continue in our efforts to unravel the physiological importance of endothelin.

ACKNOWLEDGMENTS

Portions of the work described herein were supported by research grant HL31543 from the National Institutes of Health, USPHS.

REFERENCES

1. AGRICOLA, K., G. RUBANYI, R. J. PAUL, and R. F. HIGHSMITH. Characterization of a potent coronary artery vasoconstrictor produced by endothelial cells in culture. *Federation Proc.* 43: 899, 1984.
2. DEMEY, J. G., and P. M. VANHOUTTE. Heterogeneous behavior of the canine arterial and venous wall. *Circ. Res.* 51:439–447, 1982.
3. DEMEY, J. G., and P. M. VANHOUTTE. Anoxia and endothelium-dependent reactivity of the canine femoral artery. *J. Physiol. (Lond.)* 335: 65–74, 1983.
4. FURCHGOTT, R. F., and J. V. ZAWADZKI. The obligatory role of endothelial cells in the relaxation of arterial smooth muscle by acetylcholine. *Nature* 288: 373–376, 1980.
5. GILLESPIE, M. N., J. O. OWASOYO, I. F. MCMURTRY, and R. F. O'BRIEN. Sustained coronary vasoconstriction provoked by a peptidergic substance released from endothelial cells in culture. *J. Pharmacol. Exp. Ther.* 236: 339–343, 1986.
6. HICKEY, K. A., G. RUBANYI, R. J. PAUL, and R. F. HIGHSMITH. Characterization of a coronary vasoconstrictor produced by cultured endothelial cells. *Am. J. Physiol.* 248 (*Cell Physiol.* 17): C550–C556, 1985.
7. HIGHSMITH, R. F., D. C. PANG, and R. M. RAPOPORT. Endothelial cell–derived vasoconstrictors: mechanisms of action in vascular smooth muscle. *J. Cardiovasc. Pharmacol.* 13: S36–S44, 1989.
8. HOLDEN, W. E., and E. MCCALL. Hypoxic vasoconstriction of porcine pulmonary artery strips in vitro requires an intact endothelium. *Am. Rev. Respir. Dis.* 127: 301, 1983.

9. HOLDEN, W. E., and E. MCCALL. Hypoxia-induced contractions of porcine pulmonary artery strips depend on intact endothelium. *Exp. Lung Res.* 7: 101–112, 1984.
10. HOM, G. J., R. F. HIGHSMITH, and D. C. PANG. In vivo hemodynamic effects of endothelium-derived constricting factor in the dog. *Drug Dev. Res.* 18: 145–151, 1989.
11. O'BRIEN, R. F., and I. F. MCMURTRY. Endothelial cell supernates contract bovine pulmonary artery rings. *Am. Rev. Respir. Dis.* 129: 337, 1984.
12. O'BRIEN, R. F., R. J. ROBBINS, and I. F. MCMURTRY. Endothelial cells in culture produce a vasoconstrictor substance. *J. Cell. Physiol.* 132: 263–270, 1987.
13. PANG, D. C., N. SPERELAKIS, and R. F. HIGHSMITH. Effects of endothelium-derived constricting factor and calcium antagonists on calcium uptake into aortic vascular smooth muscle cells. *Drug Dev. Res.* 18: 153–164, 1989.
14. RAPOPORT, R. M., K. A. STAUDERMAN, and R. F. HIGHSMITH. Effects of EDCF and endothelin on phosphatidylinositol hydrolysis and contraction in rat aorta. *Am. J. Physiol.* 258 *(Cell Physiol.* 27): C122–C131, 1990.
15. RUBANYI, G., and P. M. VANHOUTTE. Hypoxia releases a vasoconstrictor substance from the coronary arterial endothelium. *Circulation* 70: 122, 1984.
16. RUBANYI, G., and P. M. VANHOUTTE. Hypoxia releases a vasoconstrictor substance from the canine vascular endothelium. *J. Physiol. (Lond.)* 364: 45–56, 1986.
17. VANHOUTTE, P. M. Effects of anoxia and glucose depletion on isolated veins of the dog. *Am. J. Physiol.* 230: 1261–1268, 1976.
18. VAN NEUTEN, J. M., and P. M. VANHOUTTE. Effect of Ca^{2+} antagonist lidoflazine on normoxic and anoxic contractions of canine coronary arterial smooth muscle. *Eur. J. Pharmacol.* 64: 173–176, 1980.
19. YANAGISAWA, M., H. KURIHARA, S. KIMURA, Y. TOMOBE, M. KOBAYASHI, Y. MITSUI, Y. YAZAKI, K. GOTO, and T. MASAKI. A novel potent vasoconstrictor peptide produced by vascular endothelial cells. *Nature* 332: 411–415, 1988.

3

Molecular Biology of Endothelins

PENNY E. PHILLIPS, CHRISTINA CADE, LYNNE H. PARKER BOTELHO AND
GABOR M. RUBANYI

In an attempt to "bioassay" endothelium-derived relaxing factor (EDRF) produced by cultured bovine aortic endothelial cells, it was discovered by Hickey et al.[7] that the conditioned medium contained a potent vasoconstrictor substance (see Chapter 2). Preliminary characterization revealed that the vasoconstrictor was a peptide, and analogous to EDRF, it was named "endothelium-derived contracting factor" (EDCF).[7] This peptidergic EDCF was later isolated by Yanagisawa et al.[20] and identified as a novel 21 amino acid peptide, endothelin. The peptide contains two intrachain disulfide bonds and is homologous to sarafotoxin S6b, a venom peptide isolated from an Israeli asp, *Atractapsis engaddensis*.[11] In our brief summary of present knowledge of the cellular synthesis of this new peptide family, we emphasize the gene structure, regulation of gene expression, and biosynthesis of mature endothelin.

THE ENDOTHELIN GENE FAMILY: DISCOVERY OF
ENDOTHELIN ISOPEPTIDES

Following the initial isolation of endothelin from culture media of porcine aortic endothelial cells, a cDNA library was constructed from the same cells.[20] This library was screened with an oligonucleotide probe encoding amino acid residues 7–20 of endothelin, and the preproendothelin cDNA for the porcine peptide was isolated, cloned, and sequenced. Applying a similar strategy, the cDNA for human endothelin was isolated from a placental cDNA library and sequenced.[10] The existence of a human gene for endothelin was confirmed when Northern blot analysis demonstrated the expression of a single preproendothelin mRNA in cultured endothelial cells derived from human umbilical vein.[10] Comparison of the cDNAs encoding human versus porcine preproendothelin-1 indicated a 79% homology at the nucleic acid level and 69% at the amino acid level.[10] The predicted amino acid sequence of the mature 21 residue peptide following cellular processing of the 203 amino acid precursor preproendothelin-1 was found to be identical for human versus porcine endothelin. The human gene encoding endothelin-1 has since been localized to chromosome 6.[2] Following the sequencing of both human and porcine cDNAs, a preproendothelin-related gene from rat was cloned and sequenced.[19] The predicted 21 amino acid peptide from rat exhibited an identical sequence pattern of cysteines compared with human or porcine endothelin but differed in amino acid composition by six residues. Sequence differences among the original three endothelin peptides from porcine, human, and rat were initially thought to be species variations of a single gene product.

Later, however, it was discovered by low hybridization stringency Southern blot analysis of human, porcine, and rat genomic DNA that there are at least three genes coding for "endothelin-like" sequences in mammalian genomes.[8] This discovery resulted in the renaming of the peptides. The original porcine endothelial cell–derived endothelin that was identical to the human endothelin was termed "endothelin-1," and the rat endothelin was designated "endothelin-3." The human endothelin-3 gene was subsequently isolated and mapped to chromosome 20.[1] The third endothelin gene sequence discovered by the Southern blot analysis of rat, human, and porcine genomic DNA[8] predicted a 21 amino acid peptide with only two amino acid differences from endothelin-1 and was named "endothelin-2." Cloning and sequence analysis of a mouse genome led to the belief that there was a fourth "endothelin-like" gene that was expressed exclusively in the intestine.[15] The peptide deduced from the gene sequence was named "endothelin-β" or "vasoactive intestinal constricting peptide" (VIC). The same gene sequence was also found in rat tissues[15] and is now believed to be an isoform of the endothelin-2 gene and not a fourth gene.[21]

PROTEIN STRUCTURE OF PROENDOTHELIN AND PROCESSING TO THE VARIOUS ENDOTHELIN ISOFORMS

The amino acid sequences and proposed disulfide bonding structure of the four known isoforms of endothelin are shown in Figure 3.1. Human and porcine endothelin-1 have identical sequences. Human and mouse endothelin-2 differ from one another by one amino acid and differ from endothelin-1 by two and

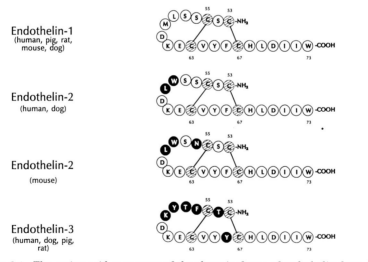

FIGURE 3.1. The amino acid sequences of the three isoforms of endothelin from different species. Cysteines believed to form disulfide bonds are represented by cross-hatched circles. The filled circles represent amino acid sequence changes compared with the endothelin-1 sequence.

MOLECULAR BIOLOGY OF ENDOTHELINS 33

three amino acids, respectively. Human and rat endothelin-3 have identical sequences, but differ from endothelin-1 by six amino acids.

The corresponding amino acid sequences and disulfide bonding structures for the four known isoforms of the immediate biological precursor of endothelin, referred to as "big endothelin" (proendothelin), are shown in Figure 3.2. The sequence of the big endothelin-1 isoforms from the human gene versus the porcine gene vary by two amino acids. The porcine isoform contains 39 amino acids, whereas the human isoform contains 38 amino acids with one additional amino acid substitution in the common 37 residues. The big endothelin-2 isoform from the human gene contains 37 residues with eight amino acid changes compared with human big endothelin-1. The big endothelin-3 isoform from both human and dog are identical to but somewhat larger than the other three big endothelin isoforms, containing 41 amino acids compared with 37–39 amino acids. There are, in addition, 12 amino acid substitutions within the first 38 amino acids of big endothelin-3 compared with human big endothelin-1. The most interesting difference among the sequences for the various proendothelin isoforms is that the proposed processing sites are different with a Trp^{21}-Ile dipeptide in big endothelin-3 versus a Trp^{21}-Val dipeptide for the others.

ENDOTHELIN GENE STRUCTURE

The genomic clones for human endothelin-1 and endothelin-3 have been isolated[1,2,9] and partially sequenced (Fig. 3.3). The endothelin-1 gene has

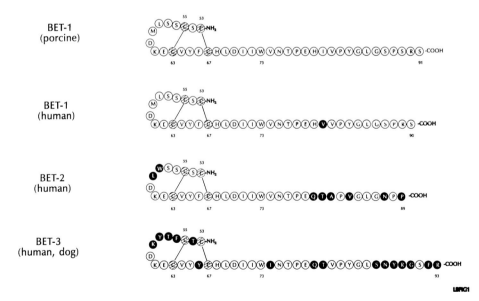

FIGURE 3.2. The amino acid sequences of the three isoforms of the immediate biological precursor of endothelin, big endothelin, from different species. Cysteines believed to form disulfide bonds are represented by cross-hatched circles. The filled circles represent amino acid sequence changes compared with the big endothelin-1 sequence.

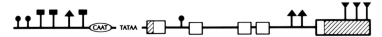

FIGURE 3.3. The proposed structure of the human preproendothelin-1 gene. In the 6.8 kb structure, exons are indicated by open boxes, the 5' and 3' untranslated sequences are indicated by cross-hatched areas, and potential regulatory sequences are indicated by ●, acute phase regulatory element; ■, AP-1/JUN elements; ▲, NF-1–binding sites; and ▼, AUUA sequences. See text for further details.

been more thoroughly characterized than has the gene for endothelin-3, but both genes contain five exons in approximately 6.8 kb.[1,2,9] Exon 1 encodes the first 21 amino acids of preproendothelin, exon 2 encodes the mature endothelin-1 peptide and the first four amino acids of big endothelin, exon 3 encodes the "endothelin-like" peptide, exon 4 encodes residues 131–178 of preproendothelin, and exon 5 contains the sequence coding for the last 34 amino acids and the 3' untranslated region of preproendothelin.[2,9] The endothelin-1 transcription start site, mapped by S1 nuclease protection, is located 31 bp 3' of a TATAA box,[9] and a CAAT box is located 65 bp 5' to the TATAA box.[2,9] Analysis of the endothelin-1 genomic sequence suggests several possible regulatory elements located 5' to the coding sequence and within introns 1 and 4. Two copies of the hexanucleotide "acute regulatory phase" element, thought to mediate induction of message under acute physiological stress in vivo, are located at nucleotides −2868 and −1425.[9] Three sequence repeats are found at nucleotides −657, −367, and −109 that match at least seven bases of the consensus octanucleotide sequence of an AP-1/JUN–binding sequence element involved in phorbol ester induction.[9] A 10 bp sequence idential to the core of a 20 bp (NF-1) element shown in other systems to confer transforming growth factor (TGF-β) induction is located at the 3' end of the first exon.[2] Similar sequences are found in intron 4 and in the 3' flanking region.[9] Two additional copies of the acute regulatory phase element also have been identified in introns 1 and 4.[9] The 3' region of the endothelin-1 gene contains AU-repeat sequences that are commonly thought to be involved in highly specific translation-dependent destabilization of mRNA[2,9] providing an additional regulatory mechanism by which the cellular level of endothelin mRNA may be controlled by post-transcriptional message degradation.

REGULATION OF ENDOTHELIN GENE EXPRESSION

Endothelin gene expression and translation into an active protein product has been studied most extensively in cultured endothelial cells. Little or no mature endothelin or big endothelin has been found intracellularly, and the rates of release into the cell culture media in the absence of serum over a 24 h period are linear, suggesting that the release is constitutive.

Studies conducted in our laboratory[3] to determine the rates of "basal" proendothelin-1 and endothelin-1 release into the cell culture media in the absence of serum over a 24 h period imply that the total amount of proendothelin-1 and endothelin-1 as well as the ratio of the two vary among different cell preparations from different species and tissue sources (Table 3.1). It was also

TABLE 3.1. Basal rates of endothelin release from endothelial cells[a]

Cell Line	Species	Tissue source	ET-1[b]	BET-1	Total
CKEND-1	Bovine	Atrial endocardium	0.3–6.5	0.06–4.2	5.2 ± 0.6 (6)
CKEC 1	Bovine	Pulmonary artery	2.4–5.8	0.6–1.97	4.6 ± 0.6 (6)
CKEND-2A	Bovine	Atrial endocardium	0.2–4.43	0.2–4.5	3.6 ± 0.6 (5)
BAEC	Bovine	Aorta	2.4–2.9	1.2–1.5	4.0 ± 0.4 (3)
PAEC	Porcine	Thoracic aorta	0.02–0.83	5.7–6.5	6.1 ± 0.4 (3)
CKEND 2V	Bovine	Ventricular endocardium	Trace amounts	—[c]	—[c]

[a]Basal rates of release are reported as pmol/10^6 cells/24 h. C. Cade, unpublished data. ET-1, endothelin-1; BET-1, big ET-1. Values in parentheses are number of experiments.
[b]Because of cross-reactivity of the antisera employed, endothelin-1 values (pmol/10^6 cells/24 h) may represent a mixture of endothelin-1 and endothelin-2 isoforms.
[c]Levels of big endothelin-1, if any, were below the detection limit of assay used.

found that both the total amount and the ratio of proendothelin-1 to endothelin-1 may vary significantly, depending on the confluency of the endothelial cells (Fig. 3.4) and the composition and the pH of the cell culture media (Fig. 3.5).[3] However, a cultured cell system might not accurately reflect the in vivo situation, and the constitutive "basal" release of endothelin may only represent release from "activated" cells or endothelial cells that lack their physiological environment (e.g., presence of smooth muscle, extracellular matrix, and so forth).

The induction of endothelin-1 mRNA (Table 3.2) and the rate of peptide release (Table 3.3) have been found to be increased by a growing number of agents or mechanical stimuli, such as thrombin, TGF-β, angiotensin II, vasopressin, hemodynamic shear stress, interleukin-1 (IL-1), phorbol esters, and calcium ionophores.[5,9,12–14,16–18,21–23] Yoshizumi et al.[22] demonstrated time- and dose-dependent induction by interleukins of endothelin-1 mRNA with concomitant appearance of elevated concentrations of the peptide in the media from cultured porcine endothelial cells. Human recombinant IL-1β was less potent than IL-1α at both levels of gene expression. Many of the reported inducers of endothelin production are known to promote intracellular calcium accumula-

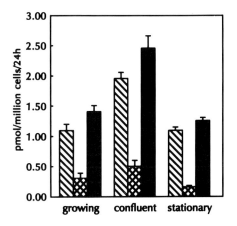

FIGURE 3.4. The release of endothelin-1 and big endothelin-1 from cultured bovine pulmonary artery endothelial cells at various stages of growth. Radioimmunoassays were done on culture medium exposed to a monolayer of cells for 24 h. Endothelin-1 levels (pmol/million cells/24 h) are shown in the diagonally hatched bars, big endothelin-1 levels are shown in the cross-hatched bars, and the total amount of endothelin and big endothelin is shown in the vertically hatched bars. The concentrations of endothelin-1 and big endothelin-1 were normalized by cell number. The values are the mean ± SEM of four individual experiments with duplicate samples per experiment.

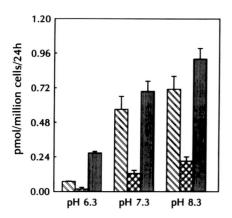

FIGURE 3.5. The release of endothelin-1 and big endothelin-1 by cultured endothelial cells and the dependence on the pH of the cell culture media. Endothelial cells prepared from bovine pulmonary arteries were grown to confluence in medium 199 plus 10% serum. Medium was replaced by serum-free 10 mM HEPES-buffered medium 199 at pHs 6.3, 7.3, and 8.3 for 24 h at 37°C in a humid non-CO_2 incubator. The medium samples were collected after 24 h and radioimmunoassayed for endothelin-1 and big endothelin-1. Quantitation of endothelin-1 (pmol/million cells/24 h) is represented by the diagonally hatched bars, the quantitation of big endothelin-1 (pmol/million cells/24 h) is represented by the cross-hatched bar, and the total endothelin-1 and big endothelin-1 is represented by the vertically hatched bars. The values represent the mean ± SEM for four separate experiments with duplicate samples per experiment.

tion and/or protein kinase C activation, which may act at the level of transcription and/or translation. The increase in mRNA induced by agents such as phorbol esters and TGF-β has been suggested to be due to increased transcription of the endothelin-1 gene, possibly via interaction of intracellular mediators with *cis*-nucleotide sequences.

CELLULAR PROCESSING OF PREPROENDOTHELIN TO
MATURE ENDOTHELIN

The cDNA sequence of the gene encoding for endothelin-1 implied that the 21 amino acid, biologically active peptide was synthesized as a much larger preproprotein precursor that required biochemical processing for conversion to the active component (Fig. 3.6).[20] The original preproendothelin-1 was postulated to be a 203 amino acid peptide[20] containing the endothelin sequence in residues 53–73 and a second "endothelin-like" sequence in residues 110–130. Subsequent isolation and sequencing of the genes for endothelin-2 and endothelin-3 revealed that these genes encoded isoforms of similar size with significant sequence homology.[10,11,20] The peptide derived from the "endothelin-like" sequence in the endothelin-1 gene has been shown to be inactive in several biological assays.[4]

Endothelin-1 (Cys^{53}-Trp^{73}) and proendothelin-1 (Cys^{53}-Ser^{92}), referred to as "big endothelin-1," have been shown to accumulate in conditioned medium from endothelial cells.[20] This discovery led to a postulated cellular processing scheme that involved cleavage of the 203 preproendothelin to the 39 amino acid proendothelin and then to the 21 amino acid endothelin. This scheme invokes N-terminal cleavage of amino acids 53 and 94 by dibasic endopeptidases, with subsequent carboxypeptidase trimming of residues 93 and 94 to give big endothelin. A novel protease, putatively named "endothelin-converting en-

TABLE 3.2. Modulators of endothelin gene transcription: relative increase or decrease in endothelin-1 mRNA[a]

Agent	Concentration	Cells	Species	mRNA levels[b]	Reference
A23187	1 μM	Aorta ECs	Porcine	+	20
ADP	1 μM	Aorta ECs	Porcine	0	12
Adrenaline	1 μM	Aorta ECs	Porcine	+	20
EGF	15 ng ml^{-1}	Amnion cells	Human	+	17
EGF	15 ng ml^{-1}	Umbilical vein ECs	Human	0	17
H7	250 μM	Retinal microvessel ECs	Human	−	13
IL-1α	800 pg ml^{-1}	Amnion cells	Human	+	17
IL-1α	800 pg ml^{-1}	Umbilical vein ECs	Human	0	17
IL-1α	10–100 ng ml^{-1}	Aorta ECs	Porcine	+	22
IL-1β	10–300 ng ml^{-1}	Aorta ECs	Porcine	+	22
PDGF	8 ng/ml	Aorta ECs	Porcine	0	12
Phenylephrine	1 μM	Retinal microvessel ECs	Human	−	13
Phorbol[d]	1 μM	Retinal microvessel ECs	Human	+	13
Serotonin	10 μM	Aorta ECs	Porcine	0	12
Shear stress	Low	Aorta ECs	Porcine	+	22
TGF-β	10–300 pM	Aorta ECs	Porcine	+	12
TGF-β	5 ng ml^{-1}	Retinal microvessel ECs	Human	+	13
TPA	0.5 μM	Umbilical vein ECs	Human	+	9
Thrombin	2 U ml^{-1}	Aorta ECs	Porcine	+	20
Thromboxane analog[e]	0.1 μM	Aorta ECs	Porcine	0	12

[a]ADP, adenosine diphosphate; EGF, epidermal growth factor; ECs, endothelial cells; IL-1, interleukin-1; PDGF, platelet-derived growth factor; TFG-β, transforming growth factor; TPA, 12-O-tetradecanoylphorbol-13-acetate.
[b]Cultured cells were exposed to various concentrations of test agents. For Northern blot analysis, total cellular RNA was size fractionated and analyzed for mRNA using cDNA probes for human preproendothelin-1. The data are presented as a relative increase (+), decrease (−), or no change (0) in mRNA detected.
[c]1-(5-Isoquinolinesulfonyl)-2-methylpiperazine.
[d]Phorbol-12,13-dibutyrate.
[e]9,11-Epithio-12-methanothromboxane A$_2$.

TABLE 3.3. Inducers of endothelin production[a]

Agent	Concentration	Cells	Species	Isoform[b]	Relative increase or decrease[c]	Reference
Calf serum	10%	Adrenal cortex capillary ECs	Bovine	ET	+	6
BSA	0.1%	Adrenal cortex capillary ECs	Bovine	ET	+	6
Thrombin	2 U/ml^{-1}	Aorta ECs	Porcine	ET	+	18
	2 U/ml^{-1}	Aorta ECs	Porcine	BET	+	18
	2 U/ml^{-1}	Hep G2 cells	Human	BET	+	18
	2 U/ml^{-1}	Hep G2 cells	Human	ET	−	18
	3 U/ml^{-1}	Aorta ECs	Porcine	ET	+	16
	10 U/ml^{-1}	Aorta ECs	Porcine	ET	++	16
[Phe^2, Orn^8]vasopressin	1–100 nM	Endometrial cells	Rabbit	ET	+	14
[Tyr^4, Gly^7]oxytocin	1–100 nM	Endometrial cells	Rabbit	ET	+++	14
TGF-β	2.5 ng/ml^{-1}	Aorta ECs	Porcine	ET	+	18
	2.5 ng/ml^{-1}	Aorta ECs	Porcine	ET	+	18
	2.5 ng/ml^{-1}	Hep G2 cells	Human	ET	+	18
	2.5 ng/ml^{-1}	Hep G2 cells	Human	BET	+	18
	10–300 pM	Aorta ECs	Porcine	ET	+	12
	10 nM	Adrenal cortex capillary ECs	Bovine	ET	++	6
Shear stress		Aorta ECs	Porcine	ET	+	23
TPA	1 μM	Carotid artery ECs	Bovine	ET	+	15
Ionomycin	1 μM	Carotid artery ECs	Bovine	ET	++	15
TPA and ionomycin	1 μM	Carotid artery ECs	Bovine	ET	+++	15
Angiotensin II	1 μM	Carotid artery ECs	Bovine	ET	++	15
	0.1 μM	Carotid artery ECs	Bovine	ET	+	15
	0.01 μM	Carotid artery ECs	Bovine	ET	+++	15
Arginine vasopressin	10 μM	Carotid artery ECs	Bovine	ET	++	15
	1 μM	Carotid artery ECs	Bovine	ET	+	15
	0.1 μM	Carotid artery ECs	Bovine	ET	+	15
IL-1α	10–100 ng/ml^{-1}	Aorta EC	Porcine	ET	++	22
IL-1β	10–300 ng/ml^{-1}	Aorta EC	Porcine	ET	+	22

[a]Confluent monolayers of cultured cells were exposed to various concentrations of the test agents. Conditioned media were subsequently collected and assayed for immunoreactive endothelin by ELISA or RIA. BET, big endothelin; BSA, bovine serum albumin; ECs, endothelial cells; ET, endothelin; TGF-β, transforming growth factor; TPA, 12-O-tetradecanoylphorbol-13-acetate; IL-1, interleukin-1.

[b]In many cases endothelin-1 may represent a mixture of endothelin-1 and/or endothelin-2 and/or endothelin-3, because in most cases cross-reactivity of antiserum with each isoform was not assessed.

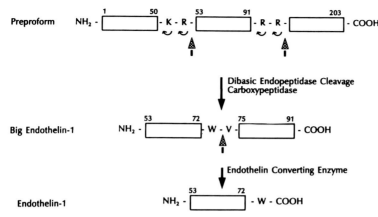

FIGURE 3.6. The proposed proteolytic processing pathway for the conversion of preproendothelin to mature endothelin. The preproform of porcine–human endothelin-1, which contains 203 amino acids, is believed to be converted to the 39 amino acid form referred to as "big endothelin-1" by dibasic endopeptidases (cross-hatched arrows) and carboxypeptidases (small curved arrows) as shown. Big endothelin-1 is then cleaved at the Trp^{73}-Val^{74} bond by a specific endopeptidase referred to as "endothelin-converting enzyme" (cross-hatched arrow). The final product is the 21 amino acid peptide endothelin, containing amino acids Cys^{53} to Trp^{73}. *Solid arrows* indicate the sequence of processing.

zyme," then cleaves Trp^{21}-Val^{22} of proendothelin-1 to form the mature endothelin peptide containing amino acids Cys^{53} to Trp^{73}. Characterization and localization of this putative endothelin-converting enzyme will be important for determining the physiological relevance of the cellular processing of preproendothelin to the biologically active endothelin.

REFERENCES

1. BLOCH, D., R. EDDY, T. SHOWS, and T. QUERTERMOUS. cDNA cloning and chromosomal assignment of the gene encoding endothelin-3. *J. Biol. Chem.* 264: 18156–18161, 1989.
2. BLOCH, D., S. FRIEDRICH, M. LEE, R. EDDY, T. SHOWS, and T. QUERTERMOUS. Structural organization and chromosomal assignment of the gene encoding endothelin. *J. Biol. Chem.* 264: 10851–10857, 1989.
3. BOTELHO PARKER, L. H. Tissue specificity of endothelin synthesis and binding. *FASEB Endothelin Symposium, Washington, D.C., April 1990.*
4. CADE, C., W. C. LUMMA, JR., R. MOHAN, G. M. RUBANYI, and L. H. PARKER BOTELHO. *Life Sci.* 47: 2097–2103, 1990.
5. EMORI, T., Y. HIRATA, K. AHTA, M. SHICHIRI, and F. MARUMO. Secretory mechanism of immunoreactive endothelin in cultured bovine endothelial cells. *Biochem. Biophys. Res. Commun.* 160: 93–100, 1989.
6. HEXUM, T., C. HOEGER, J. RIVIER, A. BAIRD, and M. BROWN. Characterization of endothelin secretion by vascular endothelial cells. *Biochem. Biophys. Res. Commun.* 167: 294–300, 1990.
7. HICKEY, K. A., G. M. RUBANYI, R. J. PAUL, and R. F. HIGHSMITH. Characterization of a coronary vasoconstrictor produced by endothelial cells in culture. *Am. J. Physiol.* 248 (*Cell Physiol.* 17): C550–C556, 1985.
8. INOUE, A., M. YANAGISAWA, S. KIMURA, Y. KASUYA, T. MIYAUCHI, K. GOTO, and T. MASAKI. The human endothelin family: three structurally and pharmacologically distinct isopeptides predicted by three separate genes. *Proc. Natl. Acad. Sci. USA* 86, 2863–2867, 1989.
9. INOUE, A., M. YANAGISAWA, Y. TAKUWA, Y. MITSUI, M. KOBAYASHI, and T. MASAKI. The human preproendothelin-1 gene. *J. Biol. Chem.* 264, 14954–14959, 1989.
10. ITOH, Y., M. YANAGISAWA, S. OHKUBO, C. KIMURA, T. KOSAKA, A. INOUE, N. ISHIDA, Y. MITSUI,

H. ONDA, M. FUJINO, and T. MASAKI. Cloning and sequence analysis of cDNA encoding the precursor of a human endothelium-derived vasoconstrictor peptide, endothelin: identity of human an porcine endothelin. *FEBS Lett.* 231, 440–444, 1988.

11. KLOOG, Y., and M. SOKOLOVSKY. Similarities in mode and sites of action of sarafotoxins and endothelins. *Trends Pharmacol. Sci.* 10: 212–214, 1989.
12. KURIHARA, H., M. YOSHIZUMI, T. SUGIYAMA, F. TAKATU, M. YANAGISAWA, T. MASAKI, M. HAMAOKI, H. KATO, and Y. YAZAKI. Transforming growth factor-β stimulates the expression of endothelin mRNA by vascular endothelial cells. *Biochem. Biophys. Res. Commun.* 159: 1435–1440, 1989.
13. MACCUMBER, M., C. ROSS, B. GLASER, and S. SNYDER. Endothelin: visualization of mRNAs by in situ hybridization provides evidence for local action. *Proc. Natl. Acad. Sci. USA* 86: 7285–7289, 1989.
14. ORLANDO, C., M. L. BRANDI, A. PERI, S. GIANNINI, G. FANTONI, E. CALABRESI, M. SERIO, and M. MAGGI. Neurohypophyseal hormone regulation of endothelin secretion from rabbit endometrial cells in primary culture. *Endocrinology* 126: 1780–1782, 1990.
15. SAIDA, K., Y. MITSUI, and N. ISHIDA. A novel peptide, vasoactive intestinal constrictor, of a new (endothelin) peptide family. *J. Biol. Chem.* 264, 14613–14616, 1989.
16. SCHINI, V. B., H. HENDRICKSON, D. M. HEUBLEIN, J. C. BURNETT, JR., and P. M. VANHOUTTE. Thrombin enhances the release of endothelin from cultured porcine aortic endothelial cells. *Eur. J. Pharmacol.* 165: 333–334, 1989.
17. SUNNERGREN, K., R. WORD, J. SAMBROOK, P. MACDONALD, and M. CASEY. Expression and regulation of endothelin precursor mRNA in avascular human amnion. *Mol. Cell. Endocrinol.* 68: R7–R14, 1990.
18. SUZUKI, N., H. MATSUMOTO, C. KITADA, S. KIMURA, and M. FUJINO. Production of endothelin-1 and big endothelin-1 by tumor cells with epithelial-like morphology *J. Biochem.* 106: 736–741, 1989.
19. YANAGISAWA, M., A. INHOUE, T. ISHIKAWA, Y. KASUYA, S. KIMURA, K. NAKAJIMA, T. X. WATANABE, S. SAKAKIBARA, K. GOTO, and T. MASAKI. Primary structure, synthesis and biological activity of rat endothelin, an endothelium-derived vasoconstrictor peptide. *Proc. Natl. Acad. Sci. USA* 85: 6964–6968, 1988.
20. YANAGISAWA, M., H. KURIHARA, S. KIMURA, Y. TOMOBE, M. KOBAYASHI, Y. MITSUI, K. GOTO, and T. MASAKI. A novel potent vasoconstrictor peptide produced by vascular endothelial cells. *Nature* 332: 441–415, 1988.
21. YANAGISAWA, M., and T. MASAKI. Endothelin, a novel endothelium-derived peptide. *Biochem. Pharmacol.* 38, 1877–1883, 1989.
22. YOSHIZUMI, M., H. KURIHARA, T. MORITA, T. YAMASHITA, Y. OH-HASHI, T. SUGIGAMA, F. TAKAKU, M. YANAGISAWA, T. MASAKI, and Y. YSZAKI. Interleukin-1 increases the production of endothelin-1 by cultural endothelial cells. *Biochem. Biophys. Res. Commun.* 166: 324–329, 1990.
23. YOSHIZUMI, M., H. KURIHARA, T. SUGIYAMA, F. TAKATU, M. YANAGISAWA, T. MASAKI, and Y. YAZAKI. Hemodynamic shear stress stimulates endothelin production by cultured endothelial cells. *Biochem. Biophys. Res. Commun.* 161: 859–864, 1989.

4

Endothelin Structure and Structure–Activity Relationships

PAUL W. ERHARDT

There are several interesting chemical features that are associated with the primary structures of the endothelins[73] and sarafotoxins (e.g., SRT-b).[33] Certain physical and spectroscopic properties of endothelin-1 are provided in order to consider these features and the conformational studies that have been conducted specifically on this molecule. A working model is developed that portrays the structural significance and functional implications of the chemical features derived during this initial analysis. The biological activity of the endothelins as it pertains to their differences in chemical structure, i.e., endothelin structure–activity relationships (SARs) is then reviewed. The latter will be accomplished by considering separately the results obtained from receptor-binding studies, in vitro studies with isolated tissues, and, finally, in vivo studies in intact animals. The data for binding and in vitro studies are standardized and tabulated to allow for their quick perusal. It should be noted at the outset that generation of endothelin SARs will be complicated by the likelihood that different tissues contain different ratios of endothelin receptor subtypes, with each subtype presumably having its own set of SARs. For the purpose of provoking further thought and experiment, a fair amount of speculation is offered at various points in this chapter. In the summary, general SAR statements are derived and assessed relative to the working model.

STRUCTURES OF THE ENDOTHELINS AND SARAFOTOXINS

The amino acid sequences for the "human endothelin family"[26] endothelin-1, endothelin-2, and endothelin-3, for the mouse vasoactive intestinal contractor (VIC, or endothelin-β)* substance,[27,54] and for the snake sarafotoxin family as represented by SRT-b are shown in Figure 4.1. This figure is color coded to highlight the presence of nonpolar, polar, and ionizable (extremely polar) amino acid substituents. A high degree of homology, 80 percent or more when the comparisons "include conservative substitutions,"[60] is apparent among the five compounds. Probably the most noticeable and interesting feature, how-

*While there seems to be some reluctance to adopt endothelin-β fully into the endothelin family, this compound should not be abandoned from a structural point of view. Biologically, it constricts isolated porcine coronary artery with about one-third less potency than endothelin-1 but has more pronounced effects than endothelin-1 on ileum preparations from the mouse and guinea pig.[27]

ever, is that these relatively small, 21 amino acid peptides are tightly constrained by the presence of two disulfide bonds. (In addition to syntheses involving spontaneous folding–coupling of all four Cys residues,[72] the location of the two disulfide bonds, as shown in Figure 4.1, has been confirmed by syntheses in which selective protection–deprotection was employed for each pair to ensure their unambiguous coupling.[25,37,48]) The exact preservation of these disulfide bonds between Cys-1 and Cys-15 and between Cys-3 and Cys-11 for all of the compounds "points to their importance as structural elements."[34] However, there is something even more intriguing about the disulfide bonds. The presence of two such bonds in this close arrangement seems redundant for simply achieving the resulting conformational fold. It would appear that the Cys-1, Cys-15 bond could, in singular fashion, provide approximately the same net conformation. This prompts the question as to what other role the Cys-3, Cys-11 bond might be playing. One can speculate that during anabolism, because they are in closer sequential proximity, the Cys-3 and Cys-11 sulfhydryl groups may react first and thereby begin to fold the molecule. In this scenario the Cys-3, Cys-11 bond is present to assist the subsequent formation of the more remote and important Cys-1, Cys-15 bond. Alternatively, one can speculate that because of the propensity for disulfides to react with nucleophiles, the Cys-3, Cys-11 bond itself plays some active role at the receptor during binding or activation. This situation could proceed by reaction with a Cys sulfhydryl group that is present as part of the receptor surface. Depending on the dynamics (reversibility) and associated kinetics of such a process, one would imagine that if the second possibility were true the endothelins might be relatively slow to dissociate from their receptors and would be rather difficult to wash out during biochemical and in vitro experiments.

The carboxy-terminal tail from His-16 to Trp-21 represents a second interesting structural feature that is highly conserved in all of the compounds. Some investigators have referred to this region as the "hydrophobic tail."[34,60] However, the His, Asp- and carboxy-terminal group all prefer to be ionized at physiological pH. Therefore, it also seems appropriate to regard this tail as having three distinct areas of very high polarity or dipoles that are insulated by adjacent nonpolar moieties. Speculatively, the neighboring nonpolar substituents by virtue of their inherent hydrophobicity may repel water molecules from the three key dipoles and, perhaps, force them to interact with one another. This would actually tend to ensure the polar nature of these groups as they approach the receptor in a manner unhindered by solvating water molecules. The interesting arrangement and possible significance of these three polar sites will be discussed further in the section on endothelin-1 theoretical conformation.

The amino-terminus, and in particular region 4 through 7, represents the area where there is considerable variation among the endothelins and SRT-b. This has been noted previously[34] and can be expected to cause variation in the biological activity observed for these compounds. The present analysis emphasizes instead the familial relationship of these compounds and attempts to discern their similarities in this region. In this regard, endothelin-1 and endothelin-2 are essentially identical because the substitutions at positions 6 and 7 in endothelin-2 still represent nonpolar amino acids (conservative substitu-

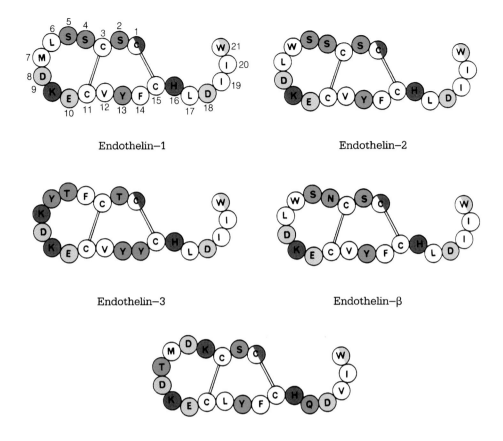

FIGURE 4.1. Amino acid sequences for the endothelins[27,54,73] and sarafotoxin-b.[2] Yellow lines indicate disulfide bonds. Clear circles indicate nonpolar amino acids; green indicates polar, uncharged amino acids; blue indicates amino acids that would tend to be anions at physiological pH; and red indicates amino acids that would tend to be cations. Endothelin-1 is associated with human, porcine, dog, and rat; endothelin-2 with human; endothelin-3 with human and rat; endothelin-β with mouse; and SRT-b represents the family of snake venom toxins from *A. engaddensis*.

tions). Taken together, endothelin-1 and endothelin-2 provide a common theme of moderate polarity for this region in that several polar amino acids are dispersed with several nonpolar amino acids. Endothelin-β is also encompassed by this general assessment because it is identical to endothelin-2 except for substitution at position 4 with a somewhat less polar amino acid while the neighboring Ser (polar) is still retained. Using the theme of moderate polarity for this region, it can be further argued that the basic Lys-7, unique to endothelin-3, could be hydrogen bonded as a proton acceptor, with its immediate neighbor Tyr-6 serving as the proton donor. The latter is also unique to endothelin-3 and, because it contains a weakly acidic phenolic hydroxy group, represents an ideal partner for such an interaction, especially when situated in an adjacent position. The net result of this interaction would be to buffer or attenuate the dramatic polarity that an ionized Lys-7 would otherwise bestow to this region of endothelin-3. As perceived by solvent or by a receptor surface, the net polarity of this pair of amino acids would thus tend to be lower because of the formation of an internal hydrogen bond or salt. Note that the same argument can be applied to the ionizable Lys-4 and Asp-5 that are unique to SRT-b and again, perhaps more than just coincidentally, are located in an adjacent relationship in the sequence. Thus, despite the high degree of amino acid variation in the 4–7 region, in a general sense all five molecules can (arguably) be regarded as having a somewhat similar, moderately polar amino-terminus sequence.

Like the carboxy-terminus, the two remaining chemically distinct regions spanning positions 8–10 and 11–15 are highly conserved within all of the compounds. The alternating charges associated with Asp-8, Lys-9, and Glu-10 represent a very interesting arrangement. There is a net charge of -1 for this trio of amino acids and for the overall charges of the entire endothelin-1, endothelin-2, endothelin-β, and SRT-b molecules as well. It seems reasonable to assume that the pronounced polarities across the 8–10 positions have a counterpart at endothelin receptors. Speculatively, the receptors could bear a positive charge in this area to greet and favorably interact with this region of the ligand. Furthermore, such a dramatic, energetically favorable interaction (e.g., salt formation) may help to provide for initial recognition and/or binding of the endothelins as they are sought from the biological milieu by their receptors. Finally, the sequence Cys-11 to Cys-15 represents a relatively nonpolar region in all of the molecules. Interestingly, however, a moderately polar Tyr is always conserved at position 13. As suggested previously, the nonpolar amino acids immediately flanking the Tyr-13 (or Tyr-13,14 pair as present in endothelin-3) may help to prohibit water molecules from masking this moderately polar amino acid. Tyr-13 could then better interact with the receptor as a hydrogen bond donor, acceptor, or perhaps as a weak acid.

PHYSICAL AND SPECTROSCOPIC PROPERTIES OF ENDOTHELIN-1

We had initial concerns about being able to obtain samples of endothelin-1 that contained only the desired disulfide linkages, especially because commercial syntheses typically rely on simultaneous formation of both bonds subsequent to a solid-phase synthesis. With time, however, high-performance liquid chromatography (HPLC) examinations* of endothelin-1 samples from five different suppliers and among different lots from the same suppliers have established that this is not a problem. Likewise, aqueous solutions of endothelin-1 are stable throughout the course of a day's experiments when the solutions are maintained at ~4°C.

Unfortunately, it appears that solid endothelin-1 does not readily lend itself toward crystallization in a manner that is suitable for X-ray (solid-state conformational) analysis. Probably contributing to this situation is its "amphipathic nature."[57] Despite the significant regions of hydrophobicity, these molecules, other than endothelin-3, which is neutral, bear a net negative charge[34] and as previously discussed also have several regions of considerable local polarity (positive- and negative-charged groups). A hydrophobicity index, which was used to average values for hexapeptide units along the human and porcine big endothelin sequences,[28] also demonstrates this amphipathic character for endothelin-1. Values ranging from -1 to 0 to $+1$ were found through the endothelin-1 portion that is common to both big endothelin molecules.[28] Exploiting the amphipathic character, some investigators have made initial progress by growing crystals at the interface between a two-phase solvent system consisting of water and ether.[57]

Efforts to derive conformational information while the endothelin-1 molecule is in the solution state have employed spectroscopic techniques and have been more successful. Particularly noteworthy are the results from nuclear magnetic resonance (NMR) experiments in which there seems to be general agreement among at least five different groups of investigators.[3,8,16,35,55] These studies suggest that in endothelin-1 the region Lys-9 to His-16 exists in an arrangement that approaches an α-helix while the carboxy-terminus appears to exhibit considerable conformational flexibility. Furthermore, preliminary circular dichroism (CD) studies[3] have suggested that endothelin-1 may be able to form a helical structure that is stabilized by electrostatic interactions. Our initial work in the CD area* tends to support the reported studies.

---

*An HPLC retention time of ~14.8 min is obtained for natural endothelin-1 after a 10 μl injection of a 0.5 mg/ml solution on a Whatman Partisil-5-ODS-3 (250 × 4.6 mm) column having a flow rate of 1.5 ml/min, using a gradient mobile phase comprised of (A) 0.1% trifluoroacetic acid (TFA) in H_2O and (B) 0.1% TFA in CH_3CN, and following the program 80:20 (A:B) to 30:70 (A:B) over 30 min. Detection was by UV absorbance at 220 nm. Separation of natural endothelin-1 (1,15; 3,11) from its 1,11; 3,15 disulfide isomer can be accomplished with an isocratic system comprised of 30% CH_3CN in 0.1% TFA. Other analytical specifications for endothelin-1 samples included appropriate amino acid analysis; peptide content typically greater than 80%; and appropriate FAB MS with $(M + H)^+$ at 2491 and 2492 (^{13}C contribution). While the latter technique has been used[43] to characterize disulfide arrangements, at this point we have not been able to make unambiguous assignments employing only this technique for the endothelin system. We have also initiated studies that employ capillary zone electrophoresis (CZE)[12] to evaluate the integrity of endothelin-related materials and samples. The electrophoretic mobility of endothelin-1 is 1.64×10^{-4} cm^2/V s after its 2 s vacuum injection onto a 72 cm capillary (50 cm to a 200 nM UV detector) buffered with sodium citrate (20 mM, pH 2.5) at 35°C while under a voltage of 20,000 V supplied by an ABI Model 270a CZE instrument. Finally, our preliminary results employing circular dichroism (CD) to study the conformation of endothelin-1 indicate that endothelin-1 has a distinct negative CD band at 209 nM when in CH_3OH and two negative bands, at 207 and 224 nM, when in CF_3CO_2H where the latter solvent is thought to "induce helix" formation.[3]

THEORETICAL CONFORMATION STUDIES OF ENDOTHELIN-1

A preliminary report[15] that employed molecular dynamic (MD) studies (gas-phase conformations) to search the entire conformational space accessible to endothelin-1 determined that there are four conformational families that merit serious consideration. One of these families contains a "helical structure extending from residue 9 through 13."[15] Its lowest energy conformer is within 15 Kcal of the low-energy conformers representing the other families. Based on this work and on the aforementioned NMR and CD solution conformation studies, we have conducted MD experiments beginning with the Lys-9 to His-16 residues constrained in an α-helix. (The details of these studies are in preparation for publication elsewhere by S. Topiol et al). As shown in Figure 4.2, the global minimum structure, even after relaxing the initial constraints, maintains the α-helix. In addition, the His-16 and Asp-18 substituents are shown to be in very close proximity (1.95 Å) such that a hydrogen bond is likely to be present between these two functional groups. Furthermore, the aromatic ring present in the Trp-21 residue also resides close to the His-16, Asp-18 pair such that it may stabilize the donation–transfer of a proton between these residues. The situation persists even when the calculations are repeated on variously charged versions of endothelin-1. Speculatively, the His-16, Asp-18, and Trp-21 residues appear to represent a key triad of amino acids that may have important functional (as well as recognitional) significance[63] at endothelin receptors. For example, the proton transfer from the Asp-18 to His-16 may be part of a longer proton relay system extending through the His-16 to a proton acceptor located on the receptor surface. That hydrogen bonding may occur between the His-16 and Asp-18 residues has also been suggested by previous workers,[57] although other aspects of their theoretical work do not coincide with the helix and floppy carboxy-terminus themes found by us and by other workers cited in this section.

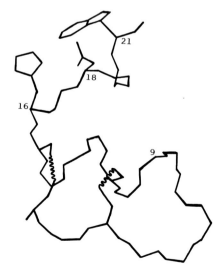

FIGURE 4.2. Molecular dynamics–derived (S. Topiol et al., in preparation) minimum energy conformation for endothelin-1 after starting from an entry structure where residues 9–16 were initially constrained in an α-helix. Note that the α-helix is retained and that the His-16, Asp-18, and Trp-21 substituents have formed a close triad where the His-16 and Asp-18 are probably internally hydrogen bonded (see text for discussion). The two disulfide bonds are shown by wavy lines.

WORKING MODEL OF THE ENDOTHELIN STRUCTURES

The key chemical features discussed individually in the preceeding sections are combined in Figure 4.3 to provide an overall working model for this interesting class of compounds. The structural significance of the key features are portrayed in this figure, and their potential functional character is further illustrated in Figure 4.4. The working model can also be used to compare the results from SAR studies. The latter will be reviewed briefly in the following sections by employing tables that have been standardized for their ready perusal.

STRUCTURE–ACTIVITY RELATIONSHIPS

Receptor-Binding Studies

Employing ^{125}I-labeled endothelin-1, endothelin receptors have been identified in a variety of tissues. They are present on cultured rat vascular smooth muscle cells,[20] on porcine aortic membranes,[29] on the atrium and in the CNS of the rat,[1] on human placenta,[9] on the atrioventricular node of the porcine heart,[71] and on cultured renal epithelial cells.[47] Autoradiographic studies have also demonstrated their presence in the cardiovascular and respiratory systems, intestine, kidney, and CNS from a variety of mammals, including monkeys and humans.[4,23,36,51] Furthermore, as the endothelin family has grown to four members and with most of these materials now available as radiolabeled ligands, ratios for endothelin-receptor subpopulations are beginning to be characterized in various tissues.[39,49,66] Finally, efforts to solubilize and isolate endothelin-receptors are also underway.[2] At this point, general conclusions about SARs should be regarded as preliminary or tentative, because the overall data set has been derived from a variety of tissues with undefined ratios of endothelin-receptor subpopulations. Interestingly, some of the early work in this area showed that "bound ^{125}I-endothelin-1 was resistant to dissociate"[21] and that it can be difficult to wash-out endothelin-1 from the receptor milieu. The results from several binding studies that have examined analogs of endothelin-1 are provided in Table 4.1.

Within the endothelin family, it appears that binding potency is only moderately reduced when comparing endothelin-2 to endothelin-1. In contrast, a striking reduction in binding occurs for endothelin-3. Likewise, the porcine endothelin-1 precursor big endothelin appears to be only weakly active as a receptor ligand. It should be noted that a small amount of conversion to endothelin during binding experiments with porcine big endothelin will complicate any analyses of the latter compound. SRT-b appears to be similar to endothelin-2 in its ability to bind to endothelin receptors and thereby displace endothelin-1. Thus, and intended only as a very general guide, the abilities of the endothelin compounds to bind with their various receptor populations, as present in the indicated tissues or cells, can be ranked as follows: endothelin-1 ⩾ endothelin-2 ⩾ SRT-b >> endothelin-3 ~ porcine big endothelin. While this ranking, to a large extent, is probably associated with endothelin-1 recep-

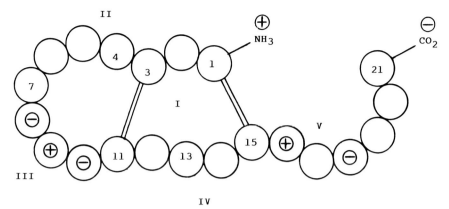

FIGURE 4.3. Working structural model of the endothelins. Key chemical features include the following. *I:* Disulfide bonds. The 1,15 bond is critical for maintaining the entire folded or loop conformation that is important for activity. The 3,11 bond may assist in the initial folding after anabolism and/or play an active role at the receptor such as reversible chemical interaction with some receptor nucleophile (see Figure 4.4A). *II:* Moderately polar amino-terminus. In a general manner this region (1–7) can be characterized in this fashion. However, while residues 1–3 are highly conserved, residues 4–7 display considerable variation among the family members. As biological assays become more discriminative, this region will probably be shown to be critical for establishing endothelin-receptor subtype specificity. *III:* Highly polar region bearing a net negative charge. Because of the net negative charge, this region (−, +, −) should be very important in establishing a strong binding relationship with the receptor. The latter probably contains a correspondingly polar region with an actual cationic site in the receptor pocket (see Fig. 4.4B). This interaction could be one of the early events associated with recognition and initial docking of the endothelin ligands. *IV:* Nonpolar region bearing a Tyr residue. In addition to hydrophobic interactions with the receptor, the nonpolar groups in this region may enhance the ability of the key Tyr residue (13) to interact with the receptor as a hydrogen bond donor, acceptor, or weak acid (see Fig. 4.4C). *V:* Alternating polar, nonpolar, carboxy-terminus. The nonpolar insulating groups may enhance the ability of the three key amino acids His-16 (+), Asp-18 (−), and Trp-21 to form an internal hydrogen bond relationship that is also part of a proton relay system involving, for example, an Asp carboxy-group located in the receptor pocket (see Fig. 4.4D). This event may occur later in the binding scheme time sequence and could play a functional role[63] toward activation of the endothelin receptors.

tors, considerable additional experiments are required to determine whether there are distinct endothelin-receptor subpopulations and then to establish more precisely the definitive potency rankings in each case.

When either of the 1–15 or 3–11 Cys pairs is left alkylated (after synthesis), the resulting monodisulfide versions of endothelin-1 (compound numbers **6** and **7**) are essentially inactive. The crossed 1–11, 3–15 bisdisulfide compound **8,** however, demonstrates weak activity. Similarly, Ala replacements at each Cys pair (compounds **9** and **10**) retain weak activity, while replacement of both pairs results in an inactive compound (**11**). Based on a comparison of compounds **9** and **10,** it can also be noted that the disulfide bridge linking positions 1 and 15 appears to be more important for binding activity than the bridge between 3 and 11.

Oxidation of the Met at position 7 (compound **12**) moderately decreases binding activity, whereas elongation of the amino-terminus with Lys-Arg (compound **13**) dramatically reduces activity. Removal of Trp-21 abolishes activity (compound **14**) as does dividing the molecule between positions 15 and 16 to

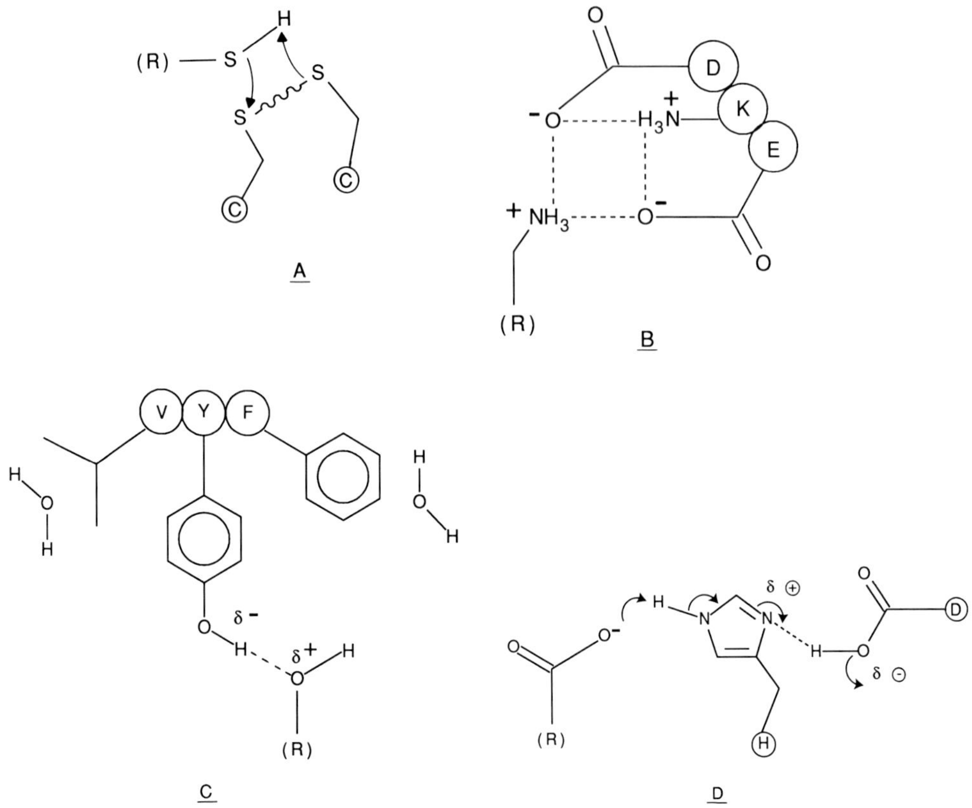

FIGURE 4.4. Possible ligand–receptor (R) interactions implied by the endothelin structural working model. *A:* reaction of the Cys-3, Cys-11 disulfide bridge with a receptor-based nucleophile. *B:* interaction of region 3 (Fig. 4.3) to form a salt relationship with the receptor. *C:* hydrogen bonding of the Tyr-13 residue with the receptor acting as a proton acceptor. The exclusion of water molecules by neighboring hydrophobic substituents is also shown. Alternatively, (not depicted), the Tyr proton could actually transfer to some receptor-based nucleophile (or anion), or via its electron pair, the Tyr oxygen could act as a hydrogen acceptor in a similar hydrogen-bonding relationship with an appropriate donor at the receptor. Finally, the Tyr may simply interact with the receptor by virtue of its aromatic ring. *D:* proton transfer from Asp-18 to His-16 with relay to a receptor based nucleophile.[63] In this case an actual anion is depicted at the receptor. Note that electron flow (arrows) is in the direction opposite to the proton transfer.

produce the two fragments represented by compounds **15** and **16**. Finally, the second "endothelin-like" sequence (see footnote *i* in Table 4–1), as either its free carboxy-terminal or NH_2 amide (compounds **17** and **18,** respectively), does not appear to bind to endothelin receptors located on canine heart membranes.

In Vitro Studies

Analogous to the widespread distribution of endothelin receptors, endothelin-1 has been found to be a potent vasoconstrictor in a variety of isolated vascular tissues and to be a potent inotropic agent on cardiac tissues isolated from several species. Dramatic constriction accompanying application of endothelin-1 has been demonstrated for renal arteries from both Wistar-Kyoto and sponta-

TABLE 4.1. Competitive receptor-binding studies between [^{125}I]endothelin-1 and related peptides[a]

Peptide[b]	RVSMC[c]	MFC/HFC[d]	PASMM[e]	CHM[f]	RC[g]	SRC[h]
1 ET-1	100	100	100	100	100	100
2 ET-2	—	39	—	8.0	—	100
3 ET-3	—	0.09, 0.07	—	NR	—	56
4 SRT-b	—	—	—	6.7	—	83
5 PBET	1.0	—	0.05	—	—	—
6 [Cys(Acm)^{3,11}]	NR	—	—	—	—	—
7 [Cys(Acm)^{1,15}]	NR	—	—	—	—	—
8 [Cys^{1,11},Cys^{3,15}]	1.0	—	1.0	—	—	—
9 [Ala^{3,11}]	—	—	—	—	2.8	—
10 [Ala^{1,15}]	—	—	—	—	0.7	—
11 [Ala^{1,3,11,15}]	—	—	—	—	NR	—
12 [Met(O)^7]	30	—	—	—	—	—
13 Lys-Arg–ET-1	0.15	—	—	—	—	—
14 (1–20)	NR	—	—	—	—	—
15 (1–15)	NR	—	—	—	—	—
16 (16–21)	NR	—	—	—	—	—
17 PPET(110–130)[i]	—	—	—	NR	—	—
18 PPET(110–130)–NH$_2$[i]	—	—	—	NR	—	—

[a] In general, individual studies employed concentrations of [^{125}I]endothelin-1 in the range of 10^{-9} to 10^{-11} M and report IC$_{50}$ data for the various related peptides. These data have been further normalized relative to endothelin-1 (which was assigned a value of 100). Smaller numbers indicate weaker binding with the endothelin receptor. Approximate IC$_{50}$ data for endothelin-1 in each of the studies are provided in the corresponding footnotes. NR indicates that an IC$_{50}$ was not reached at test concentrations as high as 10^{-6} M.

[b] ET, endothelin; PBET, porcine big endothelin-1 (1–39); SRT, sarafotoxin. Unless indicated otherwise, the tabulated modifications pertain to changes of endothelin-1.

[c] Rat vascular smooth muscle cells.[18,19] Endothelin IC$_{50}$ = 3 × 10^{-11} M.[19] The value listed for PBET was determined from Hirata et al.[18]

[d] Murine fibroblast 3T3 cells and human fibroblast FS-4 cells.[49] Endothelin IC$_{50}$ = 1.4 and 1.5 × 10^{-10} M, respectively.

[e] Porcine aortic smooth muscle membranes.[13] Endothelin IC$_{50}$ = 3.5 × 10^{-9} M.

[f] Canine heart membranes (details for this study are in preparation for publication elsewhere by C. Cade et al.) Endothelin IC$_{50}$ = 2 × 10^{-9} M.

[g] Rat cerebellum.[50] Endothelin IC$_{50}$ = 7 × 10^{-10} M.

[h] Solubilized endothelin receptors from rat cerebellum.[2] In this case, [^{125}I]SRT-b was used in place of [^{125}I]endothelin during competition studies, and K_i values were reported. Endothelin K_i = 2.5 × 10^{-9} M.

[i] These peptides represent the second "endothelin-like" sequence present between residues 110 and 124 in the 203 amino acid endothelin prepropeptide. The –NH$_2$ version signifies a capped carboxy-terminus (as the NH$_2$ amide). The exact sequence is Cys-Gln-Cys-Ala-Ser-Gln-Lys-Asp-Lys-Lys-Cys-Trp-Ser-Phe-Cys-Gln-Ala-Gly-Lys-Glu-Ile. It is also thought to have two disulfide bonds.

neous hypertensive rats,[62] for both guinea pig tracheal and human bronchial smooth muscle,[65] for rat thoracic aorta and mesenteric beds, for rabbit mesenteric artery and portal veins,[7] for rat lymphatic vessels,[11] for rat pulmonary vasculature,[68] and for cat[30] and human[5] cerebral arteries. Positive inotropic effects have been demonstrated on cardiac tissue from ferrets[56] and guinea pigs.[61] Additional in vitro studies have also demonstrated contractile responses from the guinea pig ileum, esophageal musculature, and uterus,[7] as well as an ability to inhibit both the release of renin from rat glomeruli[52] and the adrenergic neuroeffector transmission in the guinea pig femoral artery.[70] Analogous to the binding studies, reported SAR studies have used a variety of tissue types that lack an accurate description of the endothelin-receptor subpopulation ratios such that general conclusions from these studies should also be regarded as preliminary. The results from several in vitro SAR studies are summarized in Table 4.2.

A similar potency ranking for the endothelin family is obtained from the in vitro studies with isolated tissues: endothelin-1 ⩾ endothelin-2 ~ SRT-b >> endothelin-3 ~ porcine big endothelin. Likewise, while low potency is retained after removal of only the 3,11 disulfide bridge (compound **9**), removal of the 1,15 disulfide (compound **10**) and, in particular, removal of both disulfides (compounds **11, 21, 27,** and **28**) essentially abolishes activity. Further emphasizing the importance of the conformational fold or large loop associated with an intact 1,15 disulfide bridge are the dramatic reductions in potency upon crossing the two disulfide bonds (compound **11**), opening the ring after Lys-9 (compound **20**), or retaining only one of the two smaller rings present in the larger loop (compounds **25** and **26**).

As also predicted from the binding studies, both the amino- and carboxytermini seem to be very important for activity in the in vitro tissue perfusion studies. Capping the amino-terminus as an amide (compounds **13** and **30**) essentially abolishes activity. While capping the carboxy-terminus yields weakly active compounds (e.g., **24** and **31**), nearly a complete loss of activity occurs

TABLE 4.2. Vasoconstrictor potencies for endothelin-1 and related peptides in various in vitro tissue preparations[a]

Peptide[b]		PCA[c]	RA[d]	RPA[e]	GPB[f]	RMB[g]
1	ET-1	100	100	100	100	100
2	ET-2	—	—	49	—	—
3	ET-3	—	2	1.9	—	—
4	SRT-b	—	10	35.1	—	—
5	PBET	1.0, 0.7	NR	1.8	—	—
6	[Cys(Acm)^{3,11}]	—	—	NR	—	—
7	[Cys(Acm)^{1,15}]	—	—	NR	—	—
8	[Cys^{1,11},Cys^{3,15}]	—	—	0.8	—	—
9	[Ala^{3,11}]	—	7.7	29.6	—	5.3
10	[Ala^{1,15}]	—	0.5	0.4	—	NR
11	[Ala^{1,3,11,15}]	—	—	—	—	NR
12	[Met(O)^{7}]	—	—	69.1	—	—
13	Lys-Arg–ET	—	—	0.2	—	—
14	(1–20)	0.1	—	NR	—	—
15	(1–15)	—	—	—	—	—
16	(16–21)	—	NR	NR	3.0	—
17	PPET(110–130)	—	NR	—	—	—
18	PPET(110–130)–NH$_2$	—	—	—	—	—
19	[D-Trp^{21}]	0.7	—	—	—	—
20	[Lys^9 nicked]	0.7	—	—	—	—
21	[CAM^{1,3,11,15}]	NR	—	—	—	—
22	(1–19)	NR	—	—	—	—
23	(1–16)	NR	—	—	—	—
24	PBET(1–25)	2.3	—	—	—	—
25	"End ring"	—	NR	—	—	—
26	"Center ring"	—	NR	—	—	—
27	[Cys(Acm)^{1,3,11,15}]	—	—	NR	—	—
28	[Asu^{1,15},Ala^{3,11}]	—	—	NR	—	—

(continued)

when this important residue is removed (compounds **14** and **22**). Likewise, when the stereochemistry of this residue is inverted, as in compound **19**, activity is essentially gone. This is very informative because in this compound the carboxylic acid group can maintain its occupancy of the same region in space while the indole substituent cannot. This implies that while both the carboxy and indole groups are important for activity, the indole may be the most critical. Note that the reduction of activity in this case (compound **19**) seems to be greater than when only the acid moiety is altered (compounds **24** and **31**). Furthermore, other aromatic amino acids when substituted for the Trp-21 residue (e.g., compounds **40** and **41**), seem to retain reasonable activity. Taken together, these results suggest that the important features at the carboxy-terminus are an aromatic substituent and then, with lesser priority, a free carboxylic acid moiety.

In all cases, smaller fragments of endothelin-1 are essentially inactive in vitro. This includes those from the amino-terminus such as compounds **15, 23,** and **29** and those from the carboxy-terminus such as compounds **16** and **42**. Likewise, the "second sequence" structure (compound **17**) is totally inactive, as was found in the binding studies.

TABLE 4.2. (*Continued*)

Peptide[b]	PCA[c]	RA[d]	RPA[e]	GPB[f]	RMB[g]
29 (1–15)–NH$_2$	—	—	NR	—	—
30 Ac–ET	—	—	0.5	—	—
31 ET–NH$_2$	—	—	5.6	—	—
32 [Ala^4]	—	—	36; 43	—	—
33 [Ala^5]	—	—	24.2	—	—
34 [Gly^6]	—	—	78.1	—	—
35 [Asn^8]	—	—	0.8	—	—
36 [Leu^9]	—	—	54.8	—	—
37 [Gln^{10}]	—	—	NR	—	—
38 [Phe^{13}]	—	—	63.3	—	—
39 [Ala^{14}]	—	—	NR	—	—
40 [Tyr^{21}]	—	—	32.8	—	—
41 [Phe^{21}]	—	—	18.4	—	—
42 (16–21)–NH$_2$	—	NR	—	NR	—

[a]In general, individual studies report EC$_{50}$ data for endothelin-1 and the related peptides that were studied. These data have been further normalized relative to endothelin-1 (which was assigned a value of 100). Smaller numbers indicate weaker potency as constrictor agents. When results from two reports are substantially different, both values are given. Approximate EC$_{50}$ data for endothelin-1 in each of the studies is provided in the corresponding footnotes. NR indicates that an EC$_{50}$ was not reached at test concentrations as high as 10^{-6} M.

[b]ET, endothelin; PBET, porcine big endothelin; SRT, sarafotoxin. Unless indicated otherwise, the tabulated modifications pertain to changes of endothelin-1. Note that the first 18 compounds are listed in an order identical to that in Table 4.1.

[c]Porcine coronary artery with 50 mM KCl as standard constrictor.[31,32] Endothelin EC$_{50}$ = 5.22 × 10^{-10} M.

[d]Rat aorta[10] with 1 μM noradrenaline[64] or 80 mM KCl[39] employed as the standard constrictor agents. Endothelin EC$_{50}$ = 1.4 × 10^{-9} M,[64] 1.7 × 10^{-9} M,[39] and 7.7 × 10^{-8} M.[10] Also note that similar results (e.g., NR) were obtained for compounds 17 and 18 in hamster aorta (see Table 4.1, footnotes *f* and *i*).

[e]Rat pulmonary artery employing 80 mM KCl as the standard constrictor agent.[44–46] Endothelin EC$_{50}$ = 6.5 × 10^{-10} M.

[f]Guinea pig bronchus employing 80 mM KCl as the standard constrictor agent.[39] Endothelin EC$_{50}$ = 6.9 × 10^{-9} M.

[g]Rat mesenteric bed using calculations associated with the changes observed from baseline perfusion pressures.[17,53] Endothelin EC$_{50}$ = 6.6 × 10^{-11} M.

The effects on activity resulting from single amino acid substitutions at several positions on endothelin-1 can be assessed from the compilation of rat pulmonary artery (RPA column) data. Substitutions at residues 4 or 5 with Ala do not cause a dramatic decrease in activity, suggesting that the moderately polar serines that are normally present in these locations are not critical for association or activation of the endothelin receptors. Substitution of the Leu-6 with Gly has even less of an effect on activity, as does oxidation of the Met-7 (compound **12**). Alternatively, when the Asp-8 carboxylic acid moiety is converted to its amide (compound **35**) there is a nearly complete loss of activity. This is also observed when the Glu-10 acid is converted to its amide (compound **37**). Speculatively, these two results imply that the departure from an overall negative charge in this region of endothelin-1 cannot be tolerated. On the other hand, enhancing the negative charge in this region by substituting a neutral Leu amino acid for the basic Lys-9 residue (protonated) has only a small influence on in vitro potency. Apparently a net negative charge of -1 is optimal for activity. In the next region, there is only a small reduction in potency when the Tyr-13 residue is replaced by Phe, suggesting that the Tyr is probably not playing a significant role as a hydrogen bond partner at the receptor. Note, however, that an interaction via its aromatic ring could still be important. Alternatively, replacement of Phe-14 with Ala leads to an inactive compound strongly suggesting some importance for this residue. Finally, it is worth noting that the carboxy-terminal chain endothelin-1 (residues 16–21); compound **16**) is able to constrict the guinea pig bronchus, albeit weakly, even though it is totally inactive in the vascular assays both in whole tissues and in those that are associated with receptor-binding studies. This finding serves to caution again about extending SAR conclusions from tissue to tissue and even within a single tissue until there is a better appreciation of the ratio of endothelin-receptor subpopulations that are present in each tissue under consideration.

In Vivo Considerations

Attempts to determine SARs in vivo are further complicated by the variety of actions elicited by endothelin. For example, while endothelin-1 is in general a potent constrictor of various vascular beds, these responses can sometimes vary depending on the dose and on the particular bed being studied.[41] Furthermore, besides its constrictor action on the vasculature and its direct inotropic and chronotropic effects on cardiac tissue, endothelin-1 has been shown to have profound effects on the endocrine system.[14] It has been suggested that endothelin-1 may serve as a "counter-regulatory hormone to the effects of endothelial-derived vasodilator agents."[40] In various settings endothelin-1 has been shown to stimulate the release of prostacyclin, thromboxane A_2, endothelium-derived relaxing factor,[6] atrial natriuretic factor,[24] and prostaglandins E_2 and I_2,[42] as well as to inhibit the release of renin.[59]

There have been relatively few endothelin in vivo SAR studies reported. It is interesting to note that when [Ala^{1,3,11,15}] endothelin-1 was studied[38] in the pithed rat, it was only five times less potent than endothelin-1 as a pressor agent. This is in sharp contrast to the complete lack of activity exhibited by this analog in receptor binding and in in vitro tissue perfusion studies (compound **11** in Tables 4.1 and 4.2). Similarly, in this preparation endothelin-1 and endothelin-3 were found to "vary little in their potency and lethality."[67] In anesthetized rats endothelin-3 was found to be only "three times less potent in causing the subsequent rise"[58] in blood pressure. Similarly, rather small differences between endothelin-1 and endothelin-3 have been reported[69] in conscious hypertensive rats (about threefold) and in Wistar-Kyoto rats (approximately ninefold). Two other analogs that lack the disulfide bridges have also been studied[22] in vivo in the rat, and they also showed surprising activity; as pressor agents, [Cys (Acm)^{1,3,11,15}] endothelin-1 (compound **27** in Table 4.2) and [Cys SSO$_3$^{1,3,11,15}] endothelin-1 had about one-third the potency of endothelin-1. Thus it appears that the large differences in activity observed for the endothelin family and its analogs in vitro may be attenuated when the comparisons are made in vivo, although the rank orders of potency seem to be maintained.

SUMMARY

The results from the SAR studies are summarized in the model depicted in Figure 4.5. To avoid the undefined variables associated with the in vivo studies, the SAR model has been derived solely from receptor-binding and in vitro tissue perfusion studies. Although not well defined, to a large extent this model probably represents SARs associated with the endothelin-1 subpopulation of endothelin receptors. The model is arranged in a fashion identical to the one constructed for the key structural and chemical features of the endoth-

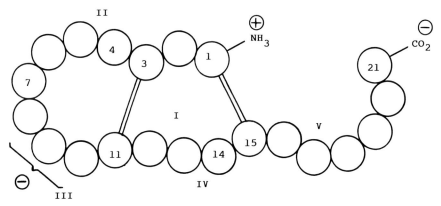

FIGURE 4.5. Structure–activity relationships model for the endothelins. Key pharmacophoric features include the following. *I:* Disulfide bonds. The presence of at least one disulfide bridge appears to be requisite for activity. The following further represents a rank ordering of their importance: 1,15 and 3,11 (as in endothelin-1) > 1,15 >> 1,11 and 3,15 ≥ 3,11 >> no disulfides (essentially inactive). *II:* Amino-terminus. A free amino group appears to be requisite for activity, whereas considerable modification can be effected at positions 4 through 7 without drastically affecting potency. *III:* Net negative charge. A composite charge of −1 in this region appears to be requisite for activity. *IV:* Nonpolar region. The Phe-14 residue appears to play an important role. *V:* Carboxy-terminus. An amino acid having an aromatic substituent and a free carboxylic acid group appears to represent the preferred pharmacophore at position 21.

elins (Fig. 4.3), thus allowing for their ready comparison. The key structural features are closely associated with the key pharmacophores, although development of the latter through SAR studies are only at a very preliminary stage. Region for region the comparison agrees quite closely except that in region IV the SAR results suggest that Phe-14 may be a very important residue while structurally Tyr-13 seems like a more prominent functional group.

Probably the most important point to emphasize, however, is one that actually pertains to future developments. Even though the entire endothelin molecule appears to be required for strong binding to its receptor(s), there will undoubtedly be much smaller fragments or mimetic structures that will be able to bind extremely well to just a few of the pharmacophoric partners or auxiliary binding areas near the receptor. These smaller molecules should then also be able to achieve status as strong ligands for this receptor family and thereby help to realize the possibility that efficient endothelin receptor antagonists can be produced. The latter will be extremely useful to characterize further both the biochemistry and the various receptors for this interesting family of compounds. Antagonists would also help to define the physiological and/or pathophysiological significance of the endothelins and aid in defining potentially new therapeutic strategies that might be derived from altering this system. Despite all the undefined variables and potential pitfalls alluded to in this chapter, the design of novel ligands via "working models" and the synthesis of analogs before a biological system is clearly understood and well defined can provide useful tools to help shed light on the new system. Toward this end it is hoped that at least some of the speculation offered herein can serve as "rationale" lamp posts for interested medicinal and peptide chemists bold enough to embark on such an excursion into the endothelin area at such an early time-point.

REFERENCES

1. AMBAR, I., Y. KLOOG, I. SCHVARTZ, E. HAZUM, and M. SOKOLOVSKY. Competitive interaction between endothelin and sarafotoxin: binding and phosphoinositides hydrolysis in rat atria and brain. *Biochem. Biophys. Res. Commun.* 158: 195, 1989.
2. AMBAR, I., Y. KLOOG, and M. SOKOLOVSKY. Solubilization of endothelin/sarafotoxin receptors in an active binding form. *Eur. J. Pharmacol.* 170: 119, 1989.
3. BROWN, S. C., M. E. DONLAN, and P. W. JEFFS. Structural studies of endothelin by CD and NMR. In: *Peptides: Chemistry, Structure and Biology*, edited by J. E. Rivier and G. R. Marshall. Leiden, The Netherlands: ESCOM, 1990, p. 595.
4. DAVENPORT, A. P., D. J. NUNEZ, J. A. HALL, A. J. KAUMANN, and J. B. MORRIS. Autoradiographical localization of binding sites for porcine [^{125}I] endothelin-1 in humans, pigs, and rats: functional relevance in humans. *J. Cardiovasc. Pharmacol.* 13(Suppl. 5): S166, 1989.
5. DEAGUILERA, E. M., A. IRURZUN, J. M. VILA, M. ALDASORO, M. S. GALEOTE, and S. LLUCH. Role of endothelin and calcium channels in endothelin-induced contraction of human cerebral arteries. *Br. J. Pharmacol.* 99: 439, 1990.
6. DENUCCI, G., R. THOMAS, P. D'ORLEANS-JUSTE, E. ANTUNES, C. WALDER, T. D. WARNER, and J. R. VANE. Pressor effects of circulating endothelin are limited by its removal in the pulmonary circulation and by the release of prostacyclin and endothelin-derived relaxing factor. *Proc. Natl. Acad. Sci. USA* 85: 9797, 1988.
7. EGLEN, R. M., A. D. MICHEL, N. A. SHARIF, S. R. SWANK, and R. L. WHITING. The pharmacological properties of the peptide, endothelin. *Br. J. Pharmacol.* 97: 1297, 1989.
8. ENDO, S., H. INOOKA, Y. ISHIBASHI, C. KITADA, E. MIZUTA, and M. FUJINO. Solution conformation of endothelin determined by nuclear magnetic resonance and distance geometry. *FEBS Lett.* 257: 149, 1989.

9. FISCHLI, W., M. CLOZEL, and C. GUILLY. Specific receptors for endothelin on membranes from human placenta: characterization and use in a binding assay. *Life Sci.* 44: 1429, 1989.
10. FOK, K. F., M. L. MICHENER, S. P. ADAMS, E. G. MCMAHON, M. A. PALOMO, and A. J. TRANPANI. Endothelin: solid-phase synthesis and structure–activity studies. In: *Peptides: Chemistry, Structure And Biology,* Edited by J. E. Rivier and G. R. Marshall. The Netherlands: ESCOM, 1990, p. 269.
11. FORTES, Z. B., R. SCIVOLETTO, and J. GARCIA-LEME. Endothelin-1 induces potent constriction of lymphatic vessels in situ. *Eur. J. Pharmacol.* 170: 69, 1989.
12. FRENZ, J., J. BATTERSBY, and W. S. HANCOCK. An examination of the potential of capillary zone electrophoresis for the analysis of polypeptide samples. In: *Peptides: Chemistry, Structure And Biology,* edited by J. E. Rivier and G. R. Marshall. Leiden, The Netherlands: ESCOM, 1990, p. 430.
13. GAETA, F. C. A., L. B. SLATER, B. R. SUNDAY, J. R. MILLER, C. L. RAMSAUR, L. GHIBAUDI, and M. CHATTERJEE. Synthesis and characterization of porcine endothelin and big endothelin. In: *Peptides: Chemistry, Structure, And Biology,* edited by J. E. Rivier and G. R. Marshall. Leiden, The Netherlands: ESCOM, 1990, p. 264.
14. GOETZ, K. L., B. C. WANG, J. B. MADWED, J. L. ZHU, and R. J. LEADLEY, JR. Cardiovascular, renal, and endocrine responses to intravenous endothelin in conscious dogs. *Am. J. Physiol.* 255 *(Regulatory Integrative Comp. Physiol.* 24): R1064, 1988.
15. HEMPEL J. C., W. A. GHOUL, J.-M. WURTZ, and A. T. HAGLER. Energy-based modeling studies of endothelin. In: *Peptides: Chemistry, Structure And Biology,* edited by J. E. Rivier and G. R. Marshall. Leiden, The Netherlands: ESCOM, 1990, p. 279.
16. HILEY, C. R. Functional studies on endothelin catch up with molecular biology [Synopsis of work by T. Kimura et al. and of work by others as presented at The First William Harvey Workshop On Endothelin, London, December 5–6, 1988]. *TIPS* 10: 47, 1989.
17. HILEY, C. R., S. A. DOUGLAS, and M. D. RANDALL. Pressor effects of endothelin-1 and some analogs in the perfused superior mesenteric arterial bed of the rat. *J. Cardiovasc. Pharmacol.* 13(Suppl. 5): S197, 1989.
18. HIRATA, Y., K. KANNO, T. X. WATANABE, S. KUMAGAYE, K. NAKAJIMA, T. KIMURA, S. SAKAKIBARA, and F. MARUMO. Receptor binding and vasoconstrictor activity of big endothelin. *Eur. J. Pharmacol.* 176, 225 (1990).
19. HIRATA, Y., H. YOSHIMI, T. EMORI, M. SHICHIRI, F. MARUMO, T. X. WATANABE, S. KUMAGAYE, K. NAKAJIMA, T. KIMURA, and S. SAKAKIBARA. Receptor binding activity and cytosolic free calcium response by synthetic endothelin analogs in cultured rat vascular smooth muscle cells. *Biochem. Biophys. Res. Commun.* 160: 228, 1989.
20. HIRATA, Y., H. YOSHIMI, S. TAKAICHI, M. YANAGISAWA, and T. MASAKI. Binding and receptor down-regulation of a novel vasoconstrictor endothelin in cultured rat vascular smooth muscle cells. *FEBS Lett.* 239: 13, 1988.
21. HIRATA, Y., H. YOSHIMI, S. TAKATA, T. X. WATANABE, S. KUMAGAI, K. NAKAJIMA, and S. SAKAKIBARA. Cellular mechanism of action by a novel vasoconstrictor endothelin in cultured rat vascular smooth muscle cells. *Biochem. Biophys. Res. Commun.* 154: 868, 1988.
22. HOEGER, C., M. R. BROWN, and J. E. RIVIER. Synthesis and biological properties of endothelin and endothelin analogs. In: *Peptides: Chemistry, Structure And Biology,* edited by J. E. Rivier and G. R. Marshall. Leiden, The Netherlands: ESCOM, 1990, p. 267.
23. HOYER, D., C. WAEBER, and J. M. PALACIOS. [^{125}I] endothelin-1 binding sites: autoradiographic studies in the brain and periphery of various species including humans. *J. Cardiovasc. Pharmacol.* 13(Suppl. 5): S162, 1989.
24. HU, J. R., U. G. BERNINGER, and R. E. LANG. Endothelin stimulates atrial natriuretic peptide (ANP) release from rat atria. *Eur. J. Pharmacol.* 158: 177, 1988.
25. IMMER, H., I. EBERLE, W. FISCHER, and E. MOSER. Solution-synthesis of endothelin. *20th European Peptide Symposium,* Tubingen, FRG, September 4–9, 1988, Abstract No. 98.
26. INOUE, A., M. YANAGISAWA, S. KIMURA, Y. KASUYA, T. MIYAUCHI, K. GOTO, and T. MASAKI. The human endothelin family: three structurally and pharmacologically distinct isopeptides predicted by three separate genes. *Proc. Natl. Acad. Sci. USA* 86: 2863, 1989.
27. ISHIDA, N., K. TSUJIOKA, M. TOMOI, K. SAIDA, and Y. MITSUI. Differential activities of two distinct endothelin family peptides on ileum and coronary artery. *FEBS Lett.* 247: 337, 1989.
28. ITOH, Y., M. YANAGISAWA, S. OHKUBO, C. KIMURA, T. KOSAKA, A. INOUE, N. ISHIDA, Y. MITSUI, H. ONDA, M. FUJINO, and T. MASAKI. Cloning and sequence analysis of cDNA encoding the precursor of a human endothelium-derived vasoconstrictor peptide, endothelin: identity of human and porcine endothelin. *FEBS Lett.* 231: 440, 1988.
29. KANSE, S. M., M. A. GHATEI, J. M. POLAK, and S. R. BLOOM. Binding and degradation of

endothelin by porcine aortic membranes. *The Bayliss And Starling Society, 8th National Science Meeting, London, September 15–16, 1988,* p. 412.
30. KAUSER, K., G. M. RUBANYI, and D. R. HARDER. Endothelium-dependent modulation of endothelin-induced vasoconstriction and membrane depolarization in cat cerebral arteries. *J. Pharmacol. Exp. Ther.* 252: 93, 1990.
31. KIMURA, S., Y. KASUYA, T. SAWAMURA, O. SHINMI, Y. SUGITA, M. YANAGISAWA, K. GOTO, and T. MASAKI. Structure–activity relationships of endothelin: importance of the C-terminal moiety. *Biochem. Biophys. Res. Commun.* 156: 1182, 1988.
32. KIMURA, S., Y. KASUYA, T. SAWAMURA, O. SHINMI, Y. SUGITA, M. YANAGISAWA, K. GOTO, and T. MASAKI. Conversion of big endothelin-1 to 21-residue endothelin-1 is essential for expression of full vasoconstrictor activity: structure–activity relationships of big endothelin-1. *J. Cardiovasc. Pharmacol.* 13(Suppl. 5), S5, 1989.
33. KLOOG, Y., I. AMBAR, M. SOKOLOVSKY, E. KOCHVA, Z. WOLLBERG, and A. BDOLAH. Sarafotoxin, a novel vasoconstrictor peptide: phosphoinositide hydrolysis in rat heart and brain. *Science,* 242: 268, 1988.
34. KLOOG, Y., and M. SOKOLOVSKY. Similarities in mode and sites of action of sarafotoxins and endothelins. *TIPS,* 10: 212, 1989.
35. KOBAYASHI, Y. Solution conformation of endothelin. In: *Peptides: Chemistry, Structure and Biology,* edited by J. E. Rivier and G. R. Marshall. Leiden, The Netherlands: ESCOM, 1990, p. 552.
36. KOSEKI, C., M. IMAI, Y. HIRATA, M. YANAGISAWA, and T. MASAKI. Binding sites for endothelin-1 in rat tissues: an autoradiographic study. *J. Cardiovasc. Pharmacol.* 13(Suppl. 5): S153, 1989.
37. LIU, W., G. H. SHIUE, and J. P. TAM. A novel strategy for the deprotection of S-acetamidomethyl containing peptides: an approach to the efficient synthesis of endothelin. In: *Peptides: Chemistry, Structure And Biology,* edited by J. E. Rivier and G. R. Marshall. Leiden, The Netherlands: ESCOM, 1990, p. 271.
38. MACLEAN, M. R., and C. R. HILEY. Blood pressure changes induced by endothelin and [Ala^{1,3,11,15}]-endothelin in pithed rats: effect of lowering ventilation volume. *Br. J. Pharmacol.* 97: 529P, 1989.
39. MAGGI, C. A., S. GIULIANI, R. PATACCHINI, P. SANTICIOLI, P. ROVERO, A. GIACHETTI, and A. MELI. The C-terminal hexapeptide, endothelin-(16-21), discriminates between different endothelin receptors. *Eur. J. Pharmacol.* 166: 121, 1989.
40. MILLER, W. L., M. M. REDFIELD, and J. C. BURNETT, JR. Integrated cardiac, renal, and endocrine actions of endothelin. *J. Clin. Invest.* 83: 317, 1989.
41. MINKES, R. K., and P. J. KADOWITZ. Differential effects of rat endothelin on regional blood flow in the cat. *Eur. J. Pharmacol.* 165: 161, 1989.
42. MIURA, K., T. YUKIMURA, Y. YAMASHITA, T. SHIMMEN, M. OKUMURA, M. IMANISHI, and K. YAMAMOTO. Endothelin stimulates the renal production of prostaglandin E_2 and I_2 in anesthetized dogs. *Eur. J. Pharmacol.* 170: 91, 1989.
43. MORRIS, H. R., and P. PUCCI. A new method for rapid assignment of S–S bridges in proteins. *Biochem. Biophys. Res. Commun.* 126: 1122, 1985.
44. NAKAJIMA, K., S. KUBO., S. KUMAGAYE, H. KURODA, H. NISHIO, M. TSUNEMI, T. INUI, N. CHINO, T. X. WATANABE, T. KIMURA, and S. SAKAKIBARA. Synthesis of some endothelin analogs and big endothelin: structure–activity relationships. In: *Peptides: Chemistry, Structure and Biology,* edited by J. E. Rivier and G. R. Marshall. Leiden, The Netherlands: ESCOM, 1990, p. 68.
45. NAKAJIMA, K., S. KUBO, S. KUMAGAYE, H. NISHIO, M. TSUNEMI, T. INUI, H. KURODA, N. CHINO, T. X. WATANABE, T. KIMURA, and S. SAKAKIBARA. Structure–activity relationship of endothelin: importance of charged groups. *Biochem. Biophys. Res. Commun.* 163: 424, 1989.
46. NAKAJIMA, K., S. KUMAGAYE, H. NISHIO, H. KURODA, T. X. WATANABE, Y. KOBAYASHI, H. TAMAOKI, T. KIMURA, and S. SAKAKIBARA. Synthesis of endothelin-1 analogues, endothelin-3, and sarafotoxin S6b: structure–activity relationships. *J. Cardiovasc. Pharmacol.* 13(Suppl. 5): S8, 1989.
47. NEUSER, D., S. ZAISS, and J.-P. STASCH. Endothelin receptors in cultured renal epithelial cells. *Eur. J. Pharmacol.* 176: 241, 1990.
48. NOMIZU, M., Y. INAGAKI, A. IWAMATSU, T. KASHIWABARA, H. OHTA, A. MORITA, K. NISHIKORI, A. OTAKA, N. FUJII, and H. YAJIMA. Application of two-step hard acid deprotection/cleavage procedures to the solid-phase synthesis of the putative precursor of human endothelin. In: *Peptides: Chemistry, Structure And Biology,* edited by J. E. Rivier and G. R. Marshall. Leiden, The Netherlands: ESCOM, 1990, p. 276.
49. OHNISHI-SUZAKI, A., K. YAMAGUCHI, M. KUSUHARA, I. ADACHI, K. ABE, and S. KIMURA. Comparison of biological activities of endothelin-1, -2 and -3 in murine and human fibroblast cell lines. *Biochem. Biophys. Res. Commun.* 166: 608, 1990.

50. PELTON, J. T., R. JONES, V. SAUDEK, and R. MILLER. Receptor binding and biophysical studies of monocyclic analogs of endothelin. In: *Peptides: Chemistry, Structure And Biology*, edited by J. E. Rivier and G. R. Marshall. Leiden, The Netherlands: ESCOM, 1990, p. 274.
51. POWER, R. F., J. WHARTON, Y. ZHAO, S. R. BLOOM, and J. M. POLAK. Autoradiographic localization of endothelin-1 binding sites in the cardiovascular and respiratory systems. *J. Cardiovasc. Pharmacol.* 13(Suppl. 5): S50, 1989.
52. RAKUGI, H., M. NAKAMARU, H. SAITO, J. HIGAKI, and T. OGIHARA. Endothelin inhibits renin release from isolated rat glomeruli. *Biochem. Biophys. Res. Commun.* 155: 1244, 1988.
53. RANDALL, M. D., S. A. DOUGLAS, and C. R. HILEY. Vascular activities of endothelin-1 and some Alanyl substituted analogues in resistance beds of the rat. *Br. J. Pharmacol.* 98: 685, 1989.
54. SAIDA, K., Y. MITSUI, and N. ISHIDA. A novel peptide, vasoactive intestinal contractor, of a new (endothelin) peptide family. *J. Biol. Chem.* 264: 14613, 1989.
55. SAUDEK, V., J. HOFLACK, and J. T. PELTON. ^{1}H-NMR study of endothelin, sequence-specific assignment of the spectrum and a solution structure. *FEBS Lett.* 257: 145, 1989.
56. SHAK, A. M., M. J. LEWIS, and A. H. HENDERSON. Inotropic effects of endothelin in ferret ventricular myocardium. *Eur. J. Pharmacol.* 163: 365, 1989.
57. SPINELLA, M. J., S. R. KRYSTEK, JR., D. H. PEAPUS, B. A. WALLACE, C. BRUNER, and T. T. ANDERSON. A proposed structural model of endothelin. *Peptide Res.* 2: 286, 1989.
58. SPOKES, R. A., M. A. GHATEI, and S. R. BLOOM. Studies with endothelin-3 and endothelin-1 on rat blood pressure and isolated tissues: evidence for multiple endothelin receptor subtypes. *J. Cardiovasc. Pharmacol.* 13(Suppl. 5): S191, 1989.
59. TAKAGI, M., H. MATSUOKA, K. ATARASHI, and S. YAGI. Endothelin: a new inhibitor of renin release. *Biochem. Biophys. Res. Commun.* 157, 1164, 1988.
60. TAKASAKI, C., M. YANAGISAWA, S. KIMURA, K. GOTO, and T. MASAKI. Similarity of endothelin to snake venom toxin. *Nature,* 335: 303, 1988.
61. TOHSE, N., Y. HATTORI, H. NAKAYA, M. ENDOU, and M. KANNO. Inability of endothelin to increase Ca^{2+} current in guinea-pig heart cells. *Br. J. Pharmacol.* 99: 437, 1990.
62. TOMOBE, Y., T. MIYAUCHI, A. SAITO, M. YANAGISAWA, S. KIMURA, K. GOTO, and T. MASAKI. Effects of endothelin on the renal artery from spontaneously hypertensive and Wistar-Kyoto rats. *Eur. J. Pharmacol.* 152: 373, 1988.
63. TOPIOL, S. The deletion model for the origin of receptors. *TIBS* 12: 419, 1987.
64. TOPOUZIS, S., J. T. PELTON, and R. C. MILLER. Effects of calcium entry blockers on contractions evoked by endothelin-1, [Ala^{3,11}] endothelin-1 and [Ala^{1,15}] endothelin-1 in rat isolated aorta. *Br. J. Pharmacol.* 98: 669, 1989.
65. UCHIDA, Y., H. NINOMIYA, M. SAOTOME, A. NOMURA, M. OHTSUKA, M. YANAGISAWA, K. GOTO, T. MASAKI, and S. HASEGAWA. Endothelin, a novel vasoconstrictor peptide, as a potent bronchoconstrictor. *Eur. J. Pharmacol.* 154: 227, 1988.
66. WAEBER, C., D. HOYER, and J.-M. PALACIOS. Similar distribution of [^{125}I]sarafotoxin-6b and [^{125}I]endothelin-1, -2, -3 binding sites in the human kidney. *Eur. J. Pharmacol.* 176: 233, 1990.
67. WALDER, C. E., G. R. THOMAS, C. THIEMERMANN, and J. R. VANE. The hemodynamic effects of endothelin-1 in the pithed rat. *J. Cardiovasc. Pharmacol.* 13(Suppl. 5): S93, 1989.
68. WANSTALL, J. C., and S. R. O'DONNELL. Endothelin and 5-hydroxytryptamine on rat pulmonary artery in pulmonary hypertension. *Eur. J. Pharmacol.* 176: 159, 1990.
69. WATANABE, T. X., S.-I. KUMAGAYE, H. NISHIO, K. NAKAJIMA, T. KIMURA, and S. SAKAKIBARA. Effects of endothelin-1 and endothelin-3 on blood pressure in conscious hypertensive rats. *J. Cardiovasc. Pharmacol.* 13(Suppl. 5): S207, 1989.
70. WIKLUND, N. P., A. OHLEN, and B. CEDERQUIST. Inhibition of adrenergic neuroeffector transmission by endothelin in the guinea-pig femoral artery. *Acta Physiol. Scand.* 134: 311, 1988.
71. YAMASAKI, H., M. NIWA, K. YAMASHITA, Y. KATAOKA, K. SHIGEMATSU, K. HASHIBA, and M. OZAKI. Specific ^{125}I-endothelin-1 binding sites in the atrioventricular node of the porcine heart. *Eur. J. Pharmacol.* 168: 247, 1989.
72. YANAGISAWA, M., A. INOUE, T. ISHIKAWA, Y. KASUYA, S. KIMURA, S.-I. KUMAGAYE, K. NAKAJIMA, T. X. WATANABE, S. SAKAKIBARA, K. GOTO, and T. MASAKI. Primary structure, synthesis, and biological activity of rat endothelin, an endothelium-derived vasoconstrictor peptide. *Proc. Natl. Acad. Sci. USA* 85: 6964, 1988.
73. YANAGISAWA, M., and T. MASAKI. Molecular biology and biochemistry of the endothelins. *TIPS* 10: 374, 1989.

5

Endothelin Receptors and Receptor Subtypes

HITOSHI MIYAZAKI, MOTOHIRO KONDOH, YASUSHI MASUDA,
HIROTOSHI WATANABE, AND KAZUO MURAKAMI

Endothelin, originally isolated from culture media of porcine aortic endothelial cells, is one of the most potent vasoactive peptide known to date and consists of 21 amino acids containing two intracellular disulfide bonds.[21] In addition to vasoconstrictor activity, it has been found that endothelin has a variety of biological actions: it increases blood pressure;[21,22] stimulates release of atrial natriuretic peptide,[4] prostacyclin, thromboxane A_2, and endothelial cell–derived relaxing factor[3]; and inhibits renin release.[12] Moreover, three isopeptides of endothelin have been identified based on the analysis of a human genomic library and named endothelin-1, endothelin-2, and endothelin-3.[5] These findings raise the question of how the endothelin isopeptides and their various biological actions are related. It is also of great interest to know whether the three endothelin isopeptides have their own specific receptors, respectively, or bind in common to a homologous population of endothelin receptors. The elucidation of these problems is essential for understanding the physiological roles of endothelin.

We describe mainly the interaction between endothelin isopeptides and the receptor subtypes after reviewing the existence of endothelin-specific receptors. It is very important to clarify that endothelin has its own specific receptors, because initially it was speculated that endothelin may be an endogenous dihydropyridine (DHP)-sensitive, voltage-dependent Ca^{2+} channel agonist.

IDENTIFICATION OF ENDOTHELIN-SPECIFIC RECEPTORS USING CHICK CARDIAC MEMBRANES

Comparison between Endothelin Receptors and Voltage-Dependent, DHP-Sensitive Ca^{2+} Channels

The amino acid sequence of endothelins shows striking homology to those of peptide neurotoxins, such as neurotoxins from scorpion venoms and bee venoms, which directly act on membrane ion channels. Endothelin-1–induced vasoconstriction is dependent on extracellular Ca^{2+} and attenuated in the presence of the voltage-dependent Ca^{2+}-channel antagonist nicardipine.[21] These data raise the possibility that endothelin directly binds to DHP-sensitive, voltage-dependent Ca^{2+} channels and opens the channels to increase intracellular

Ca^{2+} levels. To test this hypothesis, the following three experimental approaches were applied[9,10]: (1) competitive binding studies of endothelin-1 and Ca^{2+}-channel antagonists; (2) sucrose gradient sedimentation of solubilized endothelin-1–endothelin-1–binding protein and DHP–Ca^{2+}-channel complexes; (3) immunoprecipitation of solubilized endothelin-1–endothelin-1–binding protein complex with a monoclonal antibody raised against DHP-sensitive Ca^{2+} channels.

First of all, we have identified high-affinity, saturable, and specific binding sites for endothelin-1 and (+)-[^{3}H]PN200-110, a DHP derivative, on chick cardiac membranes. Figure 5.1A shows a representative saturation isotherm for [^{125}I]endothelin-1 binding to the membranes. The data yielded a curvilinear Scatchard plot, demonstrating the presence of two populations of endothelin-1–binding sites (Fig. 5.1B). The binding parameters were $K_d = 3.8 \pm 1.7$ pM, $B_{max} = 820 \pm 280$ fmol/mg protein (N = 5) and $K_d = 18 \pm 12$ pM, $B_{max} = 890 \pm 200$ fmol/mg protein (N = 5), respectively. On the other hand, binding parameters for (+)-[^{3}H]PN200-110 were $K_d = 140 \pm 74$ pM and $B_{max} = 3.0 \pm 0.4$ pmol/mg protein (N = 3).

To examine the hypothesis that endothelin may be an endogenous agonist of DHP-sensitive Ca^{2+} channels, competitive binding studies of endothelin-1 and Ca^{2+}-channel antagonists were performed. [^{125}I]endothelin-1 binding to cardiac membranes was not affected by any of the three Ca^{2+}-channel antagonists nicardipine, verapamil, and diltiazem. In contrast, unlabeled endothelin-1 inhibited binding of [^{125}I]endothelin-1 in a dose-dependent fashion. When (+)-[^{3}H]PN200-110 was incubated with cardiac membranes in the presence of nicardipine or unlabeled endothelin-1, (+)-[^{3}H]PN200-110 binding was

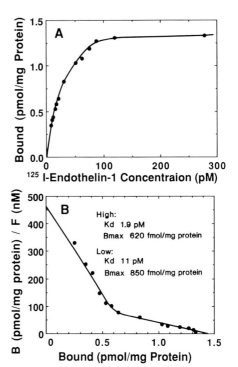

FIGURE 5.1. Saturation studies and Scatchard plot analysis of [^{125}I] endothelin-1 binding to chick cardiac membrane. A: saturation curve of [^{125}I] endothelin-1 binding. Membrane proteins (2 μg) were incubated at 25°C for 2 h in 100 μl of 10 mM Tris/HCl, pH 7.4, 10 mM MgCl$_2$, containing 0.5 mM PMSF, 10 μg/ml each of leupeptin and antipain (buffer A), with increasing concentrations of [^{125}I] endothelin-1 (10–500 pM) in the presence or absence of 100 nM unlabeled endothelin-1. B: Scatchard plot analysis. The data were transformed according to Scatchard analysis to determine the binding parameters.

displaced by nicardipine but not by unlabeled endothelin-1. Next, sucrose gradient sedimentation of (+)-[^{3}H]PN200-110–Ca^{2+} channel and of [^{125}I]endothelin-1–endothelin-1–binding protein complexes was carried out. Cardiac membranes were prelabeled with (+)-[^{3}H]PN200-110 (5 nM) and [^{125}I]endothelin-1 (250 pM), respectively, solubilized with 0.5% digitonin, and subjected to sedimentation. As shown in Figure 5–2A,C, the (+)-[^{3}H]PN200-100–Ca^{2+} channel complex produced a peak corresponding to a sedimentation (S) value of 22.3, while [^{125}I]endothelin-1–endothelin-1–binding protein complex sedimented at S = 14.1. The specificity of the radioligand binding was demonstrated by the absence of peaks when the binding reaction was carried out with 1 μM nicardipine and endothelin-1, respectively (Fig. 5.2B,D). These results are consistent with the idea that endothelin-1 does not bind to the Ca^{2+} channels. To confirm this conclusion, a monoclonal antibody raised against DHP-sensitive Ca^{2+} channels from the chick brain was used to determine whether this antibody recognizes [^{125}I]endothelin-1–endothelin-1–binding protein complex. The result was that digitonin-solubilized (+)-[^{3}H]PN200-110–Ca^{2+} channel complex was immunoprecipitated with the antibody, whereas

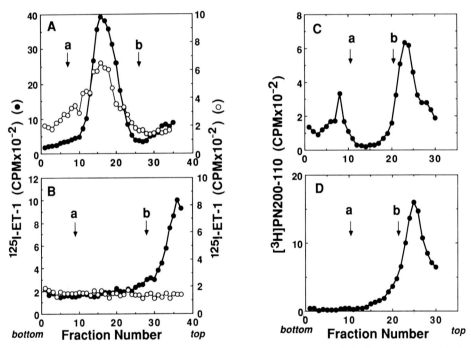

FIGURE 5.2. Sucrose gradient sedimentation of [^{125}I]endothelin-1–endothelin receptor and (+)-[^{3}H]PN200-110–Ca^{2+}-channel complexes. Cardiac membranes (300 μg/ml) were preincubated with [^{125}I] endothelin-1 (250 pM) and (+)- [^{3}H]PN200-110 (5 nM) in 0.5 ml buffer A in the absence (A) or presence (B) of 1 μM unlabeled endothelin-1 and with (D) or without (C) 1 μM nicardipine, respectively, solubilized with 0.5% digitonin, and subjected to sedimentation. After fractionation, one-half of the sample in each tube was counted for radioactivity (●) and the other half was filtrated with poly(ethyleneimine)-treated Whatmann GF/F filter to detect the radiolabeled complex. In the case of (+)-[^{3}H]PN200-110, all the sample in each tube was counted for radioactivity (○, C,D) Thin arrows indicate the position of thyroglobulin (a, 19S) and γ-immunoglobulin (b, 7S). ET, endothelin.

solubilized [^{125}I]endothelin-1–endothelin-1–binding protein complex was not (Fig. 5.3). Based on these data, it has been found that endothelin-1 has its own specific receptors and is not an endogenous agonist of DHP-sensitive, voltage-dependent Ca^{2+} channels.

Estimation of Molecular Weight of the Receptors by Affinity-Labeling Technique

To investigate the molecular weight (M_r) of the [^{125}I]endothelin-1–binding sites, chick cardiac membranes were labeled by covalent cross-linking to membrane-bound [^{125}I]endothelin-1 via the homobifunctional cross-linking agent dissuccinimidyl tartarate (DST). As shown in Figure 5.4, lane A, the autoradiogram of SDS-polyacrylamide gels of the [^{125}I]endothelin-1–labeled materials indicated one major band with an M_r of 53,000 in the presence of the disulfide-reducing reagent 2-mercaptoethanol (2-ME). Taking the M_r of endothelin-1 and DST into consideration, the band at $M_r = 53,000$ corresponds to a protein with an apparent M_r of 50,000. The addition of unlabeled endothelin-1 during the incubation of membranes with [^{125}I]endothelin-1 abolished the appearance of this band in a dose-dependent manner (Fig. 5.4, lanes B–D); at a 10 nM concentration of unlabeled endothelin-1, which is 40-fold higher than that of [^{125}I]endothelin-1 (250 pM) the intensity of the band decreased to <10% (Fig. 5.4, cf. lanes A and C). Unlike unlabeled endothelin-1, the unrelated peptides angiotensin II and substance P did not affect the appearance of the band at final 1 μM concentrations. The same autoradiographic pattern was obtained when the electrophoresis was carried out in the absence of 2-ME (Fig. 5.4, lane

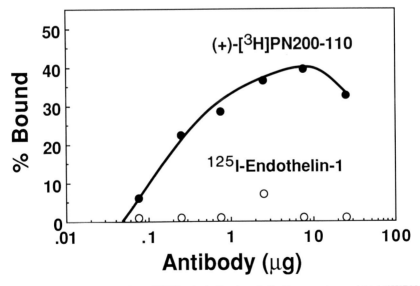

FIGURE 5.3. Immunoprecipitation of [^{125}I]endothelin-1-endothelin receptor and (+)-[^{3}H]PN200-110–Ca^{2+}-channel complexes. Solubilized [^{125}I]endothelin-1-endothelin receptor and (+)-[^{3}H]PN200-110–Ca^{2+}-channel complexes were incubated with various amounts of the monoclonal antibody. The results were expressed as the percentage of the radioactivity adsorbed by protein A–Sepharose against the original activity added into a reaction mixture.

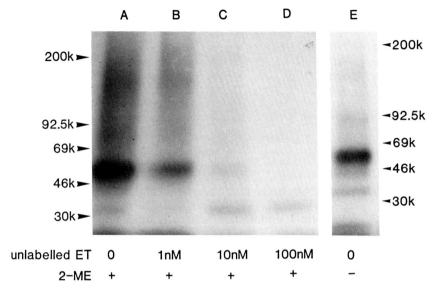

FIGURE 5.4. Affinity labeling of [^{125}I]endothelin-1 to chick cardiac membranes. The membranes (300 μg protein/ml) were incubated with 250 pM endothelin-1 in 0.5 ml buffer A in the presence or absence of indicated concentrations of unlabeled endothelin-1 and washed with 10 mM sodium phosphate, pH 7.4, 120 mM NaCl, and 5 mM EDTA (PBS). After cross-linking with 1 mM dissuccinimidyl tartarate, the samples solubilized with (+) or without (−) 2-mercaptoethanol (2-ME) were subjected to SDS-PAGE (3.5%–13%). Numbers represent M_r.

E), indicating that the affinity-labeled protein is not disulfide-linked to any other protein in cardiac membranes. Together, these data suggest that the band at $M_r = 53,000$ is not the result of artificial cross-linking between several membrane proteins and that the $M_r = 50,000$ protein represents a native endothelin receptor.

Although the equilibrium binding study indicated two populations of endothelin receptors, affinity-labeling analysis failed to substantiate the presence of the two types. These two sets of results seem contradictory, but the following explanations are possible. A curvilinear Scatchard plot results from the fact that there exist two other types of endothelin receptors, having higher affinity for endothelin-3 rather than for endothelin-1 and endothelin-2, in addition to the $M_r = 50,000$ receptor. Although these other receptors have different M_r from the $M_r = 50,000$ species, the reason why only one band appeared by affinity-labeling analysis may be due to the ratio of the concentration of [^{125}I]enothelin-1 against that of membrane proteins in the reaction: i.e., [^{125}I]endothelin-1 preferentially binds to higher affinity receptors for endothelin-1 with $M_r = 50,000$ under the conditions used here.

SUBTYPES OF ENDOTHELIN RECEPTORS: RELATION BETWEEN ENDOTHELIN ISOPEPTIDES AND THE RECEPTOR SUBTYPES

In light of recent evidence demonstrating that one kind of hormones or growth factors possesses multiple receptor forms and diversifies its physiological functions via its receptor subtypes, it is possible that endothelin might also have several different types of receptors. In the case of endothelin this possibility seems more likely because endothelin forms a family consisting of three isopeptides (endothelin-1, endothelin-2, and endothelin-3) and sarafotoxin S6b (STX),[6] a snake venom. If endothelin-receptor subtypes exist, an understanding of the interaction of endothelin isopeptides with the receptor subtypes is of great importance to elucidate the mechanism of various endothelin physiological functions. To investigate this problem, we performed competitive displacement experiments between [^{125}I]endothelins and nonradioactive endothelins on chick and rat tissue membranes and affinity labeling of the membranes by chemical cross-linking with [^{125}I]endothelins.[8,20]

Receptor Subtypes in the Chick Heart

The ability of endothelin-1, endothelin-2, endothelin-3, and STX to displace the binding of [^{125}I]endothelin-1, [^{125}I]endothelin-2, or [^{125}I]endothelin-3 to chick cardiac membranes was examined.[20] All three [^{125}I]endothelin bindings were competitively inhibited by increasing concentrations of unlabeled peptides, as shown in Figure 5.5. The order of potency in displacing the binding of [^{125}I]endothelin-1 and [^{125}I]endothelin-2 was endothelin-1 $\geq$ endothelin-2 > endothelin-3 > STX (Fig. 5.5A,B); in both cases, the potencies of endothelin-1 and endothelin-2 at the IC_{50} level were nearly equivalent to each other but four- to eightfold more potent than that of endothelin-3. In contrast, endothelin-3 was approximately five- to eightfold more effective than endothelin-1 and endothelin-2 in inhibiting [^{125}I]endothelin-3 binding to the membranes (Fig. 5.5C). It should be noted that STX acted as a competitor in every case, although the potency was weak compared with that of the other peptides. These results revealed the presence of at least two distinct types of endothelin receptors on chick cardiac membranes; one of the two has higher affinity for endothelin-1 and endothelin-2 than endothelin-3, and the other is conversely endothelin-3 preferring.

To investigate the M_r of the endothelin-receptor subtypes, the membranes were labeled by covalent cross-linking to membrane-bound [^{125}I]endothelin-1, [^{125}I]endothelin-2, or [^{125}I]endothelin-3 with the homobifunctional cross-linking agent DST.[20] Figure 5.6 shows the autoradiogram of SDS-polyacrylamide gels of [^{125}I]endothelin-labeled materials. [^{125}I]endothelin-2 was exclusively associated with a band migrating at M_r = 53,000 (lane B) that corresponds to a protein with apparent M_r = 50,000, subtracting the M_r of [^{125}I]endothelin-2 and DST. The labeling pattern obtained with [^{125}I]endothelin-2 was almost the same as that with [^{125}I]endothelin-1 (lane A), demonstrating that endothelin-1 and endothelin-2 might bind to the same type of the receptors. On the other hand, affinity labeling of the membranes with [^{125}I]endothelin-3 yielded two major and one minor bands of M_r = 34,000, 46,000, and 53,000, respectively (lane C). Specificity of these bands is indicated by the finding that the appearance of these bands was blocked by an excess amount of unlabeled endothelin-3 (10, 100 nM, lanes D and E). Under the unreducing conditions, the mobility of the M_r = 46,000 species was changed to the position at M_r = 44,000, whereas the other species migrated to the same position in both reducing and unreducing conditions, demonstrating that the affinity-labeled proteins are not disulfide linked to any other proteins on the membranes. Taking into con-

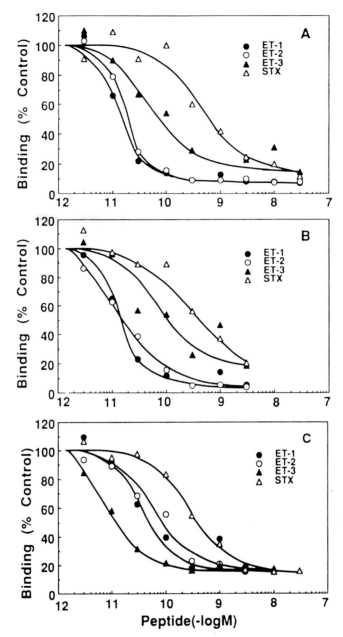

FIGURE 5.5. Competitive studies of [^{125}I]endothelins binding to chick cardiac membranes with unlabeled endothelins (ET) and sarafotoxin (STX). The membranes were incubated with 30 pM [^{125}I]endothelin-1 (A), [^{125}I]endothelin-2 (B), or [^{125}I]endothelin-3 (C) in the presence of indicated concentrations of unlabeled endothelin-1, endothelin-2, endothelin-3, and STX. Each point represents the mean of the duplicate values.

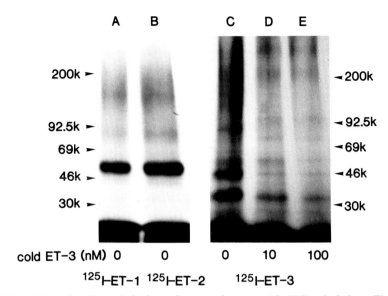

FIGURE 5.6. Affinity labeling of chick cardiac membranes with [^{125}I]endothelins. The membranes (300 μg protein/ml) were incubated in 0.5 ml buffer A with 200 pM [^{125}I]endothelin-1, [^{125}I]endothelin-2, or [^{125}I]endothelin-3 in the presence or absence of unlabeled endothelin-3. Subsequent conditions are as in Figure 5–4. ET, endothelin.

sideration the M_r of [^{125}I]endothelin-3 and DST, the M_r of the binding proteins represented by these three bands should be 31,000, 43,000, and 50,000, respectively. These results are in good agreement with the result obtained from the competitive displacement experiment. In conclusion, there exist three types of endothelin receptors having different M_r to one another on chick cardiac membranes; the $M_r = 50,000$ species exhibits higher affinity for endothelin-1 and endothelin-2 than endothelin-3, whereas the $M_r = 43,000$ and 31,000 species have a preference for endothelin-3 rather than for endothelin-1 and endothelin-2.

Receptor Subtypes in the Rat Lung

Intravenously injected endothelin-1 has been found to be distributed predominantly to the lung, kidney, and liver in rat.[13] In isolated lungs from guinea pig or rat, endothelin-1 causes the release of prostacyclin and thromboxane A_2.[3] In fact, prior to attempting characterization of endothelin receptors, we investigated the concentration of the receptors in various tissues of rat, including lung, brain, liver, and heart. Our results indicate that [^{125}I]endothelin-1 binding levels in the lung were greater than in any other tissues. These lines of information indicate that rat lung can be a suitable source for a detailed characterization of the receptor. Therefore, using rat lung membranes, we performed affinity labeling of the membranes with [^{125}I]endothelin-1, [^{125}I]endothelin-2, and [^{125}I]endothelin-3 to investigate the M_r of rat endothelin receptors and whether endothelin-receptor subtypes also exist in rat tissues.[8]

Covalent cross-linking of lung membranes with [^{125}I]endothelin-1, [^{125}I]endothelin-2, and [^{125}I]endothelin-3, respectively, via DST yielded two major bands with M_r of 45,000 and 35,000 in all cases, as shown in Figure 5–7.

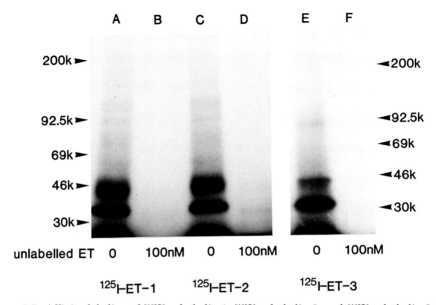

FIGURE 5.7. Affinity labeling of [^{125}I]endothelin-1, [^{125}I]endothelin-2, and [^{125}I]endothelin-3 to rat lung membranes. Rat lung membranes (600 μg/ml) were incubated with 200 pM [^{125}I]endothelin-1, [^{125}I]endothelin-2, [^{125}I]endothelin-3 in the absence (lanes A, C, and E) or presence (lanes B, D, and F) of 100 mM of unlabeled endothelin-1, endothelin-2, and endothelin-3, respectively. Subsequent conditions are the same as in Figure 5.4, except that SDS-PAGE was carried out under nonreducing conditions. ET, endothelin.

These two bands are specific for all the isopeptides of endothelin, because the presence of an excess (100 nM) of each unlabeled endothelin isopeptide during the incubation of the membranes with [^{125}I]endothelin-1, [^{125}I]endothelin-2, or [^{125}I]endothelin-3 abolished the appearance of both bands (lanes B, D, and F). The ratio of intensity of the two bands, however, was different between the labeling with [^{125}I]endothelin-1 and [^{125}I]endothelin-2 and that with [^{125}I]endothelin-3 (cf. lanes A, C, and E): i.e., the ratio of radioactivity incorporated into the two bands was approximately 1:1 when the cross-linking was performed with [^{125}I]endothelin-1 and [^{125}I]endothelin-2, while the intensity of the $M_r = 35,000$ species was fourfold greater than that of the $M_r = 45,000$ species in the case of [^{125}I]endothelin-3. This result suggests that, although all the labeled endothelin isopeptides could bind to both types of endothelin receptors, [^{125}I]endothelin-3 preferentially binds to the lower M_r type of the receptors. To confirm this finding, replacement of [^{125}I]endothelin-1 binding by three isopeptides of endothelin was examined (Fig. 5.8). When endothelin-1 and endothelin-2 were used as competitors for binding of [^{125}I]endothelin-1, the appearance of the $M_r = 45,000$ band was inhibited more effectively than was that of the $M_r = 35,000$ band (lanes A–E); in the presence of 10 nM endothelin-1 and endothelin-2, the intensity of the $M_r = 45,000$ band slightly decreased without a significant change in the intensity of the $M_r = 35,000$ band (lanes B and D). In contrast, the $M_r = 35,000$ band disappeared more effectively than did the $M_r = 45,000$ band (lanes F and G); the appearance of the $M_r = 35,000$ band was completely blocked by 100 nM unlabeled endothelin-3, but at the same concentration of endothelin-3 the $M_r = 45,000$ band did not disappear. The M_r of the two bands described above were calculated from SDS-PAGE

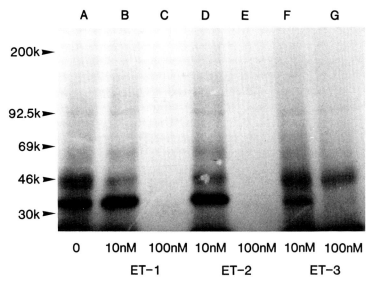

FIGURE 5.8. Competitive displacement of affinity labeling of rat lung membranes with [^{125}I]endothelin-1 by unlabeled endothelin-1, endothelin-2, and endothelin-3. The membranes labeled with [^{125}I]endothelin-1 in the absence (lane A) or presence of indicated concentrations of unlabeled endothelin-1 (lanes B and C), endothelin-2 (lanes D and E), and endothelin-3 (lanes F and G) were cross-linked with 1 mM dissuccinimidyl tartarate and subjected to SDS-PAGE (3.5%–13%) under nonreducing conditions.

without 2-ME followed by autoradiography. These two bands migrated at the positions corresponding to $M_r = 47,000$ and 35,000, respectively, under the reduced conditions, demonstrating that the M_r of the two endothelin receptors should be 44,000 and 32,000, subtracting the M_r of endothelin and DST.

In summary, two different forms of endothelin receptors are present on rat lung membranes. Although both types can bind to each endothelin isopeptide, the $M_r = 44,000$ species has a slightly higher affinity for endothelin-1 and endothelin-2, whereas endothelin-3 preferentially interacts with the $M_r = 32,000$ species.

Receptor Subtypes in Other Tissues

In addition to those in rat lung, endothelin receptors have been identified through biochemical experimentation in other tissues, organs, and several cultured cells.[7,11,15,18] Based on the M_r of the receptors analyzed by affinity-labeling technique, endothelin receptors in rat tissues could be separated into the following three groups: *(1)* $M_r = 31,000$–32,000 species in the lung, kidney, and brain[8,15]; *(2)* $M_r = 42,000$–44,000 species in the kidney, heart, lung, and brain[7,8]; *(3)* $M_r = 55,000$ species in the heart and kidney.[15] The M_r of the receptors represented here were obtained by subtracting the M_r of both endothelins and cross-linking agents from the position of affinity-labeled bands on SDS-polyacrylamide gels.

Physiological roles of endothelin in these tissues have also been investigated. Using isolated lungs from guinea pig or rat, de Nucci et al.[3] found that endothelin-1 induces the release of the potent vasodilatory substances prostacyclin and thromboxane A_2. Fukuda et al.[4] indicated that endothelin-1 inhibits the secretion of atrial natriuretic protein from cultured rat atrial cardiocytes.[4] Badr et al.[1] and Simonson et al.[14] reported multiple responses to endothelin-1 in cultured mesangial cells of the kidney such as contraction, mitogenesis, phosphatidylinositol breakdown, and so on. In the brain we identified the existence of two forms of the receptors by cross-linking technique. Yoshizawa et al.[23] suggested that endothelin is synthesized in the posterior system and may modulate the release of vasopressin or oxitocin.

Although many reports that demonstrate various endothelin physiological functions have accumulated, the relation between endothelin functions and the endothelin-receptor subtypes is not yet clear because multiple forms of the receptors are present at the same time in one tissue. On the other hand, many researchers investigated the mechanism of endothelin physiological functions, and it has been clarified that one of the biological activities of endothelin is expressed through activation of phospholipase C, which produces inositol triphosphate (IP_3) and diacylglycerol (DAG) and a subsequent increase in intracellular Ca^{2+}. Takuwa et al.[17] examined a signal transduction system of endothelin-1 in detail using cultured rat aortic vascular smooth muscle A-10 cells and cultured mouse fibroblast Swiss 3T3 cells.[16] As a result, it was found that endothelin induces mobilization of Ca^{2+} from both intra- and extracellular pools to produce a biphasic IP_3 and DAG was also observed, revealing activation of phospholipase C. Furthermore, it was indicated that the activation of phospholipase C is mediated by a pertussis toxin–insensitive guanine nucleotide-binding protein (G-protein).

Pharmacological Evidence for the Presence of Receptor Subtypes

In addition to the biochemical evidence for the presence of endothelin receptors, several lines of pharmacological evidence have also been demonstrated. Inoue et al.[5] have suggested that, although all three forms of endothelin produce strong and long-lasting pressor responses following a transient depressor response in anesthetized rats in vivo, the profiles of the actions are different from one another.[5] For example, two distinct phases of pressor effects were observed in response to endothelin-1 and endothelin-2, but in the case of endothelin-3 these two separate phases were not clearly seen. Furthermore, endothelin-3 has the most potent initial depressor effect in spite of the weakest potency in terms of presser responses. Warner et al.[19] have demonstrated, using the rat isolated mesentery, that endothelin-3 mainly produces a vasodilation effect through the release of endothelium-derived relaxing factor, while endothelin-1 exhibits mainly vasodilator activity at the low doses but acts as a vasoconstrictor agent at the higher doses. Chang et al.[2] have reported interesting results suggesting the presence of two types of endothelin receptors in neuroblastoma NG108-15 cells; i.e., although pretreatment with mouse vasoactive intestinal constrictor (VIC), which corresponds to endothelin-2, did not prevent the increase of intracellular Ca^{2+} level response to subsequent addition of endothelin-1 or VIC, respectively, the same stimulant was used as the second stimulation and the Ca^{2+} increase was completely abolished. These data cannot be explained only by the difference in the affinity of each form of endothelin to one type of endothelin receptor, but they do indicate the existence of multiple forms of endothelin receptors.

In this chapter, we mainly discussed the interactions of three forms of endothelin with several subtypes of endothelin receptors with different M_r. This issue is getting clearer, but unfortunately little is known about the relationships between endothelin-receptor subtypes, their physiological functions, and the specificity of endothelin isopeptides for a variety of biological activities of endothelin. These problems emphasize the importance of understanding the structural and functional properties of endothelin receptors. One last point to note is that one of the endothelin functions is tightly coupled to DHP-sensitive, voltage-dependent Ca^{2+} channels; this characteristic is very unique and is not observed with any other hormones and growth factors. This raises the possibility that endothelin might possess a new signal transduction system via novel type receptors.

EDITOR'S NOTE

Cloning of Two Endothelin Receptor Subtypes

Shortly after Miyazaki and colleagues submitted their manuscript, two groups independently reported the cloning of two distinct endothelin receptors.[1,2] Both receptors belong to the superfamily of rhodopsin-like receptors, they have seven transmembrane domains, and each is coupled to a G protein. The receptor cloned from the rat lung by Sakurai et al.[1] binds the three endothelin isoforms and sarafotoxins with similar affinity and is coupled through a G protein to phospholipase C. Its activation by endothelin-1 in COS-7 cells transfected with the cloned cDNA leads to transient increases in inositol phosphates and elevation of intracellular free Ca^{2+}. The messenger RNA corresponding to the cDNA is detected in many rat tissues, including brain, kidney, and lung, but not in vascular smooth muscle. These characteristics suggest that this receptor may (among others) be present on endothelial cells mediating the release of prostacyclin and EDRF (see Chapters 15 and 16).

The other receptor, cloned by Arai et al.[2] from the DNA library of bovine lung, has high selectivity for endothelin-1. The values of the inhibition constant (K_i) for endothelin-1, endothelin-2, sarafotoxin 6b, and endothelin-3 were 0.9, 7.2, 52, and 900 nM, respectively.[2] Endothelin-1 evoked fluctuating and long-lasting electric currents in transfected *Xenopus* oocytes, characteristic of the response mediated through activation of the inositol phosphate/Ca^{2+} pathway. The messenger RNA is widely distributed in bovine tissues, including brain, heart, lung, and peripheral tissues. These characteristics suggest that it may represent the vascular smooth muscle endothelin receptor. The Endothelin Receptor Nomenclature Subcommittee of IUPHAR met in Tsukuba City during the Second International Conference on Endothelins and recommended that the receptor selective for endothelin-1[2] be called ET_A and the nonselective one,[1] ET_B.

REFERENCES

1. BADR, K. E., J. J. MURRAY, M. D. BREYER, K. TAKAHASHI, T. INAGAMI, and R. C. HARRIS. Mesangial cell, glomerular and renal vascular responses to endothelin in the rat kidney: elucidation signal transduction pathways. *J. Clin. Invest.* 83: 336–342, 1989.

2. CHANG, T. F. W., N. ISHIDA, K. SAIDA, Y. MITSUI, Y. OKANO, and Y. NOZAWA. Effects of vasoactive intestinal constrictor (VIC) and endothelin on intracellular calcium level in neuroblastoma NG108-15 cells. *FEBS Lett.* 257: 351–353, 1989.
3. DE NUCCI, G., R. THOMAS, P. D'ORLEANS-JUSTE, E. ANTUNES, C. WALDER, T. D. WARNER, and J. R. VANE. Pressor effects of circulating endothelin are limited by its removal in the pulmonary circulation and by the release of prostacyclin and endothelium-derived relaxing factor. *Proc. Natl. Acad. Sci. USA* 85: 9797–9800, 1988.
4. FUKUDA, Y., Y. HIRATA, H. YOSHIMI, T. KOJIMA, Y. KOBAYASHI, M. YANAGISAWA, and T. MASAKI. Endothelin is a potent secretagogue for atrial natriuretic peptide in cultured rat atrial myocytes. *Biochem. Biophys. Res. Commun.* 155: 167–172, 1988.
5. INOUE, A., M. YANAGISAWA, S. KIMURA, Y. KASUYA, T. MIYAUCHI, K. GOTO, and T. MASAKI. The human endothelin family: three structurally and pharmacologically distinct isopeptides predicted by three separate genes. *Proc. Natl. Acad. Sci. USA* 86: 2863–2867, 1989.
6. KLOOG, Y., I. AMBAR, M. SOKOLOVSKY, E. KOCHVA, Z. WOLLBERG, and A. BDOLAH. Sarafotoxin, a novel vasoconstrictor peptide: phosphoinositide hydrolysis in rat heart and brain. *Science* 242: 268–270, 1988.
7. MARTIN, E. R., P. A. MARSDEN, B. M. BRENNER, and B. J. BALLERMANN. Identification and characterization of endothelin binding sites in rat renal papillary and glomerular membranes. *Biochem. Biophys. Res. Commun.* 162: 130–137, 1989.
8. MASUDA, M., H. MIYAZAKI, M. KONDOH, H. WATANABE, M. YANAGISAWA, T. MASAKI, and K. MURAKAMI. Two different forms of endothelin receptors in rat lung. *FEBS Lett.* 257: 208–210, 1989.
9. MIYAZAKI, H., M. KONDOH, H. WATANABE, K. MURAKAMI, M. TAKAHASHI, M. YANAGISAWA, S. KIMURA, K. GOTO, and T. MASAKI. Identification of the endothelin-1 receptor in the chick heart. *J. Cardiovasc. Pharmacol.* 13(Suppl. 5): S155–S156, 1989.
10. MIYAZAKI, H., M. KONDOH, H. WATANABE, M. MASUDA, R. MURAKANI, M. TAKAHASHI, M. YANAGISAWA, S. KIMURA, K. GOTO, and T. MASAKI. Affinity labelling of endothelin receptor and characterization of solubilized endothelin–endothelin–receptor complex. *Eur. J. Biochem.* 187: 125–129, 1990.
11. NAKAJO, S., M. SUGIURA, R. M. SNAJDAR, F. H. BOEHM, and T. INAGAMI. Solubilization and identification of human placental endothelin receptor. *Biochem. Biophys. Res. Commun.* 164: 205–211, 1989.
12. RAKUGI, H., M. NAKAMURA, H. SAITO, J. HIGAKI, and T. OGIHARA. *Biochem. Biophys. Res. Commun.* 155: 1244–1247, 1988.
13. SHIBA, R., M. YANAGISAWA, T. MIYAUCHI, Y. ISHII, S. KIMURA, Y. UCHIYAMA, T. MASAKI, and K. GOTO. Elimination of intravenously injected endothelin-1 from the circulation of the rat. *J. Cardiovasc. Pharmacol.* 13(Suppl. 5): S98–S101, 1989.
14. SIMONSON, M. S., S. WANN, P. MENE, G. R. DUBYAK, M. KESTER, Y. NAKAZATO, J. R. SEDOR, and M. J. DUNN. Endothelin stimulates phospholipase C, Na^+/H^+ exchange, c-*fos* expression, and mitogenesis in rat mesangial cells. *J. Clin. Invest.* 83: 708–712, 1989.
15. SUGIURA, M., R. M. SNAJDAR, M. SCHWARTZBERG, K. F. BADR, and T. INAGAMI. Identification of two types of specific endothelin receptors in rat mesangial cell. *Biochem. Biophys. Res. Commun.* 162: 1396–1401, 1989.
16. TAKUWA, N., Y. TAKUWA, M. YANAGISAWA, K. YAMASHITA, and T. MASAKI. A novel vasoactive peptide endothelin stimulates mitogenesis through inositol lipid turnover in Swiss 3T3. *J. Biol. Chem.* 264: 7856–7861, 1989.
17. TAKUWA, Y., Y. KASUYA, N. TAKUWA, M. KUDO, M. YANAGISAWA, K. GOTO, T. MASAKI, and K. YAMASHITA. Endothelin receptor is coupled to phospholipase C via a pertussis toxin–insensitive guanine nucleotide-binding regulatory protein in vascular smooth muscle cells. *J. Clin. Invest.* 85: 653–658, 1990.
18. WADA, K., H. TABUCHI, R. OHBA, M. SATOH, Y. TACHIBANA, N. AKIYAMA, O. HIRAOKA, A. ASAKURA, C. MIYAMOTO, and Y. FURUICHI. Purification of an endothelin receptor from human placenta. *Biochem. Biophys. Res. Commun.* 167: 251–257, 1990.
19. WARNER, T. D., G. DE NUCCI, and J. R. VANE. Rat endothelin is a vasodilator in the isolated mesentery of the rat. *Eur. J. Pharmacol.* 159: 325–326, 1989.
20. WATANABE, H., H. MIYAZAKI, M. KONDOH, Y. MASUDA, S. KIMURA, M. YANAGISAWA, T. MASAKI, and K. MURAKAMI. Two distinct types of endothelin receptors are present on chick cardiac membranes. *Biochem. Biophys. Res. Commun.* 161: 1252–1259, 1989.
21. YANAGISAWA, M., H. KURIHARA, S. KIMURA, Y. TOMOBE, M. KOBAYASHI, Y. MITSUI, Y. YAZAKI, K. GOTO, and T. MASAKI. A novel potent vasoconstrictor peptide produced by vascular endothelial cells. *Nature* 332: 411–415, 1988.
22. YANAGISAWA, M., A. INOUE, T. ISHIKAWA, Y. KASUYA, S. KIMURA, S. I. KUMAGAYE, K. NAKAJIMA, T. X. WATANABE, S. SAKAKIBARA, K. GOTO, and T. MASAKI. Primary structure, syn-

thesis, and biological activity of rat endothelin, an endothelium-derived vasoconstrictor peptide. *Proc. Natl. Acad. Sci. USA* 85: 6964–6967, 1988.
23. YOSHIZAWA, T., O. SHINMI, A. GIAID, M. YANAGISAWA, S. J. GIBSON, S. KIMURA, Y. UCHIYAMA, J. M. POLAK, T. MASAKI, and I. KANAZAWA. Endothelin: a novel peptide in the posterior pituitary system. *Science* 247: 462–464, 1990.

REFERENCES TO EDITOR'S NOTE

1. SAKURAI, T., M. YANAGISAWA, Y. TAKUWA, H. MIYAZAKI, S. KIMURA, K. GOTO, and T. MASAKI. Cloning of a cDNA encoding a non-isopeptide-selective subtype of the endothelin receptor. *Nature* 348: 732–735, 1990.
2. ARAI, H., S. HORI, I. ARAMORI, H. OHKUBO, and S. NAKANISHI. Cloning and expression of a cDNA encoding and endothelin receptor. *Nature* 348: 730–732, 1990.

6

Tissue Specificity of Endothelin Synthesis and Binding

LYNNE H. PARKER BOTELHO, CHRISTINA CADE,
PENNY E. PHILLIPS, AND GABOR M. RUBANYI

Endothelin, a potent vasoconstrictor peptide characterized by two intrachain disulfide bonds, was isolated from the cell culture media of porcine aortic endothelial cells,[96] a rich source of both contracting and relaxing factors.[22,31,57] Following the initial isolation and sequencing of the peptide by Yanagisawa et al.,[96] the same group constructed a cDNA library from porcine aortic endothelial cells and isolated, cloned, and sequenced the cDNA encoding this peptide. Based on the predicted sequence of the gene product, a putative biochemical processing pathway was proposed for conversion of a 203 amino acid preproprotein to the mature, biologically active 21 residue peptide. Applying a similar strategy for isolation of the human gene, the cDNA for human endothelin was isolated from a placental cDNA library and similarly sequenced.[10,40] The predicted amino acid sequence of the mature human peptide following cellular processing of the prepropeptide precursor was found to be identical to that of porcine endothelin. Subsequently, a preproendothelin-related gene from rat was also cloned and sequenced.[96] The 21 amino acid peptide product from the rat gene exhibited an identical sequence pattern of cysteines to that of human and porcine endothelin but differed in amino acid composition by six residues. Sequence differences among the original three endothelin peptides from porcine, human, or rat were initially thought to be species variations of a single gene product. Later, however, it was discovered that there are at least three genes coding for "endothelin-like" sequences in mammalian genomes.[38] This discovery resulted in the renaming of the peptides; the original porcine endothelial cell–derived endothelin, which was identical to the human endothelin, was termed "endothelin-1," and the rat endothelin was designated "endothelin-3." The third endothelin gene sequence[38] predicted a 21 amino acid peptide with only two amino acid differences from endothelin-1 and was named "endothelin-2." Recently, a fourth "endothelin-like" gene was identified by cloning and sequence analysis of a mouse genome.[71] The proposed gene product deduced from the sequence was named "endothelin-β" or vasoactive intestinal constricting (VIC) peptide. It has been suggested that this fourth peptide is a product of the endothelin-2 gene.[93] Each of the mature 21 amino acid endothelin isoforms have, in addition to potent contractile activity, many other interesting biological activities in vascular and nonvascular tissues.[2-4,16,88]

A large number of publications on sites of endothelin synthesis, binding, and action have appeared subsequent to the original report by Yanagisawa et al.[93] We summarize the current literature relating to the cellular sites of endothelin synthesis, including tissue- and/or organ-specific gene expression. A corollary to the question of where the various endothelin isoforms are synthesized is whether there are specific, physiologically relevant endothelin receptors for the individual endothelin isoforms and on which cells and/or tissues they are located. A further extension of this inquiry is whether endothelin acts in an autocrine, paracrine, or endocrine fashion (or some combination of the three) and whether there is functional differentiation between the isoforms. An attempt is made to address these questions as the accumulated data for endothelin synthesis and binding are reviewed.

TISSUE SPECIFICITY OF ENDOTHELIN SYNTHESIS

Endothelin Synthesis at the Cellular Level

Cellular synthesis of the big endothelin and endothelin isoforms has been assessed in cultured cells, tissue samples, and whole animals by several different techniques. The techniques cover both gene transcription as assessed by mRNA content as well as mRNA translation into a protein product.

At the cellular level, when the expression of the various genes for endothelin were examined in cultured cells using Northern blot analysis to quantitate mRNA, detectable amounts were reported for preproendothelin-1 but not for preproendothelin-2 or preproendothelin-3 (Table 6.1). The gene for endothelin-1 has been found to be constitutively expressed in cultured endothelial and epithelial cells and in various human cells of carcinoma origin.[20,39,40,52,81,96] Exposure of the cultured endothelial cells to various inducing agents, some of which increase intracellular calcium and/or activate protein kinase C, have been found to result in a severalfold increase in preproendothelin-1 mRNA levels.[38,49,52,81,96,100] Although no mRNA for endothelin-2 has been detected, it is not clear yet whether the gene for endothelin-2 is simply not expressed in the cells that have been tested or whether the tools for detecting endothelin-2 gene expression are not available yet. On the other hand, human umbilical vein endothelial cells have been specifically probed for endothelin-3 mRNA and were found to be negative.[9]

TABLE 6.1. Detection of endothelin-1 (ET-1) mRNA in cultured cells[a]

Cell type	Tissue source	Isoform	Reference
Endothelial	Porcine aorta	ET-1	96
	Bovine aorta	ET-1	20
	Human umbilical vein	ET-1	39, 40
	Human retinal microvessel	ET-1	52
Epithelial	Human amnion	Not specified[b]	81

[a]mRNA quantitation was done by Northern blot analysis. Total RNA was fractionated by size on formaldehyde–agarose gels, and transfer blots were subsequently probed with various cDNA inserts or chemically synthesized oligonucleotides.

[b]A cDNA probe for preproendothelin-1 was used for Northern analysis, but authors do not rule out the possibility that the mRNA detected is derived from endothelin-2 or endothelin-3.

The second means of determining cellular synthesis of the preproendothelin isoforms involves quantitation of accumulated protein products of mRNA translation. This is normally accomplished by immunological detection techniques such as radioimmunoassay (RIA) or ELISA of cell culture media. Currently, the endothelin-1 isoform and in some cases its protein precursor, big endothelin-1, have been detected in the culture media of endothelial, epithelial, and carcinoma cells derived from different tissues and organs from several species (Table 6.2). Inducing agents that increase the endothelin-1 mRNA levels severalfold result in smaller, but significant increases in the corresponding protein products.[20,30,64,73,82,100] The only cells to date that have been found to produce the immunologically reactive endothelin-2 peptide are monkey Cos-7 kidney cells,[47] and there are no data available on the biological precursor big endothelin-2. Similarly, an immunologically reactive endothelin-3 has been quantitated thus far exclusively in culture media from canine tracheal epithelial cells, and again there are no data available on the presence or absence of the precursor big endothelin-3.[8] Many cell lines have been investigated for endothelin isoform synthesis and have been found not to have detectable levels of any of the immunologically reactive peptides (Table 6.3).

A major problem with the quantitation of immunologically reactive endothelin products is the lack of specific antibodies that differentiate among the pre-forms and the final isoforms and, in particular, between endothelin-1 and endothelin-2. Many laboratories are using commercially available antibodies that exhibit significant cross-reactivity among the three isoforms (Table 6.4). Therefore, studies reporting quantitation of endothelin-1 that have not separated the individual isoforms by some physical separation technique such as high-performance liquid chromatography (HPLC) or that have not used a sandwich ELISA technique employing antibodies with high specificity for the different isoforms probably have not adequately distinguished among the isoforms. As a consequence, many of the values reported for endothelin-1 could theoretically be a mixture of endothelin-1 and endothelin-2. Quantitation of endothelin-3 is less of a problem, because there is less cross-reactivity with the other isoforms.

The lack of proper immunological tools for quantitation of big endothelin isoforms is also a serious problem in the interpretation of data reporting quantitation of big endothelin-1 (Table 6.2). This problem is even more complicated compared with the cross-reactivity with the endothelin isoforms, because the currently available big endothelin-1 antibody from commercial sources recognizes an unspecified sequence of the C-terminal residues (22–39) of porcine big endothelin-1 (1–39). It is known that this antibody cross-reacts with human big endothelin-1 (1–38), but it is not known what the cross-reactivity would be with big endothelin-2 or big endothelin-3 because these have not been synthesized and tested yet.

Endothelin Synthesis at the Tissue Level

Tissue production of endothelin isoforms also has been addressed at the levels of both message production and translation into protein product. In the original report, Yanagisawa et al.[96] detected endothelin-1 mRNA in porcine aortic endothelial cells, but not in tissue homogenates of porcine brain, atrium, lung,

TABLE 6.2. Basal release of endothelin from cultured cells[a]

Cell	Tissue source	Species	Endothelin (pmol/10⁶ cells/24 h)			Reference
			ET-1	Big ET-1	Total	
Vascular						
Endothelial	Carotid artery	Bovine	0.15	ND[b]	—	20
	Aorta	Porcine	2.50	0.1	2.6	20
	Pulmonary artery	Bovine	3.30	3.50	6.8	20
	Umbilical vein	Human	0.70	0.1	0.8	20
	Fetal heart	Bovine	2.30	0.20	2.5	20
	Aorta	Bovine	2.90	1.40	4.3	20
	Aorta	Porcine	2.93	ND	—	49
Capillary endothelial	Retinal	Bovine	0.01	ND	—	85
Capillary endothelial	Adrenal cortex	Bovine	3.6	3.6	7.2	30
Kidney						
COS 7	Kidney	Monkey	0.26	ND	—	47
CV-1	Kidney	Monkey	0.49	ND	—	47
Vero	Kidney	Monkey	0.11	ND	—	47
MDCK	Kidney	Canine	0.22	ND	—	47
RK-13	Kidney	Rabbit	0.05	ND	—	47
NRK-52C	Kidney	Rat	0.11	ND	—	47
BHK-21	Kidney	Hamster	0.02	ND	—	47
MDCK	Kidney	Canine	0.12	0.04	0.16	82
LLCPK	Kidney	Porcine	0.06	ND	—	76
Carcinomas						
A549	Lung	Human	0.14	0.06	0.20	82
WiDr	Colon	Human	0.03	0.002	0.03	82
LoVo	Colon	Human	NA	0.015	0.015	82
MCF-7	Breast	Human	0.014	0.001	0.015	82
HeLa-S3	Cervix	Human	0.011	0.001	0.012	82
Others						
Hep G-2	Liver	Human	0.04	0.11	0.14	82
Epithelial	Endometrium	Rabbit	11	ND	—	64
Epithelial	Trachea	Canine	0.15	ND	—	8

[a]Release rates were determined by quantitating immunologically reactive products in serum-free cell culture media. Endothelin-1 levels could in some cases represent a mixture of isoforms 1 and 2 as explained in the text.
[b]ND indicates that the levels of big endothelin were not determined.

or kidney. Similarly, Inoue and coworkers[39] reported measurable levels of endothelin-1 mRNA in human umbilical vein endothelial cells, but not in tissue homogenates of human frontal cortex, atrium, kidney, liver, spleen, stomach, placenta, testis, ovary, or uterus. In contrast, more recent investigations have identified endothelin mRNA by Northern analysis and in situ hybridization in many tissues from various species, including human and pig (Table 6.5). For

TABLE 6.3. Cultured cells that do not release endothelin[a]

Cell type	Tissue source	Species	Reference
Vascular			
A-7 VSMC	Thoracic aorta	Rat	C. C.[b]
A-10 VSMC	Thoracic aorta	Rat	C. C.
H-9 myoblast	Heart	Rat	C. C.
Neural			
B104	Neuroblastoma	Rat	C. C.
C6	Glial tumor	Rat	C. C.
Carcinoma			
HL60	Leukemia cells	Human	C. C.
S49	Lymphoma	Mouse	C. C.
K562	Leukemia	Human	81
G361	Melanoma	Human	81
C32	Melanoma	Human	81
Other			
CHO-Kl epithelial	Ovary	Chinese hamster	82
BALB/3T3	Fibroblasts	Mouse	82
Swiss 3T3	Fibroblasts	Mouse	82, C. C.
Myometrial	Uterus	Rabbit	64

[a] Conditioned medium was prepared by exposing cell monolayers to serum-free media for at least 24 h. Specific RIAs or sandwich ELISAs were used to determine presence of any immunoreactive endothelin.
[b] C. C., C. Cade, unpublished data.

TABLE 6.4. Cross-reactivity of endothelin (ET) isoforms with antibodies used in RIAs

	IC_{50} (fmol/assay)		
RIA	ET-1	ET-2	ET-3
ET-1	14 ± 2	5 ± 1	10 ± 15
ET-2	1 ± 0.3	1.7 ± 0.9	NA
ET-3	337 ± 22	190 ± 30	12 ± 3
Big ET-1	NA	NA	NA

All RIAs were carried out at 4°C for 24 h in PBS supplemented with 0.1% BSA and 0.1% Triton X-100. Antibodies to endothelin-1, endothelin-3, and big endothelin-1 were purchased from Peninsula and Peptide Institute. Unlabeled peptides were used in the range 0.1–1000 fmol per tube to determine cross-reactivity with each of the antisera used. Antibody against endothelin-2 was a gift from Dr. Sadao Kimura (Inst. Med. Sci., Univ. Tsukuba, Japan). C. Cade, (unpublished data).

TABLE 6.5. Endothelin (ET) mRNA in tissues

Tissue	Isoform	Detection method[a]	Reference
Porcine			
Aortic intima	ET-1	Northern[b]	96
Paraventricular nuclear neurons	ET-1	In situ	101

(continued)

TABLE 6.5. *(Continued)*

Tissue	Isoform	Detection method[a]	Reference
Rat, adult[c]			
Eye	ET-3	Northern, in situ	52
Intestine	ET-1	Northern	52
Liver	—[d]	Northern	52
Kidney	ET-3	Northern, in situ	52
Lung	ET-1, ET-3	Northern, in situ	52
Heart, right atrium	—	Northern	52
Cerebellum	ET-3	Northern	52
Rat, fetal (19 day)[b]			
Lung	ET-1 and/or ET-3	In situ	52
Intestine	ET-1	In situ	52
Mouse			
Brain	—	Northern	71
Kidney	—	Northern	71
Lung	—	Northern	71
Liver	—	Northern	71
Spleen	—	Northern	71
Intestine	VIC (ET-2)	Northern	71
Human amnion[e]			
Chorion laeve	Not specified	Northern	81
Villous trophoblasts	—		81
Myometrium	—		81
Human neuronal cell bodies			
Spinal cord	ET-1	In situ	24
Human dorsal root ganglia	ET-1	In situ	24
Human, fetal			
Atria	—	Northern	9
Ventricles	—	Northern	9
Adrenals	—	Northern	9
Kidney	ET-3	Northern	9
Liver	—	Northern	9
Lung	ET-1, ET-3	Northern	9
Pancreas	ET-1	Northern	9
Spleen	ET-1, ET-3	Northern	9

[a]Unless otherwise stated, for Northern analysis total RNA was extracted and size-fractionated by electrophoresis on formaldehyde–agarose gels. Transfer blots were subsequently probed with various restriction fragments of cDNA inserts. For in situ analysis, cryostat sections were analyzed with cRNA hybridization probes.

[b]For Northern analysis poly(A)+ RNAs were isolated from the various tissues probed.

[c]Two oligonucleotide probes were synthesized. Based on the published endothelin-3 sequence, one probe was predicted to hybridize strongly to endothelin-3 but weakly to endothelin-1 and endothelin-2, whereas the second probe was expected to hybridize strongly to endothelin-1 and endothelin-3 but not to endothelin-2. Different distributions of endothelin-1 and endothelin-3 mRNA were determined on the basis of hybridization to each probe and the size of the transcript as demonstrated by Northern analysis.

[d]—, mRNA was not detected in these tissues.

[e]A cDNA probe for preproendothelin-1 was used, but authors do not rule out the possibility of cross-reactivity with endothelin-2 or endothelin-3 mRNA.

example, mouse and rat intestine express a 2.5 kb transcript thought to be that of endothelin-1, whereas the kidney, brain, and eye display a 3.7 kb transcript (presumably endothelin-3) and the lung appears to possess both.[52] In addition, endothelin-1 mRNA has been localized to neuronal cell bodies of the human spinal cord and dorsal root ganglia[24] and to porcine paraventricular nuclear neurons,[101] suggesting a neuromodulatory role for endothelin. To date, Northern blot analysis has revealed endothelin-2 (VIC) gene expression only in the mouse intestine and not in other tissues.[71]

These differences may reflect species-specific expression of endothelin isoforms. Alternatively, technical difficulties in preserving endothelin message caused by the short half-life (15 min for endothelin-1 message)[39] may have lead in some tissue samples to appearently negative results. Finally, endothelin may be expressed in a developmentally regulated fashion,[9] as suggested by the finding that both endothelin-1 and endothelin-3 mRNA have been detected in a number of human fetal (lung, spleen, and pancreas) but not adult tissues.

The protein products of endothelin gene expression have also been detected in many tissue homogenates of different organ and species origin (Table 6.6). Using RIA or ELISA techniques as described above, the predominant form detected has been reported to be endothelin-1 and the levels vary from 0.02 pmol/g tissue in rat pituitary to 2.48 pmol/g tissue in porcine renal inner medulla. In almost all tissues tested, endothelin-3 was detected either at lower levels than endothelin-1 or not at all. Some of these observations may reflect cross-reactivity of endothelin-1 or endothelin-2 with the endothelin-3 antibody, but in certain tissues endothelin-3 levels were clearly too high to be explained by cross-reactivity. These tissues included rat pituitary, in which endothelin-3 levels were four times higher than were endothelin-1 levels, in rat inner medulla, in which equal amounts of endothelin-1 and endothelin-3 were detected, and in rat intestine, in which endothelin-3 levels were approximately one-half of the endothelin-1 levels. Interestingly, it was recently reported that the tissue levels of immunoreactive endothelin in the kidney medulla of the rat were significantly depressed in spontaneously hypertensive versus Wistar-Kyoto age-matched controls despite comparable levels in the lung. A significant portion of this activity was found to be endothelin-3, although exact amounts were not determined.[45]

It is important to remember that the levels of immunoreactive endothelin in tissue homogenates represent potential cellular production from a mixture of many cell types, including endothelial cells derived from capillaries and/or small arteries and venules that invariably remain during tissue preparation. Another possible source of immunoreactive endothelin in tissue homogenates could be residual peptide bound to cell surface–binding sites or synthesis by resident cells.

The question of cellular production within tissues has also been addressed with immunohistochemistry (Table 6.7) (see also Chapter 12). Using this technique to probe tissue slices derived from the vascular system with immunofluorescent antibodies, it was determined that endothelin production is localized to human endothelial cells, with no production by vascular smooth muscle cells. Rat and mouse lung tissue slices also reveal a heterogenous production of endothelin, with endothelin-1 the only isoform produced, and only in epithelial, mucous, clara, and serous cells, not in basal or ciliated cells.

TISSUE SPECIFICITY OF ENDOTHELIN SYNTHESIS AND BINDING

TABLE 6.6. Tissue distribution of endothelin (ET) isoforms

Origin	Tissue	Species	ET-1/ET-2 (pmol/g tissue)[a]	ET-3	Reference
Brain	Cortex	Porcine	0.23	ND[b]	43
	Spinal cord	Porcine	0.12	0.01	77
	Hypothalamus	Porcine	0.28	ND	43
	Cerebellum	Rat	0.10	0.06	54
	Cerebrum	Rat	0.16	0.04	54
	Medulla oblongata	Rat	0.09	0.02	54
	Pituitary	Rat	0.02	0.08	54
Heart	Atrium	Porcine	0.38	ND	43
	Atrium	Rat	0.36	ND	94
	Ventricle	Porcine	0.36	ND	43
	Ventricle	Rat	0.38	ND	94
Blood vessel	Aorta	Rat	0.07	0.001	54
	Umbilical artery	Human	0.90	ND	29
	Umbilical vein	Human	0.62	ND	29
Kidney	Inner medulla	Porcine	2.48	ND	43
	Outer medulla	Porcine	0.12	ND	43
	Inner medulla	Rat	1.75	1.75	45
	Cortex	Rat	0.23	ND	45
	Cortex	Porcine	0.14	ND	43
	Kidney	Rat	0.11	0.01	54
Gastrointestinal tract	Duodenum	Porcine	0.14	ND	43
	Duodenum	Rat	0.16	ND	54
	Stomach	Rat	0.22	0.004	54
	Intestine	Rat	0.31	0.18	54
	Colon	Rat	0.44	0.02	54
	Pancreas	Rat	0.16	0.004	54
Other	Lung	Porcine	0.88	ND	43
	Lung	Rat	1.36	0.0	54
	Liver	Porcine	0.10	ND	43
	Liver	Rat	0.08	0.009	54
	Spleen	Porcine	0.23	ND	43
	Spleen	Rat	0.16	0.001	54
	Adrenal gland	Rat	0.05	0.005	54
	Urinary bladder	Rat	0.06	0.001	54
	Testis	Rat	0.23	0.003	54
	Placenta	Human	~0.04	0.002	29

[a]Tissue homogenates were generally concentrated on C-18 cartridges and assayed by RIA. In some cases the endothelin antiserum used exhibited strong cross-reactivity with all three isoforms and thus may represent a mixture of endothelin-1/endothelin-2 and endothelin-3.
[b]ND indicates not determined specifically for endothelin-3.

Plasma, Urine, and Cerebrospinal Fluid Concentrations of Endothelin

Endothelin synthesis has been shown to occur in vascular endothelial cells that line the blood vessels. Significant quantities of endothelin are detected in circulating plasma in several species (Table 6.8), implying that the release of endothelin is not restricted to the abluminal surface in that endothelin diffuses only from the site of synthesis (endothelium) toward underlying smooth mus-

TABLE 6.7. Immunohistochemistry[a]

Cell/tissue	Species	Endothelin-1[b]	Reference
Vascular			
Endothelial cells	Human	+	35
Vascular smooth muscle cells	Human	−	35
Adventitial fibroblasts	Human	−	29
Umbilical cord vein tissue	Human	+	29
Umbilical cord vein endothelial cells	Human	+	29
Neural			
Glial cells	Rat	+	13
Neurons	Rat	+	13
Dorsal horn spinal cord	Porcine	+	99
Motorneurons	Porcine	+	77
Dorsal horn neurons	Porcine	+	77
Dorsal horn of spinal cord	Porcine	+ +	77
Neurons or spinal cord	Human	+	24
Neurons of dorsal root ganglia	Human	+	24
Lung			
Epithelial cells	Rat/mouse	+	70
Mucous cells	Rat/mouse	+	70
Serous cells	Rat/mouse	+	70
Clara cells	Rat/mouse	+	70
Basal cells	Rat/mouse	−	70
Ciliated cells	Rat/mouse	−	70

[a]For demonstration of endothelin immunoreactivity, tissue sections were incubated with antiendothelin-1, antiserum and subsequently probed with a second antibody conjugated to FITC using indirect immunofluorescence histochemistry. Alternatively, biotinylated second antibody was employed and the presence of endothelin demonstrated using the biotin–avidin enzyme complex method. In all cases the specificity of the antiserum was confirmed by incubation of sections with antiserum preabsorbed with synthetic peptide (see also Chapter 12).

[b]Endothelin-1 levels in some cases could represent a mixture of the three isoforms because of cross-reactivity of the antiserum used.

TABLE 6.8. Plasma levels of endothelin-1 in various species[a]

Species	Endothelin-1	Reference
Porcine	10.8 pM	67
Human[b]	0.10–7.5 pM	97
Rat[b]	0.7–1.6 pM	72, 83
Sheep	<8 pM[c]	58
Dog	25 pM	11

[a]Big endothelin has been detected in both human and rat plasma.

[b]Plasma samples were concentrated on C-18 Sep-Pak cartridges and assayed for immunoreactive endothelin by RIA or ELISA.

[c]In this study the detection limit of the RIA employed was 8 pM.

cle cells. This nonpolarized release or secretion is in line with the data showing that endothelin synthesis is constitutive and that endothelin is not stored in secretory granules. The circulating concentrations of endothelin-1 are in the low picomolar range, which could imply a possible endocrine action of endothelin.

The studies to date measured exclusively circulating levels of endothelin-1.[11,58,65,72,83,98] It was discovered in our laboratory that dog plasma contains a significantly higher concentration of endothelin-3 (213 ± 31 pM, N = 8) compared with endothelin-1 (25 ± 4 pM, N = 8). Because the antibodies employed in the RIAs do not distinguish between endothelin-1 and endothelin-2, it is feasible that the endothelin-1 value is a composite of both isoforms. No big endothelin-1 was detected (N = 8). The relatively high circulating concentrations of endothelin-3 in dog plasma were surprising (although none of the studies have been carried out on canine endothelial cells), because there have been no reports to date of endothelin-3 production in cultured endothelial cells. There is a single report of endothelin-3 production in dog tracheal epithelial cells,[8] but this would not explain the high circulating plasma concentrations. The origin, the physiological significance, and the potential endocrine function of these relatively high circulating levels of endothelin-3 in dog plasma are unknown at present.

In addition to the circulating peptide, immunoreactive endothelin has been detected in human urine.[7] It was found that concentrations of endothelin were on average sixfold higher in urine samples than in plasma. Similarly, endothelin is present in normal human cerebrospinal fluid (CSF) at levels that are significantly greater (about sevenfold) than in plasma.[33,36] Endothelin has also been quantitated in bronchial lavage fluid where the levels were elevated during the bronchospastic phase of an asthma attack and returned to basal levels after recovery.[62]

TISSUE SPECIFICITY OF ENDOTHELIN BINDING

Since the original discovery of endothelin numerous papers have been published on binding of the various isoforms to cell surface receptors. The currently available binding data have been obtained by a number of different approaches using intact cultured cells, isolated tissue membranes, and cryostat sections of whole tissues and whole animals in which tissue distribution was determined in situ.

Cellular Binding

The first reports on endothelin binding to specific cellular receptors was done in a rat thoracic aorta vascular smooth muscle cell line.[32] Since that time, there have been many reports of specific binding to cultured cells (Table 6.9). The binding of endothelin to cell surface receptors is rapid, specific and saturable, but the dissociation rate is very slow with a half-life of days and in most cases, the dissociation is only partial.

The binding studies compiled in Table 6.9 were all conducted with cultured cells, but under various experimental conditions. Many of the studies

TABLE 6.9. Endothelin (ET) isoform binding to cultured cells[a]

Cell	Isoform	$K_{d,\text{apparent}}$ (pM)[b]	B_{max} (Sites/cell)	Reference
Nonvascular				
Retinal capillary pericytes (bovine)	ET-1	1300	$1-2 \times 10^5$	50
Retinal pericytes (bovine)	ET-1	140	1.5×10^5	85
Adrenal zona glomerulosa (calf)	ET-1	100	6×10^4	17
Glomerular mesangial cells (rat)	ET-1	—[c]	6.8×10^2	80
	ET-1	—	27.6×10^2	80
Swiss 3T3 fibroblasts (mouse)	ET-1	180	3.8×10^5	87
	ET-1	500	1×10^5	63
	ET-2	520	1.2×10^5	63
MDCK renal epithelial (canine)	ET-1	40	5×10^3	61
LLC/PK$_1$ renal epithelial (porcine)	ET-1	50	5×10^3	61
Vascular				
Smooth muscle, aorta (rat)	ET-1	60	2.9×10^3	14
	ET-1	300	1.3×10^4	34
	ET-1	141	7.3×10^2	68
Smooth muscle, umbilical vein (human)	ET-1	126	1×10^4	15
Cardiocytes (rat)	ET-1	600–900	$5-8 \times 10^4$	32

[a]Binding studies were performed with [^{125}I]endothelin-1 or [^{125}I]endothelin-2 on confluent cell monolayers. Scatchard analysis of the binding data was used to calculate the apparent $K_{d,\text{apparent}}$ and B_{max} values.
[b]Because many of these studies were carried out at 37°C, and K_d values in these studies more accurately represent estimates of apparent half-maximal saturation constants (see text for details).
[c]—, Data were not available.

were carried out at 37°C or at room temperature instead of at 4°C. The use of higher temperatures complicates the interpretation, because the values documented probably represent internalization of ligand as well as cell surface binding. In addition, most studies involve a Scatchard analysis of the saturation binding data from which a K_d value and a maximal binding concentration have been calculated. Because the binding of endothelin-1 is not rapidly and completely reversible, this derivation of a K_d value is at best an estimate of an apparent half-maximal saturation concentration ($K_{d,\text{apparent}}$).

The values for $K_{d,\text{apparent}}$ obtained from the various studies on cells range from 40 to 1300 pM, which agrees with the concentrations required for production of a biochemical signal in isolated cells. As a specific example, a Scatchard analysis of the saturation binding data for endothelin-2 binding to 3T3 fibroblasts (Fig. 6.1) yields a $K_{d,\text{apparent}}$ of 24.2 ± 3.3 pM. Endothelin-2 increases intracellular free calcium in these cells (measured as Fura-2 fluorescence) from a resting level of ~100 nM to a peak level of ~800 nM in a concentration-dependent manner, with an EC$_{50}$ of ~1 nM (Fig. 6.2).

Endothelin-1 receptor autoradiography using cryostat sections has also been performed, and a diverse tissue distribution and density of binding sites have been demonstrated in rat, human, rabbit, monkey, and canine tissues as shown in Table 6.10. Binding studies using isolated membrane preparations have been conducted, and Scatchard analyses of saturation binding curves

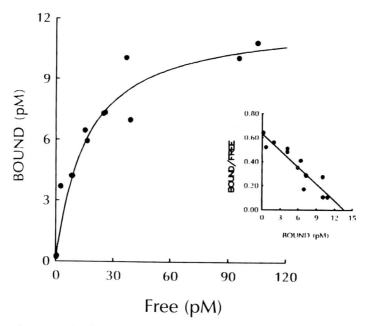

FIGURE 6.1. Saturation binding curve and Scatchard analysis of endothelin-2 binding to Swiss 3T3 fibroblasts. Increasing concentrations from 0.1 to 100 pM of [^{125}I]endothelin-2 (2200 Ci/mmol) were incubated with confluent monolayers in 60 mm tissue culture dishes for 3 h at 4°C. Cells were harvested by addition of Triton X-100 (0.5 ml, 0.5% v/v) to the monolayers for at least 5 min to solubilize the cells, and the cell-associated radioactivity was counted. Nonspecific binding was obtained in the presence of 10 μM endothelin-1. These data were obtained from three experiments in which individual data points were obtained in triplicate.

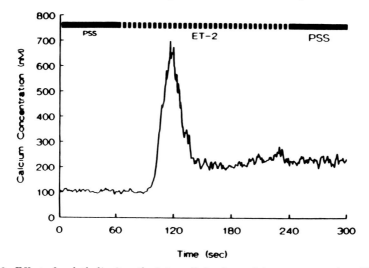

FIGURE 6.2. Effect of endothelin-2 on the intracellular free calcium concentration of Swiss 3T3 cells. A monolayer of porcine pulmonary artery endothelial cells, on a 22 × 22 mm glass coverslip, was loaded with Fura-2 and mounted in a flowthrough spectrofluorometer. A physiological salt solution (PSS) was passed through the cell chamber (5 ml) at 2 ml/min. The solution bathing the cells was changed to one containing 100 nM endothelin-2 (ET-2) at 60 s, and the cells were returned to PSS after 3 min. The intracellular calcium concentration was calculated according to the method of Grynkiewicz et al.[26]

have yielded $K_{d,\,apparent}$ and B_{max} values as shown in Table 6.11. A representative saturation binding experiment conducted in our laboratory on dog heart membranes is shown in Figure 6.3. The binding of endothelin-1 to these membrane receptors is rapid, saturable, and specific, but not rapidly or completely reversible. The lack of reversibility has been demonstrated by two techniques: addition of excess cold ligand (Fig. 6.4) and infinite dilution of the ligand–receptor complex (Fig. 6.5). As is demonstrated in these figures, the dissociation rate of bound endothelin-1 appears to have two components: a rapidly

TABLE 6.10. Densities of [^{125}I]endothelin-1 binding sites[a]

Tissue	Species	Density (fmol/mg^{-1} protein)[b]	References
Cardiac			
Ventricle	Rat	580	48
Neural			
Telencephalon	Rat	600	48
Mesencephalon	Rat	880	48
Metencephalon	Rat	2860	48
Cerebellum	Rat	882	42
Hypothalamus	Rat	721	42
Thalamus	Rat	517	42
Olfactory bulbs	Rat	493	42
Hippocampus	Rat	372	42
Caudate putamen	Rat	255	42
Cerebral cortex	Rat	193	42
Caudale putamen	Rat	330	42
Caudate (body)	Human	874	42
Cerebellum molecular	Rat	1660	42
Cerebellum molecular	Human	872	42
Cerebellum granular	Rat	3219	42
Cerebellum granular	Human	1377	42
Pituitary anterior	Rat	509	42
Pituitary anterior	Rat	3414	42
Pituitary posterior	Rat	117	42
Spinal cord grey	Human	197	42
Spinal cord white	Human	130	42
Hippocampus str. molec.	Rat	1776	42
Hippocampus str. molec.	Human	1830	42
Choroid plexus	Rat	2934	42
Medial thalamus	Rat	1075	42
Frontoparietal cortex I	Rat	253	42
III	Rat	416	42
VI	Rat	533	42

(*continued*)

dissociating component that is 20%–30% of the bound ligand and a slower component with an estimated half-life of days.

In addition to the concern described above regarding the derivation of K_d values where the ligand binds irreversibly within the time-frame of the experiment, another source of confusion regarding literature values for K_d is the result of competitive ligand-binding experiments mistakenly referred to as "ligand displacement experiments." This type of experiment yields data on the concentration of ligand required to inhibit the binding of labeled ligand by 50%

TABLE 6.10. (Continued)

Tissue	Species	Density (fmol/mg^{-1} protein)[b]	References
Temporal cortex I–VI	Human	276	42
VI	Human	670	42
Inferior olive	Rat	1825	42
Lateral geniculate	Rat	1719	42
Pretectal nucleus	Rat	2167	42
Habenular nucleus	Rat	2689	42
Fornix	Rat	619	42
Renal			
Kidney	Rat	1060	48
Glomeruli	Rat	6672	41
Medulla	Rat	823	41
Renal papilla	Rat	2276	41
Glomeruli	Monkey	783	41
Medulla	Monkey	2387	41
Glomeruli	Human	701	41
Medulla	Rabbit	330	41
Cortex	Rabbit	1–2	41
Glomeruli	Canine	2306	41
Medulla	Canine	1170	41
Other			
Lung	Rat	2730	48
Colon	Rat	160	48
Intestine	Rat	520	48
Stomach	Rat	400	48
Spleen	Rat	290	48
Liver	Rat	370	48
Skeletal muscle	Rat	50	48
Adrenal gland	Rat	840	48

[a]Receptor autoradiography was carried out in vitro using cryostat sections from various tissue sources and [^{125}I]labeled endothelin-1. Nonspecific binding was assessed by coincubating an adjacent section with excess unlabeled ligand.
[b]Binding to the section was quantitated densitometrically, and values were corrected for receptor occupancy (using predetermined affinities) and for complete occupancy.

TABLE 6.11. Endothelin-1 binding to tissue membranes[a]

Tissue membrane	$K_{d, apparent}$ (pM)[b]	B_{max} (fmol/mg¹)[b]	Reference
Heart, left ventricle (rat)	227	89	28
Heart (chick)	3.8, 18	820, 890	56
Heart (rat)	200	94	28
Heart (mouse)	108	132	60
Heart, atria (porcine)	760	2720	86
Heart, ventriculi (porcine)	630	520	86
Coronary artery (porcine)	60	790	86
Coronary vein (porcine)	120	450	86
Jugular vein (porcine)	110	270	86
Basilar vein (porcine)	120	350	86
Pia mater vessels (porcine)	520	1260	86
Aorta (porcine)	470, 140	1250, 390	55, 86
Aorta (human)	130	150	86
Pulmonary artery (porcine)	90	750	86
Kidney, medulla (human)	—	105	91
Kidney, glomeruli (human)	—	122	91
Kidney, glomeruli (rat)	182, 1300	420, 5583	23, 53
Kidney, cortex (human)	—	57	91
Kidney, blood vessels (human)	—	88	91
Renal vein (porcine)	100	440	86
Renal artery (porcine)	110	1390	86
Renal papillary (rat)	662	7666	53
Placenta (human)	24, 760	240, 1800	21, 59
Brain (rat)	120, 3000	130, 2200	27, 74

[a]Membranes were prepared from various tissues and species, and saturation binding studies were conducted using increasing concentrations of [^{125}I]endothelin-1. Nonspecific binding was determined by addition of excess unlabeled endothelin-1 to the assay.
[b]Binding parameters were calculated by Scatchard analysis of equilibrium saturation binding curves. The K_d and B_{max} values were derived by linear regression of the Scatchard plots.

(IC$_{50}$). The IC$_{50}$ value can differ substantially from a true or apparent K_d, as illustrated by data from our laboratory on endothelin-1 binding to membranes prepared from dog hearts. As illustrated in Figure 6.3, a saturation binding experiment carried out with various concentrations of [^{125}I]endothelin-1 incubated at 4°C for 3 h with a 40,000g membrane pellet resulted in a $K_{d, apparent}$ of 0.5 nM and a B_{max} of 400 fmol mg^{-1} as determined by Scatchard analysis. A competitive binding experiment carried out under identical conditions with the apparent half-maximal saturating concentration of [^{125}I] endothelin-1 (0.5 nM) resulted in an IC$_{50}$ of 1.0 nM for unlabeled endothelin-1 following a 3 h incubation at 4°C. An experiment designed to determine the concentration of endothelin-1 required to displace 50% of the rapidly reversibly bound ligand is shown in Figure 6.6. In this experiment, 0.5 nM [^{125}I]endothelin-1 was incubated with dog heart membranes for 3 h at 4°C at which time various concentrations of the unlabeled ligand were added and the incubation was carried out

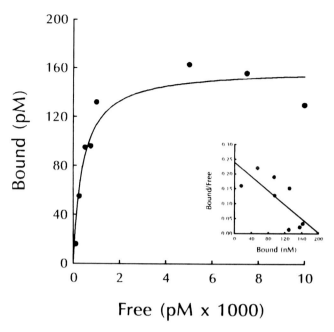

FIGURE 6.3. Saturation binding curve and Scatchard analysis of endothelin-1 binding to dog heart membranes. Increasing concentrations from 50pM to 10 nM of [^{125}I]endothelin-1 (2200 Ci/mmol) were incubated for 3 h at 4°C. Nonspecific binding was determined in the presence of 10 μM unlabeled endothelin-1. This is a representative experiment in which individual data points were obtained in triplicate and the experiment was repeated four times.

for an additional 21 h at 4°C. The data shown are for the 21 h displacement point; this did not differ significantly from the 3 h point. The percentage of displacement of bound ligand was determined assuming that only 30% of the ligand was freely reversible, and a $K_{d,\ apparent}$ value of 0.06 nM was calculated. Current K_d values in the literature for endothelin binding were derived from a mixture of these techniques, none of which can accurately predict a K_d value when the ligand binding is essentially irreversible within the time-frame of

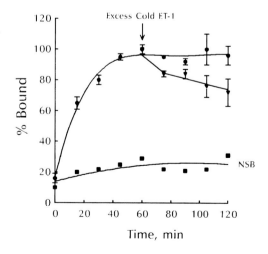

FIGURE 6.4. Association and dissociation of [^{125}I]endothelin-1 binding to dog heart membranes. Membranes prepared from dog hearts (0.1 mg protein/ml) were incubated with 0.5 nM [^{125}I]endothelin-1 for 3 h at 4°C. At this time, the reaction was continued in the presence (▼) or absence (●) of excess unlabeled (100 nM) endothelin-1. Nonspecific binding (NSB, ■) was determined by incubation in the presence of 10 μM endothelin-1. The data are the mean ± SEM of three separate membrane preparations in which each data point was obtained in triplicate. Samples were removed at specific times, and the membrane-bound ligand was separated from the free ligand by centrifugation at 100,000g for 30 min. ET, endothelin.

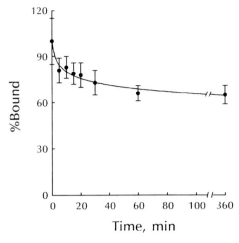

FIGURE 6.5. Dissociation of [^{125}I]endothelin-1 from dog heart membranes. Membranes prepared from dog hearts (0.1 mg protein/ml) were incubated with 0.5 nM [^{125}I]endothelin-1 for 3 h at 4°C. At this time, a 30-fold excess of ice cold buffer was added and the reaction continued for 21 h. Samples were removed at specific times, and the membrane-bound ligand was separated from the free ligand by centrifugation at 100,000g for 30 min. Nonspecific binding was determined in the presence of 10 μM endothelin-1. The data are the mean ± SEM of three separate membrane preparations in which each data point was obtained in triplicate.

the experiment. At best the different methods provide an estimate of the binding affinity of endothelin for the receptor.

Competitive inhibition binding studies have also been conducted, predominantly with [^{125}I]endothelin-1, but also with other [^{125}I]labeled isoforms. In our own laboratory we have demonstrated specific binding of [^{125}I]endothelin-1 and [^{125}I]endothelin-2 but not [^{125}I]endothelin-3 to dog heart membranes (Table 6.12). Binding of [^{125}I]endothelin-1 was competitively inhibited by increasing concentrations of unlabeled endothelin isoforms. As shown in Figure 6.7, the order of potency for the three isoforms in competing for [^{125}I]endothelin-1 binding was endothelin-2 > endothelin-1 >> endothelin-3. Other examples of competitive binding data using membranes prepared from cultured cells and from several different tissue sources from different species for the isoforms are listed in Table 6.13.

In Vitro Autoradiography of Tissue Samples

Quantitative autoradiography using [^{125}I]labeled endothelin-1 has been employed to determine the localization and density of specific high-affinity binding sites for endothelin-1 in many tissues derived from different species. Dis-

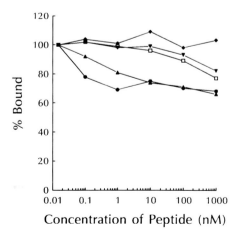

FIGURE 6.6. Competitive displacement of [^{125}I]endothelin-1 from dog heart membranes by various endothelin isoforms. [^{125}I]endothelin-1 (0.5 nM, 2200 Ci/mmol) was incubated with 100 μg of dog heart membranes for 3 h at 4°C, at which time concentrations from 0.1 to 1000 nM of endothelin-1 (●), endothelin-2 (▲), endothelin-3 (▼), big endothelin-1 (◆), and sarafotoxin S6b (□) were added. Aliquots were removed at 3 and 21 h, and membrane-bound ligand was determined on a Brandel Cell Harvester following three washes with ice cold buffer. The data in the figure are from the 21 h point and represent one experiment in which each data point was obtained in triplicate. The experiment was repeated three times.

TABLE 6.12. Endothelin (ET) isoform binding to dog heart membranes[a]

Ligand	Scatchard analysis of saturation binding data[b]	
	$K_{d, apparent}$	B_{max}
[^{125}I]ET-1	0.5 nM	400 fmol/mg
[^{125}I]ET-2	0.1 nM	12 fmol/mg
[^{125}I]ET-3	NSB[c]	NSB

[a] E. Ho and L. H. Parker Botelho (unpublished data).
[b] Saturation binding experiments were carried out with 0.010–500 nM of the three [^{125}I]labeled isoforms, and membranes were prepared from dog hearts. Incubations were carried out for 3 h at 4°C, and bound ligand was separated from free by filtration on a Brandel Cell Harvester. Nonspecific binding was determined in the presence of 10 μM of the respective unlabeled ligands.
[c] NSB, no specific binding detectable at the highest concentration of [^{125}I]endothelin-3 used (500 nM).

placeable binding sites for endothelin-1 were shown to be widely distributed in the vascular system as well as within other tissues, including the intestine, heart, lung, kidney, and brain (Table 6.14), suggesting a more extensive biological function of endothelin-1 than simply control of vascular tone. Binding sites have been identified in blood vessels of various sizes and of different anatomical origin, including the coronary artery, intrapulmonary vessels, and intrarenal and intrasplenic arteries.[66,67] A differential pattern of distribution of recognition sites within various organs has also been observed, and, similarly, there are apparent species differences in both densities and relative enrichment patterns.[18,19,37,41,42] Waeber et al.[91] recently demonstrated different total binding but similar relative binding site densities of the three endothelin isoforms in various regions of the human kidney: glomeruli > medulla > blood vessels > cortex.

In Vivo Injection of [^{125}I]Endothelin and Autoradiography of Tissue Samples

The autoradiographic localization of [^{125}I]endothelin-1 following intravenous administration has been investigated in rats. Enrichment of radioactivity was observed in both the cardiovascular system (myocardium, atrium, and smooth muscle of the thoracic aorta and coronary artery) and the nonvascular organs,

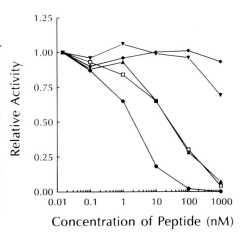

FIGURE 6.7. Competitive inhibition of [^{125}I]endothelin-1 binding to dog heart membranes by the various endothelin isoforms. [^{125}I]endothelin-1 (0.5 nM) was added simultaneously with concentrations of unlabeled endothelin-1 (●), endothelin-2 (▲), endothelin-3 (▼), big endothelin-1 (♦), and sarafotoxin S6b (□) from 0.1 to 1000 nM to dog heart membranes (100 μg) and incubated for 3 h at 4°C. Membrane-bound ligand was determined on a Brandel Cell Harvester following three washes with ice cold buffer. Each data point was determined in triplicate, and the experiment was repeated at least three times.

TABLE 6.13. Competitive inhibition of [^{125}I]labeled endothelin (ET) binding

Tissue	Species	[^{125}I]ET isoform	Competitive binding (IC$_{50}$, nM)^a			Reference
			ET-1	ET-2	ET-3	
Cardiac membranes	Chick	ET-1	0.013	0.018	0.07	92
		ET-2	0.012	0.013	0.09	92
		ET-3	0.05	0.085	0.01	92
Cardiac membranes	Rat	ET-1	0.56	—^{b}	—	51
Aortic membranes	Porcine	ET-1	~0.1	0.25	5.0	86
Adrenal glomerulosa	Bovine	ET-1	~0.9	—	100	17
Placenta	Human	ET-1	0.5	1.50	—	60
Placenta (particulate)	Human	ET-1	0.4	—	2.5	21
Placenta (solubilized)	Human	ET-1	0.08	—	—	60
Ventricular membranes	Rat	ET-1	0.16	—	—	27
Swiss 3T3 fibroblasts	Murine	ET-1	0.14	0.36	150	64
Swiss 3T3 fibroblasts	Murine	ET-2^c	0.14	0.29	250	64
FS-4 fibroblasts	Human	ET-1	0.15	0.38	215	64
FS-4 fibroblasts	Human	ET-2	0.13	0.46	220	64
Retinal pericytes	Bovine	ET-1	0.56	1.53	>1000	50
Vascular SMC^d	Rat	ET-1	0.20	—	—	69

^{a}Competitive inhibition of [^{125}I]endothelin binding was determined by addition of increasing concentrations of different unlabeled isoforms. Half-maximal inhibition (IC$_{50}$) of binding is given for various membrane preparations or for monolayers of cultured cells.
^{b}—, The respective isoform was not tested in the competitive binding assay.
^{c}Mouse endothelin-2 (VIC).
^{d}SMC, smooth muscle cells.

TABLE 6.14. Autoradiographic detection of binding sites: [^{125}I]endothelin-1 binding to cryostat sections

Tissue, species	Density of binding sites (amol/mm^2)^a	References
Heart (rat, porcine, human)	2.8–8.0	19
Vascular		
Coronary artery (rat)	2.5	19
Coronary artery (porcine)	200	66
Pulmonary artery (rat)	2.5	19
Aorta (rat)	2.2	19
Vascular smooth muscle (porcine)	250	66
Neural		
Brain (rat)	4.8–10.3	19
Nerve (porcine)	10.4	19
Spinal cord (rat)	8	19
Kidney		
Kidney (rat)	4.4–9.2	19
Renal outer medulla (rat)	No sites	19

(*continued*)

TABLE 6.14. (Continued)

Tissue, species	Density of binding sites (amol/mm^2)^a	References
Glomerulae (rat)	9.2	18
Glomerulae (porcine, human)	No sites	18
Kidney cortex (rat, porcine, human)	2.2–4.4	18
Kidney medulla (porcine, human)	9.5–11.1	18
Inner medulla (rat)	6.2	18
Outer medulla (rat)	No sites	18
Papilla (rat, porcine, human)	7.0–13.1	18
Vasa recta (rat, porcine, human)	10.1–14.6	18
Other		
Lung (rat)	9–400	19
Trachea (rat, porcine)	210–1200	66, 89
Adrenal (rat, porcine, human)	2.6–11.2	19
Cartilage (porcine)	No sites	66
Connective tissue (porcine)	Only NSB	66
Respiratory epithelium (porcine)	No sites	66
Adipocyte (rat, porcine)	No sites	19

^{a}Incubation of cryostat tissue sections with [^{125}I]endothelin-1 was performed in vitro. Nonspecific binding was assessed by exposure of an adjacent section to excess unlabeled ligand. Dried sections were subjected to autoradiography, and density of specific endothelin binding was determined using computer-assisted image analysis by comparison with a series of [^{125}I]labeled standards.

particularly the lung, kidney, brain cortex, and intestine.[44,46,48,79] Interestingly, intraaortic administration of the radioisotope labeled only circumventricular structures of the media eminence and subfornical organ in the brain, which are located outside the blood–brain barrier and choroid plexus. These findings indicate that endothelin-1 does not cross the blood–brain barrier.[48]

In similar studies, elimination of intravenously administered endothelin and subsequent tissue distribution in rat has been reported. Endothelin-1 has a short half-life in the rat bloodstream, <1 min when given via the left ventricle[1] or jugular vein[78] and ~7 min when administered into the femoral artery.[75] Within a few minutes after administration, the injected [^{125}I] endothelin-1 was found bound primarily to the parenchyma of lung (82%), but binding also occurred in the kidneys (10%) and liver (3%),[75] suggesting that circulating endothelin-1 is rapidly cleared by the pulmonary circulation. Removal of endothelin-1 isoform has been reported to be somewhat more rapid than that of endothelin-3, but the organ distribution pattern for both isoforms are comparable.[1]

AUTOCRINE, PARACRINE, AND ENDOCRINE FUNCTIONS OF ENDOTHELIN

Data Supporting an Autocrine Function of Endothelin

Initially, reports that endothelin-producing vascular endothelial cells do not have endothelin receptors[95] implied that there was no autocrine role for endothelin. Recently, however, there have been several reports of functional receptors on endothelin-producing cells, i.e., endothelial and epithelial cells. Takahashi et al.[85] reported the release of endothelin-1 (9.2 pM) by cultured bovine retinal capillary endothelial cells that also were shown to exhibit low levels of endothelin-1 binding (<2 × 10^3 binding sites per cell). A similar situation was found for bovine retinal capillary pericytes. Conditioned culture media from these cells contained 2.9 pM endothelin-1, and specific binding sites were present on these cells at a density of 1.5 × 10^5 sites per cell.[85] Recently, Vigne et al.[90] demonstrated the presence of high-affinity surface-binding sites for endothelin-1 on endothelial cells from rat brain microvessels (400 fmol/mg protein). They were, however, unable to show that these cells release the peptide into the culture media. Interestingly, endothelin-1 was found to be mitogenic in these cells, exhibiting higher potency than basic fibroblast growth factor, probably acting via phosphatidylinositol hydrolysis and intracellular calcium mobilization. Further evidence for an autocrine role for endothelin-1 was recently presented. Takagi et al.[84] reported that specific antibodies raised against endothelin-1 strongly inhibited DNA synthesis in human vascular endothelial cells in response to fetal calf serum, consistent with a role of endogenously produced endothelin in endothelial cell proliferation. Table 6.15 contains a list of binding studies done with endothelin-producing cells.

Data Supporting a Paracrine Function of Endothelin

Data supporting a paracrine action of endothelin in the vasculature are numerous. Endothelial cells lining the blood vessels, for example, release endothelin, and the underlying vascular smooth muscle cells have specific functional receptors for endothelin. A similar situation may exist in the bronchopulmonary bed such that endothelin synthesized by the epithelial cells lining the airways interact with specific, functional receptors on the underlying bronchiolar smooth muscle cells. Canine tracheal epithelial cells have been reported to synthesize and release endothelin-1 and endothelin-3,[8] and recent data have demonstrated the existence of functional receptors on guinea pig tracheal and bronchial smooth muscle preparations.[6]

Furthermore, since endothelin-1 appears to be released from the posterior pituitary during water deprivation[101] and has been shown to promote vasopressin release,[25] it has been suggested that endothelin released from nerve terminals in the pituitary may act locally to modulate release of classic neurosecretory hormones such as vasopressin and oxytocin. Recently, a paracrine function for endothelin has also been proposed in human mammary tissue.[5] Cultured breast epithelial cells were found to exhibit low constitutive expression of mRNA for endothelin-1, and release of immunoreactive peptide was enhanced in response to prolactin administration. Breast stromal cells located proximally to the epithelial cells did not express mRNA for endothelin but were found to contain specific cell surface receptors for endothelin-1. The in situ hybridzation studies of MacCumber et al.[52] also lend support to a paracrine action of endothelin, since it was found that endothelin-1 mRNA is synthesized in close proximity to endothelin-1–binding sites in several of the rat organs tested, i.e., lung, kidney, intestine, and eye.

TABLE 6.15. Endothelin-1 binding to cultured cells that synthesize endothelin

Cell	Immunoreactive endothelin/24 h[a]	Reference	$K_{d,\ apparent}$ (pM)[b]	B_{max}	Reference
Nonvascular					
Retinal capillary pericytes (bovine)	—		1300	$1-2 \times 10^5$ sites/cell	50
Retinal capillary pericytes (bovine)	1.5 pM	85	140	1.5×10^5 sites/cell	85
MDCK renal epithelial (canine)	0.22 pmol/10^6 cells	47	40	5×10^3 sites/cell	61
LLC/PK$_1$ renal epithelial (porcine)	0.06 pmol/10^6 cells	76	50	5×10^3 sites/cell	61
Vascular					
Retinal endothelial cells (bovine)	4.6 pM	85	—[c]	$<2 \times 10^3$ sites/cell	85
Brain capillary endothelial cells (rat)	0 pM	90	800 pM	400 fmol/mg protein	90

[a]Immunoreactive endothelin produced was quantitated is serum-free conditioned media derived from confluent monolayers of the cultured cells by radioimmunoassay.
[b]Dissociation constants and binding capacities were obtained by linear regression of Scatchard plots.
[c]Binding of [^{125}I]endothelin-1 was detectable but very low. Measurements of K_d and B_{max} were considered to be unreliable.

Data Supporting an Endocrine Function of Endothelin

There are fewer data and less direct evidence supporting an endocrine function of endothelin. The in situ hybridzation studies of MacCumber et al.[52] found that, in the heart and renal cortex, endothelin-1–binding sites were present in the absence of endothelin-1 mRNA. The authors felt that these data were suggestive of an endocrine action in that endothelin-1 in the bloodstream acts distant from the site of synthesis. In addition, endothelin-1 mRNA and binding sites have been identified in a number of regions in the brain, providing support for a role in neurosecretory functions, e.g., in stimulation of substance P release from the spinal cord, hypothalamus, and pituitary of the rat.[12,99] Certainly, endothelin has a wide range of actions on cardiovascular, renal, and several endocrine systems that taken together with the observed plasma levels of endothelin-1 in the low picomolar range (Table 6.8) could be consistent with the contention that endothelin(s) are circulating hormones.

Following the discovery in our laboratory of circulating endothelin-3 levels (approximately eightfold higher than endothelin-1) in dog plasma, several canine tissue membranes were tested for specific binding sites for endothelin-1 versus endothelin-3. When appropriate, the tissues from which the membranes were prepared were also tested for functional receptors. Saturation binding experiments carried out with [^{125}I]endothelin-1 or [^{125}I]endothelin-3 and membranes from dog coronary arteries showed specific, rapidly saturable receptors for endothelin-1 (Fig. 6.8), but no specific binding sites for endothelin-3 up to 5 nM. Although there are no specific endothelin-3 receptors, this ligand is capable of competing with [^{125}I]endothelin-1 in a competitive binding assay (Fig. 6.9) and is capable of eliciting a biological contractile response in intact dog coronary tissue rings (Fig. 6.10).

Similarly, no high-affinity receptors specific for endothelin-3 were found on the tissue membranes prepared from dog mesenteric vein, heart, or lung. The contractile activity data showed a less potent response to endothelin-3 than to endothelin-1, implying that the endothelin-3 effect occurs through interaction with the endothelin-1 receptor. The data from all of the dog tissues tested for comparative endothelin-1 versus endothelin-3 responses are shown in Table 6.16.

Although there are relatively high circulating levels of endothelin-3 in dog plasma, neither the site of synthesis nor the site of action of endothelin-3 is known. An "endothelin-3-like" immunoreactive material has been isolated from porcine brain homogenates,[77] leading to the speculation that endothelin-3 is a neural form of endothelin, but no similar experiments have been performed thus far in dog tissues.

SUMMARY AND CONCLUSIONS

The different isoforms of endothelin that have been identified in porcine, human, rat, and mouse tissues are the products of three separate genes that appear to be differentially regulated. The gene for endothelin-1 has been found to be constitutively expressed in cultured endothelial and epithelial cells and

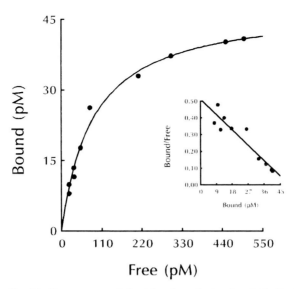

FIGURE 6.8. Saturation binding curve and Scatchard analysis of endothelin-1 binding to dog coronary artery membranes. Increasing concentrations from 1 to 500 pM of [^{125}I]endothelin-1 (2200 Ci/mmol) were incubated for 3 h at 4° C with 100 μg of membranes. Nonspecific binding was determined in the presence of 10 μM unlabeled endothelin-1. This is a representative experiment in which individual data points were obtained in triplicate and the experiment was repeated four times.

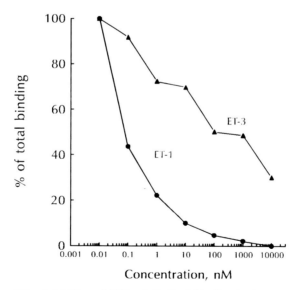

FIGURE 6.9. Competitive inhibition of [^{125}I]endothelin-1 binding to dog coronary artery membranes by the endothelin-1 and endothelin-3 isoforms. [^{125}I]endothelin-1 (0.5 nM) was added simultaneously with various concentrations of unlabeled endothelin-1 (●) or endothelin-3 (▲) (0.1 to 1000 nM) and incubated for 3 h at 4°C. Membrane-bound ligand was determined on a Brandel Cell Harvester following three washes with ice cold buffer. Each data point was determined in triplicate, and the experiment was repeated at least three times. ET, endothelin.

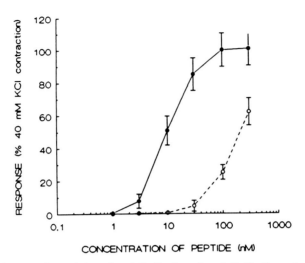

FIGURE 6.10. Contractile activity of endothelin-1 and endothelin-3 on dog coronary artery. Rings of coronary artery were set up under 4 g tension in physiological salt solution at 37°C. Cumulative log-concentration response curves of endothelin-1 (●) or endothelin-3 (○) were determined by addition of stock solutions of each isoform. The responses were normalized by expressing them as a percentage of the response to 40 mM KCl, determined before the endothelin dose–response curves. The points shown are the means ± SEM of five individual experiments.

TABLE 6.16. Dog tissue membrane binding versus biological response[a]

	Membrane receptor binding (nM)				Contractile activity (nM; EC_{50}, % 80 mM KCL)	
	Specific binding (EC_{50})[b]		Competitive binding (IC_{50})[c]			
Tissue, dog	ET-1	ET-3	ET-1	ET-3	ET-1	ET-3
Mesenteric vein	0.043	NSB[d]	0.1	600	10	60
Coronary artery	0.087	NSB	0.07	100	8	200
Lung	—	—	9.5	>10^6	—	—
Heart	1.6	NSB	1.7	2400	—	—
Right atria	—	—	0.2	—	—	—
Left atria	—	—	0.75	—	—	—
Right ventricle	—	—	0.3	—	—	—
Left ventricle	—	—	0.35	—	—	—

[a] C. Cade, P. Devesly, E. Ho, and L. H. Parker Botelho (unpublished data).

[b] [^{125}I]endothelin-1 or [^{125}I]endothelin-3 (sp. ac. 2200 Ci/mmol) was incubated with the tissue membranes at concentrations from 10 to 500 pM for 3 h at 4°C. Nonspecific binding was determined in the presence of 10 μM unlabeled ligand. The saturation binding curves were analyzed by Scatchard analysis from which a half-maximal binding concentration (EC_{50}) was determined.

[c] Competitive binding experiments were done by simultaneous incubation of [^{125}I]endothelin-1 and increasing concentrations of unlabeled endothelin-1 or endothelin-3 from 0.01 nM to 10 μM for 3 h at 4°C. A half-maximal inhibitory concentration (IC_{50}) was determined from the linear part of the inhibition curve.

[d] NSB, no specific binding sites could be detected for [^{125}I]endothelin-3 binding to the tissue membranes up to the highest concentration tested (5 nM).

in various human cells of carcinoma origin. Exposure of the cultured endothelial cells to various inducing agents has been found to result in a severalfold increase in mRNA levels and in a smaller but significant increase in the protein product. Endothelin-2 has been shown to be constitutively expressed in Cos-7 kidney cells, and similar data have been found for endothelin-3 in canine tracheal epithelial cells. The sequence of the genes for preproendothelin-2 and preproendothelin-3 have not been published, and no data have yet been reported on inducing agents that increase synthesis of these isoforms.

The mRNA for the various isoforms and the three peptides themselves have been found in many vascular tissues as well as in brain, heart, renal tissue, gastrointestinal tissue, lung, liver, spleen, adrenal gland, urinary bladder, and testes. In most cases, the levels of immunoreactive endothelin-3 are much lower than endothelin-1, and no endothelin-2 levels have been detected. The few exceptions are in rat medulla where equal quantities of immunoreactive endothelin-1 and endothelin-3 were found, rat pituitary where endothelin-3 levels are fourfold higher than endothelin-1 levels, and rat intestine where the level of endothelin-3 is one-half that of endothelin-1.

Measurable circulating endothelin-1 levels have been reported in human, porcine, rat, sheep, and canine plasma. Endothelin-3 has been found in canine plasma, and big endothelin-1 has been detected in both human and rat plasma. The plasma levels of endothelin-1 and endothelin-3 are in the low picomolar range. Experimentally, endothelin-1 levels can be increased by endotoxin infusion in pigs, rats, and sheep. Interestingly, endothelin-1 levels are sixfold higher in human urine and sevenfold higher in human CSF than in plasma.

Endothelin-1 has been shown to have specific, saturable binding sites on many cultured cells and membranes isolated from various tissue sources. Autoradiographic studies have also shown specific binding to a great variety of tissues. The binding of [^{125}I]endothelin-1 to cell membrane receptors is rapid, specific, and saturable, but not rapidly reversible. In most systems the ligand is only partially displaced, and the half-life can be days. The nonreversibility of endothelin binding has resulted in inaccurate calculations of $K_{d,\ apparent}$ values derived from Scatchard analyses of saturation binding data. Under these conditions, this type of analysis can only provide an estimate of the half-maximal saturating concentration of ligand. Additional confusion has resulted from investigators referring to competitive binding experiments as "displacement experiments" and incorrectly reporting K_d values instead of IC_{50} values. Neither of these protocols results in a K_d value, because the ligand binding is essentially irreversible within the time-frame of the experiment. Despite the confusion caused by inaccurately reported kinetic constants for endothelin binding, the reported $K_{d,\ apparent}$ values range from 40 to 1300 pM and are in agreement with the concentration ranges required to obtain a biochemical signal.

In vivo binding has been confirmed by quantitative autoradiography of tissue slices following intravenous administration of [^{125}I]endothelin-1.

Recent data showing both endothelin synthesis by and specific binding sites in endothelial cells imply a possible autocrine function for this peptide. Examples of endothelin functioning in a paracrine role are numerous, with the best example being synthesis by the vascular endothelium and specific binding and action in the underlying vascular smooth muscle cells. Detectable circulating plasma concentrations of endothelium in the low picomolar range in conjunction with tissue diversity of receptor distribution is consistent with a possible endocrine role, but no definitive data are available yet.

REFERENCES

1. ANGGARD, E., S. GALTON, G. RAE, R. THOMAS, L. MCLOUGHLIN, G. NUCCI, and J. VANE. The fate of radioiodinated endothelin-1 and endothelin-3 in the rat. *J. Cardiol. Pharmacol.* 13(Suppl. 5): S46–S49, 1989.
2. ASANO, T., I. IKEGAKE, Y. SUZUKI, S. SATOH, and M. SHIBUYA. Endothelin and the production of cerebral vasospasm in dogs. *Biochem. Biophys. Res. Commun.* 159: 1345–1351, 1989.
3. AUGUET, M., S. DELAFLOTTE, P. CHABRIER, E. PIROTZKY, F. CLOSTRE, and P. BRAQUET. Endothelin and Ca^{++} agonist bay K 8644: different vasoconstrictive properties. *Biochem. Biophys. Res. Commun.* 156: 186–192, 1988.
4. BADR, K., J. MURRAY, M. BREYER, K. TAKAHASHI, T. INAGAMI and R. HARRIS. Mesangial cell, glomerular and renal vascular responses to endothelin in the rat kidney. *J. Clin. Invest.* 83: 336–342, 1989.
5. BALEY, P. A., T. J. RESINK, U. EPPENBERGER and A. W. HAHN. Endothelin messenger RNA and receptors are differentially expressed in cultured human breast epithelial and stromal cells. *J. Clin. Invest.* 85: 1320–1323, 1990.
6. BATTISTINI, B., J. FILEP and P. SIROIS. Potent thromboxane-mediated in vitro bronchoconstrictor effect of endothelin in the guinea pig. *Eur. J. Pharmacol.* 178: 141–142, 1990.
7. BERBINSCHI, A., and J. M. KETELSLEGERS. Endothelin in urine. *Lancet* 2: 46, 1989.
8. BLACK, P. N., M. A. GHATEI, K. TAKAHASHI, D. BRETHERTON-WATT, T. KRAUSZ, C. T. DOLLERY and S. R. BLOOM. Formation of endothelin by cultured airway epithelial cells. *FEBS Lett.* 255: 129–132, 1989.
9. BLOCH, K., R. EDDY, T. SHOWS, and T. QUERTERMOUS. cDNA cloning and chromosomal assignment of the gene encoding endothelin-3. *J. Biol. Chem.* 264: 18156–18161, 1989.
10. BLOCH, K., S. FRIEDRICH, M. LEE, R. EDDY, T. SHOWS and T. QUERTERMOUS. Structural organization and chromosomal assignment of the gene encoding endothelin. *J. Biol. Chem.* 264: 10851–10857, 1989.
11. BOTELHO PARKER, L. H. Tissue Specificity of Endothelin Synthesis and Binding. Presented at the *FASEB Endothelin Symp., Washington, DC, April 1990.*
12. CALVO, J. J., R. GONZALEZ, L. F. DECARVALHO, K. TAKAHASHI, S. M. KANSE, G. R. HART, M. A. GHATEI and S. R. BLOOM. Release of substance P from rat hypothalamus and pituitary by endothelin. *Endocrinology* 126: 2288–2295, 1990.
13. CINTRA, A., K. FUXE, E. ANGGARD, B. TINNER, W. STAINES and L. AGNATI. Increased endothelin-like immunoreactivity in ibotenic acid–lesioned hippocampal formation of the rat brain. *Acta Physiol. Scand.* 137: 557–558, 1989.
14. CLOZEL, M. Endothelin sensitivity and receptor binding in the aorta of spontaneously hypertensive rats. *J. Hypertens.* 7: 913–917, 1989.
15. CLOZEL, M., W. FISCHLI and C. GUILLY. Specific binding of endothelin on human vascular smooth muscle cells in culture. *J. Clin. Invest.* 83: 1758–1761, 1989.
16. COCEANI, F., C. ARMSTRONG, and L. KELSEY. Endothelin is a potent constrictor of the lamb ductus arteriosus. *Can. J. Physiol. Pharmacol.* 67: 902–904, 1989.
17. COZZA, E., C. GOMEZ-SANCHEZ, M. FOECKING and S. CHIOU. Endothelin binding to cultured calf adrenal zona glomerulose cells and stimulation of aldosterone secretion. *J. Clin. Invest.* 84: 1032–1035, 1989.
18. DAVENPORT, A. P., D. NUNEZ, M. BROWN. Binding sites for [^{125}I]-labeled endothelin-1 in the kidneys: differential distribution in rat, pig and man demonstrated by using quantitative autoradiography. *Clin. Sci.* 77: 129–131, 1989.
19. DAVENPORT, A. P., D. J. NUNEZ, J. A. HALL, A. J. KAUMANN, and M. J. BROWN. Autoradiographical localization of binding sites for porcine [^{125}I]endothelin-1 in humans, pigs and rats: functional relevance in human. *J. Cardiovasc. Pharmacol.* 13: S166–S170, 1989.
20. EMORI, T., Y. HIRATA, K. OHTA, M. SHICHIRI, and F. MARUMO. Secretory mechanism of immunoreactive endothelin in cultured bovine endothelial cells. *Biochem. Biophys. Res. Commun.* 160: 93–100, 1989.
21. FISCHLI, W., M. CLOZEL, and C. GUILLY. Specific receptors for endothelin on membranes from human placenta: characterization and use in a binding assay. *Life Sci.* 44: 1429–1436, 1989.

22. FURCHGOTT, R. F., and J. V. ZAWADSKI. The obligatory role of endothelial cells in the relaxation of arterial smooth muscle by acetylcholine. *Nature* 288: 373–376, 1980.
23. GAUQUELIN, G., G. THIBAULT, and R. GARCIA. Characterization of renal glomerular endothelin receptors in the rat. *Biochem. Biophys. Res. Commun.* 164: 54–57, 1989.
24. GIAID, A., S. GIBSON, N. IBRAHIM, S. LEGON, S. BLOOM, M. YANAGISAWA, T. MASAKI, I. VARNDELL and J. POLAK. Endothelin-1, an endothelium-derived peptide, is expressed in neurons of the human spinal cord and dorsal root ganglia. *Proc. Natl. Acad. Sci. USA* 86; 7634–7638, 1989.
25. GOETZ, K., B. WANG, J. MADWED, J. ZHU, and R. LEADLEY. Cardiovascular, renal, and endocrine responses to intravenous endothelin in conscious dogs. *Am. J. Physiol.* 255 (*Regulatory Integrative Comp. Physiol.* 24): R1064–R1068, 1988.
26. GRYNKIEWICZ, G., M. POENIE, and R. Y. TSIEN. A new generation of Ca^{2+} indicators with greatly improved fluorescence properties. *J. Biol. Chem.* 260:3440–3450, 1985.
27. GU, X., D. CASLEY, M. CINCOTTA, and W. NAYLER. [^{125}I]Endothelin-1 binding to brain and cardiac membranes from normotensive and spontaneously hypertensive rats. *Eur. J. Pharmacol.* 177: 205–209, 1990.
28. GU, X. H., D. CASLEY, and W. NAYLER. Specific high-affinity binding sites for [^{125}I]labelled porcine endothelin in rat cardiac membranes. *Eur. J. Pharmacol.* 167: 281–190, 1989.
29. HAEGERSTRAND, A., A. HEMSEN, C. GILLIS, O. LARSSON, and J. LUNDBERG. Endothelin: presence in human umbilical vessels, high levels in fetal blood and potent constrictor effect. *Acta Physiol. Scand.* 137: 541–542, 1989.
30. HEXUM, T., C. HOEGER, J. RIVIER, A. BAIRD, and M. BROWN. Characterization of endothelin secretion by vascular endothelial cells. *Biochem. Biophys. Res. Commun.* 167: 294–300, 1990.
31. HICKEY, K. A., G. RUBANYI, R. J. PAUL, and R. F. HIGHSMITH. Characterization of a coronary vasoconstrictor produced by endothelial cells in culture. *Am. J. Physiol.* 248 (*Cell Physiol.* 17): C550–C556, 1985.
32. HIRATA, Y., Y. FUKUDA, H. YOSHIMI, T. EMORI, M. SHICHIRI, and F. MARUMO. Specific receptor for endothelin in cultured rat cardiocytes. *Biochem. Biophys. Res. Commun.* 160: 1438–1444, 1989.
33. HIRATA, Y., T. MATSUNAGA, K. ANDO, T. FURUKAWA, H. TSUKAGOSHI, and F. MARUMO. Presence of endothelin-1–like immunoreactivity in human cerebrospinal fluid. *Biochem. Biophys. Res. Commun.* 166: 1274–1278, 1990.
34. HIRATA, Y., H. YOSHIMI, S. TAKAICHI, M. YANAGISAWA, and T. MASAKI. Binding and receptor down-regulation of a novel vasoconstrictor endothelin in cultured rat vascular smooth muscle cells. *FEBS Lett.* 239: 13–17, 1988.
35. HIROE, M., Y. HIRATA, F. MARUMO, M. NAGATA, T. TOYOZAKI, M. HASUMI, Y. OHTA, T. HORIE, and M. SEKIGUCHI. Immunohistochemical localization of endothelin in human vascular endothelial cells. *Peptides* 10: 1281–1282, 1989.
36. HOFFMAN, A., H. R. KEISER, E. GROSSMAN, D. S. GOLDSTEIN, P. W. GOLD, and M. KLING. Endothelin concentrations in cerebrospinal fluid in depressive patients. *Lancet* 2: 1519, 1989.
37. HOYER, D., C. WAEBER, and J. PALACIOS. [^{125}I]endothelin-1 binding sites: autoradiographic studies in the brain and periphery of various species including humans. *J. Cardiol. Pharmacol.* 13(Suppl. 5): S162–S165, 1989.
38. INOUE, A., M. YANAGISAWA, S. KIMURA, Y. KASUYA, T. MIYAUCHI, K. GOTO, and T. MASAKI. The human endothelin family: three structurally and pharmacologically distinct isopeptides predicted by three separate genes. *Proc. Natl. Acad. Sci. USA* 86: 2863–2867, 1989.
39. INOUE, A., M. YANAGISAWA, Y. TAKUWA, Y. MITSUI, M. KOBAYASHI, and T. MASAKI. The human preproendothelin-1 gene. *J. Biol. Chem.* 264: 14954–14959, 1989.
40. ITOH, Y., M. YANAGISAWA, S. OHKUBO, C. KIMURA, T. KOSAKA, A. INOUE, N. ISHIDA, Y. MITSUI, H. ONDA, M. FUJINO, and T. MASAKI. Cloning and sequence analysis of cDNA encoding the precursor of a human endothelin-derived vasoconstrictor peptide, endothelin: identity of human and porcine endothelin. *FEBS Lett.* 231: 440–444, 1988.
41. JONES C., C. HILEY, J. PELTON, and R. MILLER. Autoradiographic localization of endothelin binding sites in kidney. *Eur. J. Pharmacol.* 163: 379–382, 1989.
42. JONES, C., C. HILEY, J. PELTON, M. MOHR. Autoradiographic visualization of the binding sites for [^{125}I]-endothelin in rat and human brain. *Neurosci. Lett.* 97: 276–279, 1989.
43. KITAMURA, K., T. TANAKA, J. KATO, T. ETO, and K. TANAKA. Regional distribution of immunoreactive endothelin in porcine tissue: abundance in inner medulla of kidney. *Biochem. Biophys. Res. Commun.* 161: 348–352, 1989.
44. KITAMURA, K., T. TANAKA, J. KATO, T. ETO, and K. TANAKA. Chromatographic characterization of immunoreactive endothelin in rat lung. *Life Sci.* 46: 405–409, 1990.
45. KITAMURA, K., T. TANAKA, J. KATO, T. OGAWA, T. ETO, and K. TANAKA. Immunoreactive

endothelin in rat kidney inner medulla: marked decrease in spontaneously hypertensive rats. *Biochem. Biophys. Res. Commun.* 162: 38–44, 1989.
46. KOHZUKI, M., C. JOHNSTON, S. CHAI, D. CASLEY, F. ROGERSON, and F. MENDELSOHN. Endothelin receptors in rat adrenal gland visualized by quantitative autoradiography. *Clin. Exp. Pharmacol. Phys.* 16: 239–242, 1989.
47. KOSAKA, T., N. SUZUKI, H. MATSUMOTO, Y. IOTH, T. YASUHARA, H. ONDA, and M. FUJINO. Synthesis of the vasoconstrictor peptide endothelin in kidney cells. *FEBS Lett.* 249: 42–46, 1989.
48. KOSEKI, C., M. IMAI, Y. HIRATA, M. YANAGISAWA, and T. MASAKI. Autoradiographic distribution in rat tissues of binding sites for endothelin: a neuropeptide? *Am. J. Physiol.* 256 (*Rural Fluid Electrolyte Physiol.* 25): R858–868, 1989.
49. KURIHARA, H., M. YOSHIZUMI, T. SUGIYAMA, F. TAKAKU, M. YANAGISAWA, T. MASAKI, M. HAMAOKI, H. KATO, and Y. YAZAKI. Transforming growth factor-β stimulates the expression of endothelin mRNA by vascular endothelial cells. *Biochem. Biophys. Res. Commun.* 159: 1435–1440, 1989.
50. LEE, T., K. HU, T. CHAO, and G. KING. Characterization of endothelin receptors and effects of endothelin on diacylglycerol and protein kinase C in retinal capillary pericytes. *Diabetes* 38: 1643–1646, 1989.
51. LIU, J., D. CASLEY, and W. NAYLER. Ischaemia causes externalization of endothelin-1 binding sites in rat cardiac membranes. *Biochem. Biophys. Res. Commun.* 164: 1220–1225, 1989.
52. MACCUMBER, M., C. ROSS, B. GLASER, and S. SYNDER. Endothelin: visualization of mRNAs by *in situ* hybridization provides evidence for local action. *Proc. Natl. Acad. Sci. USA* 86: 7285–7289, 1989.
53. MARTIN, E., P. MARSDEN, B. BRENNER, and B. BELLERMANN. Identification and characterization of endothelin binding sites in rat renal papillary and glomerular membranes. *Biochem. Biophys. Res. Commun.* 162: 130–137, 1989.
54. MATSUMOTO, H., N. SUZUKI, H. ONDA, and M. FUJINO. Abundance of endothelin-3 in rat intestine, pituitary gland and brain. *Biochem. Biophys. Res. Commun.* 164: 74–80, 1989.
55. MATSUMURA Y., R. IKEGAWA, M. TAKAOKA, and S. MORIMOTO. Conversion of porcine big endothelin to endothelin by an extract from the porcine aortic endothelial cells. *Biochem. Biophys. Res. Commun.* 167: 203–210, 1990.
56. MIYAZAKI, H., M. KNODOH, H. WATANABE, Y. MASUDA, K. MURAKAMI, M. TAKAHASHI, M. YANAGISAWA, S. KIMURA, K. GOTO, and T. MASAKIN. Affinity labelling of endothelin receptor and characterization of solubilized endothelin receptor complex. *Eur. J. Biochem.* 187: 125–129, 1990.
57. MONCADA, S., A. G., HERMAN, E. A. HIGGS, and J. R. BAIN. Differential formation of prostacyclin (PGX or PG12) by layers of the arterial wall: an explanation for the antithrombotic properties of vascular endothelium. *Thromb. Res.* 11: 323–44, 1977.
58. MOREL, D. R., J. S. LaCROIX, A. HEMSEN, D. A. STEINIG, J. PITTET, and J. M. LUNDBERG. Increased plasma and pulmonary lymph levels of endothelin during endotoxin shock. *Eur. J. Pharmacol.* 167: 427–428, 1989.
59. NAKAJO, S., M. SUGIURA, and T. INAGAMI. Native form of endothelin receptor in human placental membranes. *Biochem. Biophys. Res. Commun.* 167: 280–286, 1990.
60. NAYLER, W., X. GU, D. CASLEY, S. PANAGIOTOPOULOS, J. LIU, and P. MOTTRAM. Cyclosporine increases endothelin-1 binding site density in cardiac cell membranes. *Biochem. Biophys. Res. Commun.* 163: 1270–1274, 1989.
61. NEUSER, D., S. ZAISS, and J. STASCH. Endothelin receptors in cultured renal epithelial cells. *Eur. J. Pharmacol.* 176: 241–243, 1990.
62. NOMURA, A., Y. UCHIDA, M. KAMBYAMA, M. SAOTOME, K. OKI, and S. HASEGAWA. Endothelin and bronchial asthma. *Lancet* 2: 747–748, 1989.
63. OHNISHI-SUZAKI, A., K. YAMAGUCHI, M. KUSUHARA, I. ADACHI, K. ABE, and S. KIMURA. Comparison of biological activities of endothelin-1, -2 and -3 in murine and human fibroblast cell lines. *Biochem. Biophys. Res. Commun.* 166: 608–614, 1990.
64. ORLANDO, C., M. L., BRANDI, A. PERI, S. GIANNINI, G. FANTONI, E. CALABRESI, M. SERIO, and M. MAGGI. Neurohypophyseal hormone regulation of endothelin secretion from rabbit endometrial cells in primary culture. *Endocrinology* 126: 1780–1782, 1990.
65. PERNOW, J., and J. LUNDBERG. Endothelin-like immunoreactivity is released from the pig spleen during asphyxia. *Acta Physiol. Scand.* 137: 553–554, 1989.
66. POWER, R., J. WHARTON, S. SALAS, S. KANSE, M. GHATEI, S. BLOOM, and J. POLAK. Autoradiographic localization of endothelin binding sites in human and porcine coronary arteries. *Eur. J. Pharmacol.* 160: 199–200, 1989.
67. POWER, R. F., J. WHARTON, Y. ZHAO, S. R. BLOOM, and J. M. POLAK. Autoradiographic lo-

calization of endothelin-1 binding sites in the cardiovascular and respiratory systems. *J. Cardiovasc. Pharmacol.* 13: S50–S56, 1989.
68. RESINK, T. J., T. SCOTT-BURDEN, E. WEBER, and F. R. BUHLER. Phorbol ester promotes a sustained down-regulation of endothelin receptors and cellular responses to endothelin. *Biochem. Biophys. Res. Commun.* 166: 1213–1219, 1989.
69. ROUBERT, P., V. GILLARD, P. PLAS, P. E. CHABRIER, and P. BRAQUET. Presence of a specific binding site of endothelin on cultured smooth muscle cells. *Arch. Mal. Coeur* 82(7): 1261–1263, 1989.
70. ROZENGURT, N., D. SPRINGALL, and J. POLAK. Localization of endothelin-like immunoreactivity in airway epithelium of rats and mice. *J. Pathol.* 160: 5–8, 1990.
71. SAIDA, K., Y. MITSUI and N. ISHIDA. A novel peptide, vasoactive intestinal contractor, of a new (endothelin) peptide family. *J. Biol. Chem.* 264: 14613–14616, 1989.
72. SAITO, Y., K. NAKAO, G. SHIRAKAMI, M. HOUGASAKI, T. YANAKA, H. ITOH, M. MUKOYAN-MA, H. ARAI, K. HOSODA, S. SUGA, Y. OGAWA, and H. IMURA. Detection and characterization of endothelin-1–like immunoreactivity in rat plasma. *Biochem. Biophys. Res. Commun.* 163: 1512–1516, 1989.
73. SCHINI, V. B., H. HENDRICKSON, D. M. HEUBLEIN, J. C. BURNETT, JR., and P. M. VANHOUTTE. Thrombin enhances the release of endothelin from cultured porcine aortic endothelial cells. *Eur. J. Pharmacol.* 165: 333–334, 1989.
74. SCHVARTZ, I., O. and ITTOOP, E. HAZUM. Identification of endothelin receptors by chemical cross-linking. *Endocrinology* 126: 1829–1833, 1990.
75. SHIBA, R., M. YANAGISAWA, T. MIYAUCHI, Y. ISHII, S. KIMURA, Y. UCHIYAMA, T. MASAKI, and K. GOTO. Elimination of intravenously injected endothelin-1 from the circulation of the rat. *J. Cardiol. Pharmacol.* 13(Suppl. 5): S98–S101, 1989.
76. SHICHIRI, M., Y. HIRATA, T. EMORI, K. OHTA, T. NAKAJIMA, K. SATO, A. SATO, and F. MARUMO. Secretion of endothelin and related peptides from renal epithelial cell lines. *FEBS Lett.* 253: 203–206, 1989.
77. SHINMI, O., S. KIMURA, T. SAWAMURA, Y. SUGITA, T. YOSHIZAWA, Y. UCHIYAMA, M. YANAGISAWA, K. GOTO, T. MASAKI, and I. KANAZAWA. Endothelin-3 is a novel neuropeptide: isolation and sequence determination of endothelin-1 and endothelin-3 in porcine brain. *Biochem. Biophys. Res. Commun.* 164: 587–593, 1989.
78. SIRVIO, M. L., K. METSARINNE, O. SAIJONMAA, and F. FYHRQUIST Tissue distribution and half-life of [^{125}I]endothelin in the rat: importance of pulmonary clearance. *Biochem. Biophys. Res. Commun.* 167: 1191–1195, 1990.
79. STASCH, J., W. STEINKE, S. KAZDA, and D. NEUSER. Autoradiographic localization of [^{125}I]endothelin in rat tissues. *Drug Res.* 39: 59–61, 1989.
80. SUGIURA, M., R. SNAJDAR, M. SCHWARTZBERG, K. BADR, and T. INAGAMI. Identification of two types of specific endothelin receptors in rat mesangial cell. *Biochem. Biophys. Res. Commun.* 162: 1396–1401, 1989.
81. SUNNERGREN, K., R. WORD, J. SAMBROOK, P. MACDONALD, and M. CASEY. Expression and regulation of endothelin precursor mRNA in avascular human amnion. *Mol. Cell. Endocrinol.* 68: R7–R14, 1990.
82. SUZUKI, N., H. MATSUMOTO, C. KITADA, S. KIMURA, and M. FUJINO. Production of endothelin-1 and big endothelin-1 by tumor cells with epithelial-like morphology. *J. Biochem.* 106: 736–741, 1989.
83. SUZUKI, N., T. MIYAUCHI, Y. TOMOBE, H. MATSUMOTO, K. GOTO, T. MASAKI, and M. FUJINO. Plasma concentration of endothelin-1 in spontaneously hypertensive rats and DOCA-salt hypertensive rats. *Biochem. Biophys. Res. Commun.* 167: 941–947, 1990.
84. TAKAGI, Y., M. FUKASE, S. TAKATA, H. YOSHIMI, O. TOKUNAGA, and T. FUJITA. Autocrine effect of endothelin on DNA synthesis in human vascular endothelial cells. *Biochem. Biophys. Res. Commun.* 168: 537–543, 1990.
85. TAKAHASHI, K., R. BROOKS, S. KANSE, M. GHATEI, E. KOHNER, and S. BLOOM. Production of endothelin-1 by cultured bovine retinal endothelial cells and presence of endothelin receptors on associated pericytes. *Diabetes* 38: 1200–1202, 1989.
86. TAKAYANAGI, R., T. HASHIGUCHI, M. OHASHI, and H. NAWATA. Regional distribution of endothelin receptors in porcine cardiovascular tissues. *Regulatory Peptides* 27: 247–255, 1990.
87. TAKUWA, N., Y. TAKUWA, M. YANAGISAWA, K. YAMASHITA, and T. MASAKI A novel vasoactive peptide endothelin stimulates mitogenesis through inositol lipid turnover in Swiss 3T3 fibroblasts. *J. Biol. Chem.* 264: 7856–7861, 1989.
88. TOMOBE, Y., T. MIYAUCHI, A. SAITO, M. YANAGISAWA, S. KIMURA, K. GOTO, and T. MASAKI. Effects of endothelin on the renal artery from spontaneously hypertensive and Wistar-Kyoto rats. *Eur. J. Pharmacol.* 152: 373–374, 1988.
89. TURNER, N., R. POWER, J. POLAK, S. BLOOM, and C. DOLLERY. Endothelin-induced contrac-

tions of tracheal smooth muscle and identification of specific endothelin binding sites in the trachea of the rat. *Br. J. Pharmacol.* 98: 361–366, 1989.
90. VIGNE, P., R. MARSAULT, J. P. BREITTMAYER, and C. FRELIN. Endothelin stimulates phosphatidylinositol hydrolysis and DNA synthesis in brain capillary endothelial cells. *Biochem. J.* 266: 415–420, 1990.
91. WAEBER, C., D. HOYER, and J. PALACIOS. Similar distribution of ^{125}I sarafotoxin-6b and [^{125}I]endothelin-1, -2, -3 binding sites in the human kidney. *Eur. J. Pharmacol.* 176: 233–236, 1990.
92. WATANABE, H., H. MIYAZAKI, M. KONDOH, Y. MASUDA, S. KIMURA, M. YANAGISAWA, T. MASAKI, and K. MURAKAMI. Two distinct types of endothelin receptors are present on chick cardiac membranes. *Biochem. Biophys. Res. Commun.* 161: 1252–1259, 1989.
93. YANAGISAWA, M. Molecular Biology of Endothelins. Presented at the *FASEB Endothelin Symp., Washington, DC, April 1990.*
94. YANAGISAWA, M., A. INOUE, T. ISHIKAWA, Y. KASUYA, S. KIMURA, S. KUMAGAYE, K. NAKAJIMA, T. X. WATANABE, S. SAKAKIBARA, K. GOTO, and T. MASAKI. Primary structure, synthesis and biological activity of rat endothelin, an endothelin-derived vasoconstrictor peptide. *Proc. Natl. Acad. Sci. USA* 85: 6964–6968, 1988.
95. YANAGISAWA, M., H. KURIHARA, S. KIMURA, K. GOTO and T. MASAKI. A novel peptide vasoconstrictor, endothelin, is produced by vascular endothelium and modulates smooth muscle Ca^{2+} channels. *J. Hypertens.* 6 (Suppl 4): S188–S191, 1988.
96. YANAGISAWA, M., H. KURIHARA, S. KIMURA, Y. TOMOBE, M. KOBAYASHI, Y. MITSUI, Y. YAZAKI, K. GOTO and T. MASAKI. A novel potent vasoconstrictor peptide produced by vascular endothelial cells. *Nature* 332:411–415, 1988.
97. YANAGISAWA, M., and T. MASAKI. Molecular biology and biochemistry of the endothelins. *Trends Pharmacol. Sci.* 10: 374–378, 1989.
98. YOSHIMI, H., Y. HIRATA, Y. FUKUDA, Y. KAWANO, T. EMORI, M. KURAMOCHI, T. OMAE, and F. MARUMO. Regional distribution of immunoreactive endothelin in rats. *Peptides* 10: 805–808, 1989.
99. YOSHIZAWA, T., S. KIMURA, I. KANAZAWA, Y. UCHIYAMA, M. YANAGISAWA, and T. MASAKI. Endothelin localizes in the dorsal horn and acts on the spinal neurones: possible involvement of dihydropyridine-sensitive calcium channels and substance P release. *Neurosci. Lett.* 102: 179–184, 1989.
100. YOSHIZUMI, M., H. KURIHARA, T. SUGIYAMA, F. TAKATU, M. YANAGISAWA, T. MASAKI, and Y. YAZAKI. Hemodynamic shear stress stimulates endothelin production by cultured endothelial cells. *Biochem. Biophys. Res. Commun.* 161: 859–864, 1989.
101. YOSHIZAWA, T., O. SHINMI, A. GIAID, M. YANAGISAWA, S. GIBSON, S. KIMURA, Y. UCHIGAMA, J. M. POLAK, T. MASAKI, and I. KANAZAWA. Endothelin: a novel peptide in the posterior pituitary system. *Science* 247: 462–464, 1990.

7

Cellular Actions of Endothelin in Vascular Smooth Muscle

TOMMY A. BROCK AND N. RAJU DANTHULURI

Vascular endothelium is an essential component of the blood vessel wall, important in the maintenance of normal vascular structure and function. By virtue of their location at the blood–tissue interface in vivo, endothelial cells are constantly exposed to a wide diversity of physical and chemical stimuli. It has long been recognized that the vascular endothelium plays a pivotal role in the regulation of blood vessel tone by producing vasoactive substances that act on underlying smooth muscle cells.[33] Endothelin is the most recent addition to the list of endothelium-derived vasoactive molecules.[92] Consistent with the initial observation that endothelin is by far one of the most powerful vasoconstrictors, it has now been widely documented that endothelin is a potent stimulator of both vascular and nonvascular smooth muscle contraction.[50,77,93] It is now recognized that the human endothelin "family" contains three distinct isoforms (endothelin-1, endothelin-2, and endothelin-3) that are products of three separate genes.[93]

As illustrated in Figure 7.1, endothelin-1–induced contractions are more slowly developing, are maintained for longer periods of time, and are more resistant to agonist removal when compared with contractions in response to classical contractile agonists, such as angiotensin II or the α_1-receptor agonist phenylephrine. The contractile pattern of isolated arteries consists of a rapid, initial phase and of a slowly developing, tonic phase. It is universally accepted that an increase in cytosolic free Ca^{2+} ($[Ca^{2+}]_i$) leads to an increase in the activity of myosin light chain kinase, a Ca^{2+}–calmodulin-dependent enzyme[28,80] during the rapid contractile response. This enzyme complex initiates contraction by stimulating myosin light chain phosphorylation, thus permitting stimulation of mysosin ATPase activity by actin and leading to cross-bridge formation. Although the mechanisms underlying tonic contraction are not as well understood,[24,67,80] recent observations suggest that localized Ca^{2+}–protein kinase C interactions adjacent to the plasmalemma may be an important stimulus for the maintenance of contraction.[66,67] Experimental observations supporting this hypothesis are (1) the $[Ca^{2+}]_i$ increase in response to agonist stimulation is transient, returning to basal or near basal levels within 2–3 min while the contraction is maintained near maximum for longer periods of time[55]; (2) the tonic phase of agonist-induced contraction is associated with sustained sn-1,2-diacylglycerol (DAG) formation and protein kinase C stimulation in intact arteries[17,28–30,67]; (3) phorbol esters (potent activators of

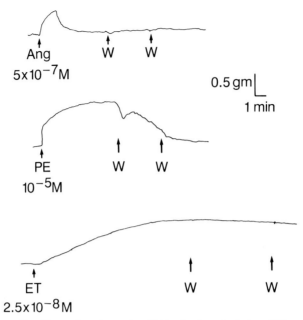

FIGURE 7.1. Contractile effects of angiotension II (Ang), phenylephrine (PE), and endothelin-1 (ET-1) in isolated rat aorta. Isometric contractions were recorded as described.[15] W, wash.

protein kinase C) cause prolonged contractions that are associated with small or undetectable $[Ca^{2+}]_i$ increases in swine carotid arteries[68] or rat aorta,[40] respectively (however, phorbol ester–induced contractions appear to be dependent on external Ca^{2+})[4,16,24,39]; and *(4)* phorbol esters can cause contraction of skinned muscle fibers when $[Ca^{2+}]_i$ is low.[11] Thus there is good evidence that $[Ca^{2+}]_i$ is not the sole determinant of vascular contractility. The slowly developing, persistent time-course of endothelin-1–stimulated contractions has a striking similarity to that induced by phorbol esters.

We summarize what is known about the cellular mechanisms underlying endothelin-induced contractile responses in vascular smooth muscle cells (VSMC), including phosphoinositide-derived second messengers, cytosolic $[Ca^{2+}]_i$, and cytosolic pH (pH_i), and the role of protein kinase C in mediating endothelin-1–stimulated VSMC contraction. Because vascular endothelial cells appear only to produce endothelin-1, and not endothelin-2 or endothelin-3,[93] our discussion focuses on VSMC actions of endothelin-1.

CA^{2+} SIGNALING PATHWAYS IN VSMC

Intracellular Ca^{2+} Mobilization

Agonist-stimulated $[Ca^{2+}]_i$ transients and contractions in VSMC have been shown to consist of two components: *(1)* a rapid phase that is initiated by releasing Ca^{2+} from an intracellular compartment, presumably the sarcoplasmic reticulum; and *(2)* a sustained phase that has been shown to be dependent on external Ca^{2+}. Numerous studies utilizing both intact arteries and cultured VSMC have documented that, in the absence of external Ca^{2+}, $[Ca^{2+}]_i$ transients and contractile responses to norepinephrine[19,76] and angiotensin II,[10,19,72] as well as to endothelin-1[4,5,15,27,42,44,47,51,54,61,84,92], are more transient and return to baseline more rapidly than are those in the presence of external Ca^{2+}. Figure 7.2 illustrates the ability of endothelin-1 at two different doses to increase $[Ca^{2+}]_i$ in cultured A-10 VSMC (a cell line derived from neonatal rat aorta) in the presence and absence of extracellular Ca^{2+}. Exposure of VSMC to either 50 or 0.5 nM endothelin-1 induced a rapid increase in $[Ca^{2+}]_i$. When VSMC were bathed in Ca^{2+}-free EGTA (2 mM)–physiological salt solution (PSS) for 5 min, the initial $[Ca^{2+}]_i$ rise caused by endothelin-1 remained intact; however, the response was more transient compared with that in the presence of extracellular Ca^{2+}. Subsequent exposure of endothelin-1 (50 nM)–treated A10 VSMC to 1.5 mM Ca^{2+} resulted in a $[Ca^{2+}]_i$ overshoot compared with that seen in untreated controls (Fig. 7–2, middle trace). These observations suggest that the endothelin-induced $[Ca^{2+}]_i$ increase is due to both Ca^{2+} release from an intracellular compartment, presumably the sarcoplasmic reticulum, and Ca^{2+} influx across the sarcolemma. Kai et al.[42] and Kodama et al.[47] have recently reported that the endothelin-1–sensitive Ca^{2+} store overlaps the caffeine-sensitive Ca^{2+} store in VSMC. Consistent with the hypothesis that endothelin-1 mobilizes an internal store of Ca^{2+}, we[51] and others[8,53] have reported that endothelin-1 induces a rapid increase in $[Ca^{2+}]_i$ and in $^{45}Ca^{2+}$ efflux in the absence of extracellular calcium in rat and rabbit aortic VSMC. Furthermore, endothelin-1 also induced a sustained decrease (up to 20 min) in total exchangeable cell ^{45}Ca content in A-10 VSMC,[15] suggesting that this Ca^{2+} pool

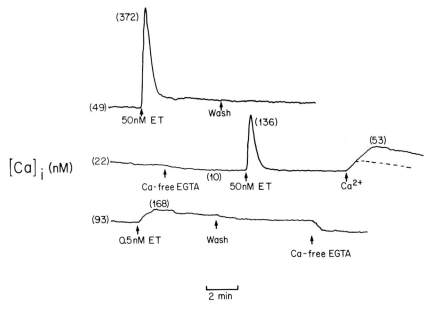

FIGURE 7.2. Endothelin effects on $[Ca^{2+}]_i$ in cultured A-10 VSMC in the presence and absence of external Ca^{2+}. Cultures of VSMC were grown on circular glass coverslips and loaded with the fluorescent Ca^{2+} indicator Fura-2 and exposed to endothelin-1 (0.5 or 50 nM) in the presence or absence (2 mM EGTA) of Ca^{2+}. In certain experiments, Ca^{2+} was readmitted to the bath. ET, endothelin. (From Danthuluri and Brock.[15])

does not refill during prolonged endothelin-1 exposure. Taken together, then, these observations suggest that endothelin-1 receptor stimulation leads to intracellular Ca^{2+} mobilization and to sustained transmembrane Ca^{2+} flux.

Ca^{2+} Influx Pathway: Receptor-Operated vs. Potential-Operated Channels

It is believed that VSMC contain at least two distinct Ca^{2+} influx pathways, i.e., voltage-gated channels (VGC) and receptor-operated channels (ROC).[45] In general, membrane VGC in VSMC are sensitive to organic Ca^{2+}-channel antagonists, whereas ROC are relatively insensitive to this group of compounds. Based on the observations that endothelin-1 shared significant sequence homology to α-scorpiotoxins, e.g., sarafotoxin (known to bind directly and to activate voltage-dependent Na^+ channels), and that endothelin-1–induced contractions of porcine coronary artery strips were sensitive to extracellular Ca^{2+} removal, as well as nicardipine, a Ca^{2+}-channel antagonist, endothelin-1 was first proposed to be an endogenous, direct activator of VSMC voltage-gated Ca^{2+} channels.[92] However, there has been no direct evidence supporting this mechanism to date. The extracellular Ca^{2+} dependency, as well as the Ca^{2+} channel antagonist sensitivity of the endothelin-1–induced contraction, appears to be vessel and species dependent. For example, preincubation with nicardipine caused a 100-fold increase in EC_{50} for endothelin-1–induced contraction in porcine coronary artery,[43] but not in rabbit aorta.[61] Likewise, diltiazem, D600, nifedipine, and verapamil are weak inhibitors of endothelin-1–induced contractile responses in rat[4,5,20] and rabbit,[61] whereas these Ca^{2+}-channel antagonists are known to inhibit potently KCl-induced contractions. Kim et al.[46] have reported that BRL34915 (cromakalin), a K^+-channel opener, and valinomycin, a specific K^+ ionophore, inhibit endothelin-1–stimulated contractions of rat aorta. Both of these latter compounds increase K^+ conductance and cause membrane hyperpolarization, which decreases Ca^{2+} entry, resulting in vascular relaxation.

As discussed above, it is clear that the sustained contractile actions of endothelin-1 are dependent on external Ca^{2+}.[5,44,47,61,92] Furthermore, numerous studies[43,51,63,90] have indicated that endothelin-1 does stimulate ^{45}Ca influx in VSMC. However, the pathway(s) and underlying mechanism(s) for transmembrane Ca^{2+} influx remain unclear. Goto et al.,[27] using the whole-cell patch-clamp technique, have shown that endothelin-1 stimulates an L-type Ca^{2+}-channel current in porcine coronary artery VSMC that can be blocked by nifedipine (Fig. 7.3). Likewise, Silberberg et al.[78] have shown that endothelin-1, when added to the bath, induces a slow activation of Ca^{2+} channels in the cell-attached patch-clamp recording mode in porcine coronary artery VSMC (Fig. 7.3). Endothelin-1 also increases the action potential plateau in atrial cells[38] that is produced by the Ca^{2+}-dependent, slow inward current sensitive to dihydropyridines. By contrast, Van Renterghem et al.[89] have shown that endothelin-1 does not activate this L-type Ca^{2+} channel in cultured A_7r_5 VSMC (a cell line derived from neonatal rat aorta). Although PN200-110 inhibited endothelin-1–stimulated contractions in rat aorta, endothelin-1 actually induced a small inhibition of the dihydropyridine PN200-110–sensitive Ca^{2+} current in voltage-clamped VSMC. Initially, endothelin-1 caused a transient hyperpolar-

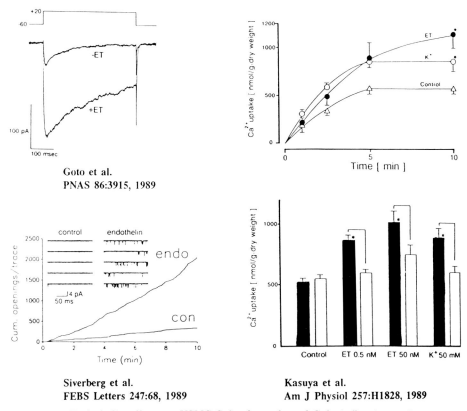

FIGURE 7.3. Endothelin effects on VSMC Ca^{2+} channels and Ca^{2+} influx in porcine coronary artery VSMC. Data in the top left are from Goto et al.[27]; bottom left, Silberberg et al.[78]; and bottom right, Kasuyu et al.[43] ET, endothelin; endo, endothelin; con, control: *open bars,* + nicardipine; *solid bars,* − nicardipine. (See corresponding references for further description.)

ization followed by a sustained depolarization that resulted from opening of a nonselective cation channel that was permeable to Ca^{2+}. This depolarization subsequently activated voltage-gated Ca^{2+} channels, leading to increased PN200-100–sensitive Ca^{2+} influx and contraction. Thus the initial effects of endothelin-1 on Ca^{2+} current in rat VSMC may be due to the opening of a Ca^{2+}-permeable, nonselective cation channel that is insensitive to organic Ca^{2+}-channel blockers. The ability of Ca^{2+}-channel blockers to inhibit endothelin-1–induced transmembrane Ca^{2+} fluxes is also inconsistent. A majority of studies,[45,51,63,90] but not all,[8] have shown that endothelin-1 stimulates ^{45}Ca influx in different VSMC preparations. Nicardipine inhibits endothelin-1–stimulated ^{45}Ca uptake in porcine coronary artery (Fig. 7.3),[44] whereas nifedipine has been reported to inhibit ^{45}Ca uptake in cultured rabbit aortic VSMC.[51] In contrast, the sustained plateau of the endothelin-1–stimulated $[Ca^{2+}]_i$ transient is not blocked by nicardipine or nifedipine in cultured rat VSMC,[54,87] but is inhibited by external Ca^{2+} removal or by Ni^{2+}, an inorganic Ca^{2+}-entry blocker. Therefore, the differences in the ability of endothelin-1 to activate VGC and ROC appear to be related to species and/or vascular bed. More detailed comparative studies on different vascular beds are clearly warranted.

Despite the differences in Ca^{2+} entry mechanisms in different VSMC, the observation that endothelin-1 induces ion channel activation in both whole-cell and cell-attached patch-clamp recording modes[27,78] supports the hypothesis that endothelin-exerts its action via a diffusible second messenger. The nature of the second messenger remains to be defined. Direct protein kinase C stimulation using phorbol esters has been shown to activate Ca^{2+} channels in several cell types,[41] including L-type Ca^{2+} channels in A_7r_5 VSMC[22] and to increase nifedipine-sensitive ^{45}Ca influx[82] in A_7r_5 VSMC. To date, studies examining the ability of protein kinase C inhibitors to inhibit endothelin-1–induced ion channel activation have not been performed. Protein kinase C–mediated Ca^{2+} channel activation in response to endothelin-1 would represent a positive feedback for supplying the Ca^{2+} required to stimulate persistent protein kinase C activity. This is consistent with the observation that nicardipine, at a concentration that blocks the endothelin-1–induced Ca^{2+} current, only shifts the contraction dose–response of endothelin-1 to the right in intact aortic strips without decreasing the maximal tension achieved[27].

RECEPTOR SIGNAL TRANSDUCTION MECHANISMS

Receptor Characteristics in VSMC

High-affinity binding sites for endothelin-1 have been reported in porcine coronary artery[43] and in different types of cultured VSMC derived from neonatal[87] and adult rat aorta,[36,37,75] rabbit aorta,[51] human omental vessels,[70] and human umbilical vein.[13] In general, [^{125}I]endothelin-1 binding either to intact VSMC or to plasma membrane fractions has been shown to be time dependent, saturable, and specific. Scatchard analysis of [^{125}I]endothelin-1 binding data is linear, showing a single class of high-affinity binding site (K_d = 0.06–2.12 nM) and maximal binding capacity ranging between 5,500 to 67,000 sites per cell, depending on the vascular tissue of origin. In all of the vascular tissues that have been studied [^{125}I]endothelin-1 binding is not displaced either by structurally unrelated peptides[37,51,75] or by different Ca^{2+}-channel ligands, including D-600, diltiazem, nicardipine, and nifedipine.[13,37,51] Likewise, endothelin-1 does not displace [^{125}I]iodopine, a Ca^{2+}-channel ligand.[43] Recently, Takuwa et al.[87] affinity cross-linked human [^{125}I]endothelin-1 to a putative receptor in A-10 VSMC membranes and have identified a single major band of approximately M_r = 65,000–75,000. [^{125}I]endothelin-1 binding to plasma membranes is inhibited by guanosine 5'-O-(thiotriphosphate) (GTPγS), a nonhydrolyzable GTP analog, suggesting that the receptor may be regulated via a guanine nucleotide–regulatory protein (see later under Guanine Nucleotide–Regulatory Proteins).

Interestingly, Roubert et al.[75] reported that preincubation of cultured rat aortic VSMC with angiotensin II for 18 h resulted in a dose-dependent "down-regulation" of endothelin-1–binding capacity without a change in binding affinity, suggesting that there may be significant "cross talk" between angiotensin II and endothelin-1 receptors. Although the underlying mechanisms are not fully understood, this phenomenon may be related to protein kinase C activation because the angiotensin II effect could be mimicked by the phorbol ester phorbol 12,13-dibutyrate.[70,75] The effects of phorbol ester to downregulate specific [^{125}I]endothelin-1 binding in cultured human umbilical vein VSMC are rapid ($t_{1/2} \approx 10$ min) and complete within 2 h.[70] This phenomenon may partially explain the ability of phorbol esters to inhibit endothelin-1–stimulated phosphoinositide hydrolysis and arachidonic acid release.

Phospholipid Signaling Mechanisms

Considerable attention in recent years has focused on the relationship between receptor-mediated phosphatidylinositol 4,5-bisphosphate (PIP_2) hydrolysis and Ca^{2+} mobilization.[7] A wide variety of hormones, neurotransmitters, and vasoactive agonists are believed to induce blood vessel contraction by activating phosphoinositide-specific phospholipase C.[28] In support of this hypothesis, it has been documented that vasoconstrictor agonists such as norepinephrine, angiotensin II, vasopressin, and serotonin can stimulate phospholipase C–mediated 1,4,5-inositol trisphosphate (IP_3) formation in VSMC. In this scheme of events, agonist-receptor binding leads to phospholipase C stimulation, which catalyzes the breakdown of PIP_2, leading to the generation of IP_3 and DAG. Ca^{2+} release from an intracellular site, presumably the sarcoplasmic reticulum, can be stimulated directly by IP_3 in cultured[91] or intact[81] VSMC that have been permeabilized. There are now numerous reports indicating that endothelin-1 increases phospholipase C–mediated IP_3 formation in cultured VSMC derived from rat[2,15,56,69,84,88] and rabbit[3,51,85] aorta, as well as in intact rat aorta[65] and dog coronary artery.[63] The time-course of endothelin-1–stimulated IP_3 formation in cultured A-10 VSMC is illustrated in Figure 7.4. Endothelin-1 (50 nM) induces a rapid and transient increase in [^{3}H]IP_3 that is kinetically similar to the rapid [Ca^{2+}]$_i$ increase shown in Figure 7.2. Endothelin-1 also induces a rapid accumulation of 1,3,4,5-inositol tetrakisphosphate (IP_4) in cultured A-10 VSMC cells,[49] most likely via the action of a specific IP_3-3-kinase.[7] IP_4 may play a role in mediating Ca^{2+} entry into the IP_3-sensitive pool.[35]

Although endothelin-1 causes a transient elevation in VSMC IP_3, it induces a biphasic and sustained accumulation of DAG in [^{3}H]arachidonic acid–labeled VSMC (Fig. 7.4). This pattern is similar to that described for angiotensin II in rat aortic VSMC.[30] Other studies with different cultured VSMC types have also reported biphasic,[31,49,85] as well as monophasic,[56,69,87] increases in DAG in response to endothelin-1. In addition, increases in cellular DAG mass have been detected in cultured rabbit aortic[85] and bovine aortic[49] VSMC using an enzymatic assay to measure DAG mass. While IP_3 can be formed only from phospholipase C hydrolysis of PIP_2, DAG formation in stimulated cells can occur via several distinct pathways. In addition to PIP_2, it has been suggested that phosphatidylinositol or phosphatidylcholine can also serve as a source of DAG formation during sustained angiotensin II exposure in VSMC.[28,30] Sunako et al.[85] have found that the dose–response curves for early (30 s) and late (5 min) phases of endothelin-1–stimulated DAG formation in cultured rabbit aortic VSMC are significantly different, with the early response being more sensitive. This result coupled with their observation that the EC_{50} value for late-phase endothelin-1–stimulated DAG formation did not correlate with the EC_{50} value for endothelin-1–stimulated IP_3 formation provides evidence that phos-

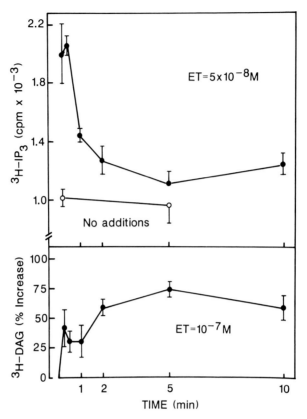

FIGURE 7.4. Time-course of 1,4,5-inositol trisphosphate (IP_3) and sn-1,2-diacylglycerol (DAG) formation in endothelin-stimulated cultured A-10 VSMC. Cells were prelabeled with either [^{3}H]myoinositol or [^{3}H]arachidonic acid and exposed to endothelin-1 (ET) for the indicated times. [^{3}H]inositol phosphates or [^{3}H]DAG were separated and quantitated as described. (From Danthuluri and Brock.[15])

phoinositides may not be the sole DAG source in endothelin-1–stimulated VSMC.

It has been shown that both phospholipases C and D contribute to phosphatidylcholine hydrolysis in different cell types, although phospholipase D is responsible for this effect in most cells.[21] Whereas phospholipase C action on phosphatidylcholine would generate DAG directly, phospholipase D acts to form phosphatidic acid that could then be dephosphorylated by phosphatidic acid phosphohydrolase to produce DAG. We have found that endothelin-1 stimulates a significantly greater increase in DAG formation when [^{3}H]glycerol is used to label phospholipid precursors than when [^{3}H]arachidonic acid is used (Fig. 7.5), suggesting that other sources of DAG are indeed likely. Similar observations have been made by Lee et al.[49] using cultured bovine aortic VSMC. Moreover, when VSMC were incubated with [^{3}H]choline to label phosphatidylcholine, the addition of endothelin-1 stimulated a rapid accumulation of water-soluble tritiated metabolites, presumably [^{3}H]choline and [^{3}H]phosphocholine (Fig. 7.5). Although the contributions of different phospholipid sources, as well as of phospholipases C and D, to endothelin-1–stimulated DAG formation remain to be fully understood, the data suggest that phosphatidylcholine hydro-

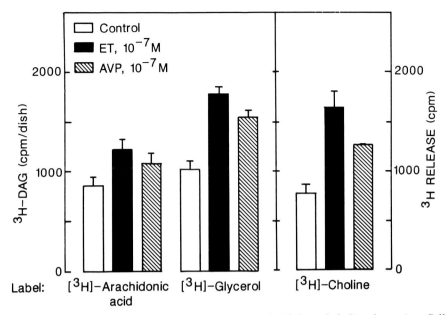

FIGURE 7.5. Endothelin-1 effects on sn-1,2-diacylglycerol (DAG) and choline formation. Cells were labeled with either [^{3}H]arachidonic acid (2 µCi, 4 h), [^{3}H]glycerol (4 µCi, 48 h) or [^{3}H]choline (1 µCi, 48 h), washed, and exposed to endothelin-1 (ET, 10^{-7} M) for 5 min. AVP, arginine vasopressin.

lysis can also contribute to sustained DAG formation in endothelin-1–stimulated VSMC. Because the cellular content of phosphatidylcholine is considerably higher than the content of phosphoinositides, phosphatidylcholine-derived DAG may be a biologically relevant stimulus for long-term protein kinase C stimulation.

Endothelin-1 also stimulates arachidonic acid release in cultured rabbit renal artery[73] and rat aortic[69a] VSMC, suggesting that phospholipase A_2 activity also increases. Brief VSMC pretreatment with phorbol esters to activate protein kinase C enhances endothelin-1–stimulated arachidonic acid release.[73] In contrast, phorbol esters inhibit endothelin-1–stimulated inositol phosphate release in cultured rabbit renal artery[73] and rat aortic[3,69a] VSMC. These observations argue against the possibility that the effects of phorbol esters can be solely attributed to their effects on endothelin-1 receptor number.[70,75] Similar effects of phorbol ester on receptor-stimulated phospholipases A_2 and C activities have also been described for other cell types.[6a] Thus it appears that phospholipases C and A_2 activities can be independently regulated by protein kinase C activation in endothelin-1–stimulated VSMC.

Guanine Nucleotide–Regulatory Proteins

Guanine nucleotide–regulatory proteins (G-proteins) are believed to play an essential role as a transducing element in the coupling of hormone receptors to adenylate cyclase, cGMP phosphodiesterase, phospholipases A_2, C, and D, ion channels, and mitogenesis.[21,26,57] G-proteins are heterotrimeric complexes consisting of α (M_r = 39,000–55,000), β (M_r = 35,000 or 36,000), and γ (M_r =

8,000–10,000) subunits.[26] The α-subunits are known to bind guanine nucleotides and to contain GTPase activity and in general are believed to activate their effector proteins. In recent years, several highly homologous, but unique G-protein α-subunits have been isolated, cloned, and sequenced using molecular biological techniques,[23,26,52] including *(1)* four splice variants of α_s-subunits involved in adenylate cyclase stimulation; *(2)* three distinct α_i-subunits (α_{i1}, α_{i2}, α_{i3}) that belong to a family of α-subunits previously thought to inhibit adenylate cyclase; *(3)* rod and cone α_t-subunits that stimulate cGMP phosphodiesterase; and *(4)* brain α_o- and α_z-subunits, whose functions are uncertain. Multiple forms of β- and γ-subunits also exist.[26,57] However, β–γ-subunits remain as a tightly coupled complex and dissociate only after denaturation.[57] Previous studies have provided evidence suggesting that the β–γ complexes activate phospholipase A_2.[57]

Nonhydrolyzable guanine nucleotides, such as GTPγS, readily stimulate IP_3 accumulation in permeabilized VSMC[79] or isolated membranes.[87] GTPγS is believed to exert its effects by binding to the α-subunit of G-proteins and promoting its dissociation from the β–γ complex.[26] However, GTPγS cannot be readily hydrolyzed by the GTPase and thus promotes persistent activation of the α-subunit–phospholipase C complex. Takuwa et al.[87] have presented evidence that the actions of endothelin-1 to induce phospholipase C activation is also mediated via a G-protein. In these studies, GTPγS stimulated a dose-dependent increase in IP_3 formation in membranes prepared from A-10 VSMC. Furthermore, GTPγS significantly augmented endothelin-1–stimulated IP_3 formation. These observations provide evidence indicating the involvement of a G-protein in regulating endothelin-1-receptor–phospholipase C interactions.

The exact nature of the G-protein coupled to endothelin-1–mediated phospholipase C activation has not been defined in VSMC. *Bordetella pertussis* toxin catalyzes the NAD-dependent, ADP ribosylation of the α-subunit of the G_i-protein family, rendering them insensitive to receptor regulation.[26] Although phospholipase C activity can be increased by guanine nucleotides, pertussis toxin does not inhibit receptor-stimulated phospholipase C in several cell types.[26] Takuwa et al.[87] recently reported that endothelin-1–stimulated IP_3 formation in intact A-10 VSMC is not blocked by pertussis toxin pretreatment, despite blocking measurable in vitro ADP-ribosylated substrate (41 kd protein) in A-10 membranes.[56,87] Consistent with this observation, pertussis toxin does not inhibit endothelin-1–stimulated $[Ca^{2+}]_i$ increases[87] or c-*fos* expression (a protooncogene that can be induced by receptor-mediated $[Ca^{2+}]_i$ increases[88]) in cultured rat aortic VSMC.[9] Interestingly, pertussis toxin does partially inhibit[9] the well-described mitogenic actions of endothelin-1 in VSMC.[9,48] Pertussis toxin partially inhibited ($\approx 30\%$) endothelin-1–stimulated inositol phosphate and arachidonic acid release in cultured rabbit renal artery VSMC.[73] Thus endothelin-1 appears to activate multiple G-proteins; however, phospholipase C activation occurs via a pertussis toxin–insensitive G-protein in certain VSMC types, suggesting that it belongs to a family of G-proteins that is different from that containing the G_i-like proteins. One likely candidate for the G-protein regulating phospholipase C may be G_z; it has been shown recently that this particular α-subunit lacks a pertussis toxin–catalyzed ADP-ribosylation site.[23,52]

PROTEIN KINASE C SIGNALING MECHANISMS

Protein kinase C is a ubiquitous Ca^{2+}- and phospholipid-dependent enzyme that can be activated by DAG and by tumor-promoting phorbol esters,[60] and it is believed to be involved in mediating sustained cellular responses to hormonal stimulation. As noted above, the slowly developing, persistent time-course of endothelin-1–stimulated blood vessel contractions is temporally similar to that induced by phorbol esters, such as phorbol 12,13-dibutyrate[4] or phorbol 12-myristate-13-acetate.[16,24,39] Indirect evidence that protein kinase C activation plays a critical role in mediating the sustained tension development in response to endothelin-1 comes from intact artery studies utilizing relatively selective inhibitors of protein kinase C, such as staurosporine and H-7 (1-[5-isoquinolinesulfonyl]-2-methylpiperazine).[4,15,61,84] Staurosporine may interact directly with the catalytic site of protein kinase C, whereas H-7 inhibits the enzyme by competition at the ATP-binding site.[60,86] The inhibitory actions of staurosporine on phenlyephrine-stimulated, phorbol 12,13-dibutyrate–stimulated, and endothelin-1–stimulated rat aortic contractions are illustrated in Figure 7.6. While the rapid phase of phenylephrine contraction was largely unaffected by staurosporine, it is clear that this protein kinase C inhibitor completely abolished the slow phase of the contractile response. By contrast,

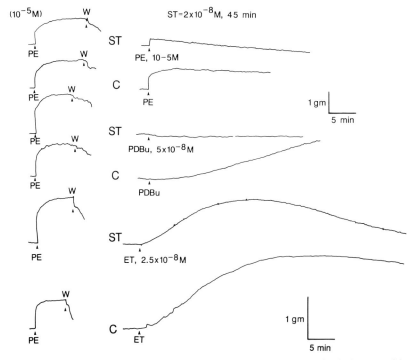

FIGURE 7.6. Staurosporine (ST) effects on phenylephrine (PE), phorbol 12,13-dibutyrate (PDBu), and endothelin-1 (ET-1) contraction. Isolated aortic rings were preincubated with 20 nM ST for 20 min and then exposed to the indicated concentration agonist. Isometric contractions were recorded as previously described.[15] C, control.

staurosporine caused a complete inhibition of phorbol 12,13-dibutyrate–stimulated tension development (middle set of tracings). The endothelin-1–stimulated contraction reached a maximum within 25 min and remained elevated for at least 60 min in the absence of staurosporine (Fig. 7.6).[15] However, addition of endothelin-1 to aortic rings that had been preexposed to staurosporine induced significantly less maximal tension and shortened the contraction duration compared with that of untreated aortic rings. It has been reported that staurosporine also antagonizes endothelin-1–stimulated contractions in rabbit aorta,[61] whereas H-7 has been shown to block endothelin-1–stimulated contractions in both rabbit[84] and rat[4] aorta.

Recent studies by Griendling et al.[31] with cultured rat aortic VSMC have demonstrated that endothelin-1 induces the rapid and sustained phosphorylation of an acidic (pI = 4.5) protein substrate (M_r = 76,000). This particular protein (Fig. 7.7) is also phosphorylated in response to phorbol 12-myristate-13-acetate, but not by the Ca^{2+} ionophore ionomycin, thus indicating that it is a protein kinase C–specific substrate. Endothelin-1 also stimulates translocation of protein kinase C from the cytosol to the membrane, resulting in sustained membranous enzymic activity in cultured bovine aortic VSMC.[49] Both of these indices of protein kinase C activation were shown to correlate strongly with the time-course of endothelin-1–stimulated DAG formation. Endothelin-1–induced, as well as phorbol 12-myristate-13-acetate–induced, protein phosphorylation was markedly reduced when protein kinase C was downregulated by pretreatment with phorbol 12,13-dibutyrate.[31] Taken together, these observations provide strong evidence suggesting that protein kinase C activation could explain the extremely potent, slowly developing phorbol ester-like contractions seen in intact arteries.

Endothelin-Induced Intracellular pH Changes

A wide variety of agonists that are linked to phospholipase C activation and to Ca^{2+} mobilization have been shown to increase pH_i via protein kinase C–mediated activation of Na^+–H^+ exchange.[32] Protein kinase C appears to activate Na^+–H^+ exchange by increasing the intracellular H^+ affinity for the

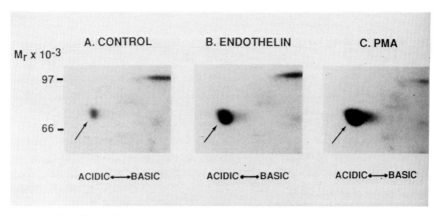

FIGURE 7.7. Two-dimensional polyacrylamide gel analysis of endothelin and phorbol 12,13-dibutyrate effects on phosphorylation of M_r = 76,000 protein in cultured rat aortic VSMC. (From Griendling et al.[30])

antiporter.[64] Numerous vasoconstrictors[6,14,34,62] have been shown to increase Na^+–H^+ antiporter activity via the DAG–protein kinase C signaling pathway leading to intracellular alkalinization. As illustrated in Figure 7.8, endothelin-1 also stimulates cytosolic alkalinization in VSMC, and this effect appears to be a consequence of protein kinase C stimulation. Endothelin-1 (50 nM) causes an alkaline pH_i shift in cultured A-10 VSMC (Fig. 7.8B). Typically, the pH_i response is quick in onset (<20 s), is maximal within 5 min, and is sustained for at least 15 min. The alkalinization time-course is qualitatively similar to that of the endothelin-1–stimulated DAG accumulation (cf. Figs. 7.4 and 7.8B). Phorbol 12,13-dibutyrate also increased pH_i in A-10 VSMC (Fig. 7.8A). However, a maximal dose of phorbol 12,13-dibutyrate did not stimulate further alkalinization when added at the peak of the endothelin-1–induced alkalinization. When cellular protein kinase C activity was reduced ("down-regulation") by pretreating VSMC with phorbol 12,13-dibutyrate (800 nM, 48 h) (Fig. 7.8C) or by pretreating VSMC with staurosporine (Fig. 7.8D), the endothelin-1–induced alkalinization response was completely blocked. These results indicate that endothelin-1–induced cytosolic alkalinization in cultured

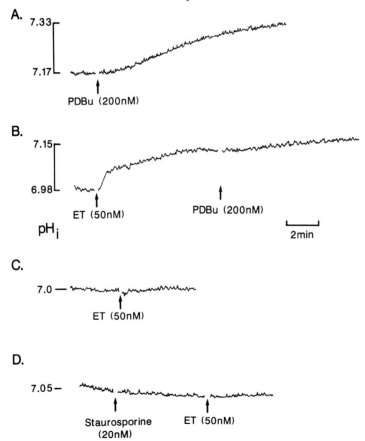

FIGURE 7.8. Endothelin (ET) and phorbol 12,13-dibutyrate PDBu) effects on pH_i in cultured A-10 VSMC. Cultures of VSMC were grown on glass coverslips and loaded with the fluorescent pH indicator 2′,7′-bis(2-carboxyethyl-5[−6]carboxyfluorescein) and exposed to endothelin-1 (0.5 or 50 nM) and/or PDBu as indicated by the arrows. In C, cells were pretreated with phorbol 12,13-dibutyrate (800 nM) for 48 h. (From Danthuluri and Brock.[15])

A-10 VSMC appears to be a protein kinase C–mediated event. The endothelin-1–induced alkalinization in VSMC is also inhibited by 5-N-ethylisopropyl-amiloride (EIPA), a specific inhibitor of the Na^+–H^+ antiporter and by external Na^+ removal, thus indicating that the pH_i increase was a consequence of increased Na^+–H^+ exchange.[15]

In A-10 VSMC monolayers that exhibit similar resting pH_i, the net alkalinization caused by 1 nM endothelin-1 does not appear to differ significantly from that induced by 50 nM endothelin-1 (Fig. 7.9), although the initial rate of pH_i increase is markedly slower. In contrast, 1 nM endothelin-1 is a submaximal dose for increasing $[Ca^{2+}]_i$ (Fig. 7.9). Because endothelin-1–stimulated Na^+–H^+ exchanger stimulation is one consequence of protein kinase C activation in A-10 VSMC, these data suggest that the endothelin-1 dose–response for protein kinase C activation may be shifted leftward to that of Ca^{2+} mobilization. However, at longer exposure times the $[Ca^{2+}]_i$ levels caused by 1 nM endothelin-1 are similar to those in response to 25 nM endothelin-1. Thus it is tempting to speculate that at any given dose endothelin-1 may differentially stimulate the Ca^{2+}-dependent DAG–protein kinase C signaling pathway more than the IP_3-dependent Ca^{2+} release events. This observation may be related to the ability of endothelin-1 to stimulate DAG formation from multiple sources (i.e., PIP_2, phosphatidylinositol, phosphatidylcholine), as well as its ability to stimulate sustained increases in Ca^{2+} influx. These biochemical events acting in concert would be expected to act as a positive feedback mech-

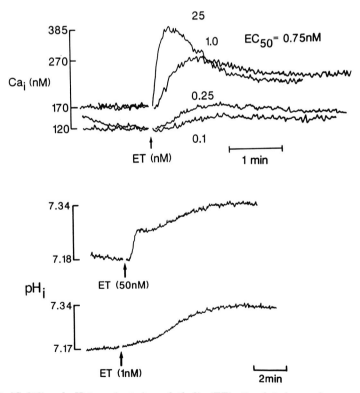

FIGURE 7.9. $[Ca^{2+}]i$ and pH_i transients in endothelin (ET)-stimulated vascular smooth muscle cells. Agonists were added as indicated by arrows. (From Danthuluri and Brock.[15])

anism and to promote persistent protein kinase C activation, ultimately resulting in the slowly developing, sustained contractions that are characteristic of endothelin-1.

It has been suggested that pH_i regulates the contractile state of VSMC.[18,28,29] In support of this hypothesis, Danthuluri and Deth[18] have used NH_4Cl to elevate pH_i directly and found that increased pH_i leads to a slowly developing contracture of rat aorta that is dependent on external Ca^{2+}. However, it should be pointed out that most studies examining the effects of agonist stimulation on pH_i have been done in the absence of $HCO_3^- - CO_2$ buffer, thus inactivating two pH_i regulating systems, Na^+-dependent Cl–HCO_3^- (acid extruder) and Na^+-independent Cl^-–HCO_3^- (alkalai extruder) exchangers (see Fig. 7.10).[25] When the latter buffering system was employed, it was found that EIPA, a Na^+–H^+ exchange inhibitor, did not inhibit contraction of rat aorta[71] and that the agonists norepinephrine and vasopressin actually decrease pH_i in rat mesenteric arteries[1] and mesangial cells,[25] respectively. In the latter study, Ganz et al.[25] showed that vasopressin stimulated all three pH_i-regulating mechanisms. However, since pH_i actually decreased during stimulation, the dominate mechanism appeared to be Na^+-independent Cl^-–HCO_3^-. Similar effects on pH_i are seen when cultured VSMC are exposed to endothelin-1 or to

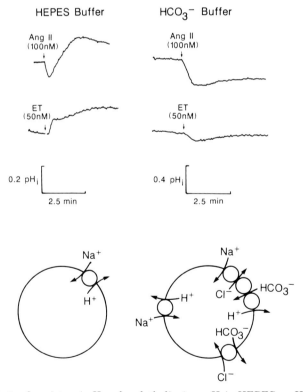

FIGURE 7.10. Effects of angiotensin II and endothelin-1 on pH_i in HEPES vs. HCO_3^+–CO_2 buffers. Cultures of rat aortic VSMC (angiotension II, 100 nM) or A-10 VSMC (endothelin-1, 100 nM) were grown on glass coverslips and loaded with the fluorescent pH indicator 2′,7′-bis(2-carboxyethyl-5[−6]carboxyfluorescein) as previously described.[15] Diagrams represent the known pH-regulating systems in VSMC that are active in each buffer system. ET, endothelin.

angiotensin II in a $HCO_3^- - CO_2$ buffer (Fig. 7.10), i.e., both agonists induce a decrease in VSMC pH_i, suggesting that endothelin-1, as well as angiotensin II, may also activate $Cl^- - HCO_3^-$ exchangers.

*Unique Nature of Endothelin Contractile Response:
Possible Mechanistic Variations from Other Vasoconstrictors*

It is evident that the interaction of endothelin with its receptor leads to phospholipase C activation, resulting in $[Ca^{2+}]_i$ increases, protein kinase C stimulation, and pH_i changes (Fig. 7.11). However, while endothelin appears to follow the signaling pathways proposed for other classic vasoconstrictors like norepinephrine and angiotensin, it is clear that the slowly developing nature of the contractile response in various smooth muscle preparations cannot be fully explained by the simple sequence of events related to IP_3, Ca^{2+} mobilization, and DAG formation. For example, why is it that even at relatively high doses endothelin-1 increases tone gradually in intact arteries, while the same doses increase rapid IP_3 formation and Ca^{2+} mobilization in cultured VSMC, as well as intact tissue? Peptide hormones such as angiotensin II show a small, but distinct lag in initiating contraction compared with that elicited by norepinephrine.[83] However, once contraction is initiated, the kinetics of angiotensin II–stimulated tension development are similar to that seen in response to norepinephrine (see Fig. 7.1). In contrast, endothelin-1–induced contractions take 10–20 min to develop fully maximal tension. The initial rapid phase of agonist-induced smooth muscle contraction is clearly associated with myosin

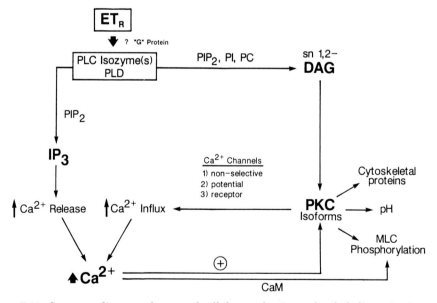

FIGURE 7.11. Summary diagram of proposed cellular mechanisms of endothelin action in vascular smooth muscle cells. ET_r, endothelin receptor; DAG, diacylglycerol; PIP_2, phosphatidylinositol 4,5-bisphosphate; IP_3, 1,4,5-inositol trisphosphate; MLC, myosin light chain; PI, phosphatidylinositol; PC, phosphatidylcholine; PKC, protein kinase C; PLC, phospholipase C; CaM, calmodulin.

light chain (MLC) phosphorylation.[28,80] Thus it may be possible that the lack of rapid endothelin-1 effects on contraction may be related to the kinetics and net amount of IP_3 generated by phospholipase C. In certain vascular tissues, endothelin-1–stimulated IP_3 may not be adequate to pulse-release threshold concentrations of Ca^{2+} required to stimulate rapid MLC phosphorylation. Support for this possibility is derived from recent data reported by Cilea et al.[12] showing that endothelin-1–stimulated MLC phosphorylation is dependent on the presence of extracellular Ca^{2+}. Several phospholipase C subtypes have now been isolated, and multiple cDNAs have been cloned from several tissue sources.[74] Thus it may be possible that endothelin-1 stimulates a different phospholipase C subtype than does either angiotensin II or norepinephrine.

Despite strong evidence implicating a role for protein kinase C in maintaining the sustained phase of vascular smooth muscle contraction, the underlying mechanism(s) are not clear at the present time. Rembold and Murphy[68] have reported that the maintenance of phorbol 12,13-dibutyrate contractions was associated with sustained increases in myosin phosphorylation, but not $[Ca^{2+}]_i$ increases. Although protein kinase C stimulates MLC phosphorylation, the net effect is to decrease the rate of phosphorylation by MLC kinase and to decrease actin-activated MgATPase activity.[58] Furthermore, protein kinase C prephosphorylation of MLC kinase with no bound calmodulin also decreases MLC kinase activity.[59] Alternatively, Somlyo and Himpens[80] have postulated that the potent contractile effects of phorbol esters in intact ateries may be explained by protein kinase C–stimulated phosphorylation of a protein phosphatase inhibitor, thereby limiting MLC phosphatase activity. All of these events would be expected to favor slower contractions. Multiple protein kinase C isozymes, as well as cDNAs, have been isolated,[60] thus adding to the molecular diversity of receptor signaling mechanisms. One important future direction will be to identify the particular protein kinase C isoform(s), as well as specific phorphorylated protein substrate(s), that are stimulated by endothelin-1 in VSMC.

CONCLUSIONS

In summary, endothelin appears to share basic receptor signaling mechanisms in common with other vasoactive hormones, including the abilities *(1)* to stimulate contraction; *(2)* to induce phospholipid hydrolysis via multiple phospholipases (A_2 and C; D?), thereby yielding rapid IP_3 formation and sustained DAG accumulation; *(3)* to increase $[Ca^{2+}]_i$ by releasing intracellular Ca^{2+} and stimulating Ca^{2+} influx; and *(4)* to stimulate a persistent increase in protein kinase C activity. Although these mechanisms are believed to contribute to the cellular mechanism of endothelin action in VSMC, as well as in other diverse cell types, it remains to be determined how these events are temporally integrated to yield the slowly developing, sustained contraction that is characteristically different from that induced by other vasoactive agonists.

ACKNOWLEDGMENTS

This work was supported by NIH grant HL 41180 from the National Heart, Lung and Blood Institute.

REFERENCES

1. AALKJAER, C., and E. J. CRAGOE, JR. Intracellular pH regulation in resting and contracting segments of rat mesenteric resistance vessels. *J. Physiol. (Lond.)* 402:391–410, 1988.
2. AMBAR, I., Y. KLOOG, I. SCHVARTZ, E. HAZUM, and M. SOKOLOVSKY. Competitive interaction between endothelin and sarafotoxin: binding and phosphoinositides hydrolysis in rat atria and brain. *Biochem. Biophys. Res. Commun.* 158: 195–201, 1989.
3. ARAKI, S., Y. KAWAHARA, K. KARIYA, M. SUANAKO, H. FUKUSAKI, and Y. TAKAI. Stimulation of phospholipase C–mediated hydrolysis of phosphoinositides by endothelin in cultured rabbit aortic smooth muscle cells. *Biochem. Biophys. Res. Commun.* 159: 1072–1079, 1989.
4. AUGET, M., S. DELAFLOTTE, P.-E. CHABRIER, and P. BRANQUET. Comparative effects of endothelin and phorbol 12,13-dibutyrate in rat aorta. *Life Sci.* 45: 2051–2059, 1989.
5. AUGET, M., S. DELAFLOTTE, P.-E. CHABRIER, E. PIROTZKY, C. CLOSTRE, and P. BRAQUET. Endothelin and Ca^{++} agonist bay K8644: different vasoconstrictive properties. *Biochem. Biophys. Res. Commun.* 156: 186–192, 1988.
6. BERK, B. C., T. A. BROCK, M. A. GIMBRONE, JR., and R. W. ALEXANDER. Early agonist-mediated ionic events in cultured vascular smooth muscle cells: calcium mobilization is associated with intracellular acidification. *J. Biol. Chem.* 262: 5065–5072, 1987.
6a. BERRIDGE, M. J. Inositol trisphosphate and diacylglycerol: two interacting second messengers. *Annu. Rev. Biochem.* 56: 159–193, 1987.
7. BERRIDGE, M. J., and R. F. IRVINE. Inositol phosphates and cell signalling. *Nature* 341: 197–205, 1989.
8. BIALECKI, R. A., N. J. IZZO, JR., and W. S. COLUCCI. Endothelin-1 increases intracellular calcium mobilization but not calcium uptake in rabbit vascular smooth muscle cells. *Biochem. Biophys. Res. Commun.* 164: 474–479, 1989.
9. BOBIK, A., A. GROOMS, J. A. MILLAR, A. MITCHELL, and S. GRINPUKEL. Growth factor activity of endothelin on vascular smooth muscle. *Am. J. Physiol.* 258 (*Cell Physiol.* 27): C408–C415, 1990.
10. BROCK, T. A., R. W. ALEXANDER, L. S. EKSTEIN, W. J. ATKINSON, and M. A. GIMBRONE, JR. Angiotensin increases cytosolic free calcium in cultured vascular smooth muscle cells. *Hypertension* 7(Suppl. 1): I105–I109, 1985.
11. CHATTERJEE, M., and M. TEJADA. Phorbol ester–induced contraction in chemically skinned vascular smooth muscle. *Am. J. Physiol.* 251(*Cell Physiol.* 20): C356–C361, 1986.
12. CILEA, J., S. MORELAND, and R. S. MORELAND. Maintenance of endothelin contractions of swine carotid artery requires extracellular calcium and protein kinase C activation, abstracted. *FASEB J.* 4: A332, 1990.
13. CLOZEL, M., W. FISCHLI, and C. GUILLY. Specific binding of endothelin on human vascular smooth muscle cells in culture. *J. Clin. Invest.* 83: 1758–1761, 1989.
14. DANTHULURI, N. R., B. C. BERK, T. A. BROCK, E. J. CRAGOE, JR., R. C. DETH. Protein kinase C–mediated intracellular alkalinization in rat and rabbit aortic smooth muscle cells. *Eur. J. Pharmacol.* 141: 503–506, 1987.
15. DANTHULURI, N. R., and T. A. BROCK. Endothelin receptor–coupling mechanisms in vascular smooth muscle: a role for protein kinase C. *J. Pharmacol. Exp. Ther.* 254: 393–399, 1990.
16. DANTHULURI, N. R., and R. C. DETH. Phorbol ester–induced contraction of arterial smooth muscle and inhibition of α-adrenergic response. *Biochem. Biophys. Res. Commun.* 125: 1103–1109, 1984.
17. DANTHULURI, N. R., and R. C. DETH. Acute desensitization to angiotensin II: evidence for a requirement of agonist-induced diacylglycerol production during tonic contraction of rat aorta. *Eur. J. Pharmacol.* 126: 135–139, 1986.
18. DANTHULURI, N. R., and R. C. DETH. Effects of intracellular alkalinization on resting and agonist-induced vascular tone. *Am. J. Physiol.* 256(*Heart Circ. Physiol.* 25): H867–H875, 1989.
19. DETH, R., and C. VAN BREEMEN. Relative contributions of Ca^{2+} influx and cellular Ca^{2+} release during drug-induced activation of the rabbit aorta. *Pflugers Arch.* 348: 13–22, 1976.
20. D'ORLEANS-JUSTE, P., G. DE NUCCI, and J. R. VANE. Endothelin-1 contracts isolated vessels independently of dihydropyridine-sensitive Ca^{2+} channel activation. *Eur. J. Pharmacol.* 165: 289–295, 1989.
21. EXTON, J. J. Signaling through phosphatidylcholine breakdown. *J. Biol. Chem.* 265: 1–4, 1990.
22. FISH, D. R., G. SPERTI, W. S. COLUCCI, and D. E. CLAPHAM. Phorbol ester increases the

dihydropyridine-sensitive calcium conductance in a vascular smooth muscle cell line. *Circ. Res.* 62: 1049–1054, 1988.
23. FONG, H. K. W., K. K. YOSHIMOTO, P. EVERSOLE-CIRE, and M. I. SIMON. Identification of a GTP-binding protein that lacks an apparent ADP-ribosylation site for pertussis toxin. *Proc. Natl. Acad. Sci. USA* 85: 3066–3070, 1988.
24. FORDER, J., A. SCRIABINE, and H. RASMUSSEN. Plasma membrane calcium flux, protein kinase C activation and smooth muscle contraction. *J. Pharmacol. Exp. Ther.* 235: 267–273, 1985.
25. GANZ, M. B., G. BOYARSKY, R. B. STERZEL, and W. F. BORON. Arginine vasopressin enhances pH_i regulation in the presence of HCO_3^- by stimulating three acid–base transport systems. *Nature (Lond.)* 337: 648–651, 1989.
26. GILMAN, A. G. G proteins: transducers of receptor generated signals. *Annu. Rev. Biochem.* 56: 615–49, 1987.
27. GOTO, K., Y. KASUYA, N. MATSUKI, Y. TAKUWA, H. KURIHARA, T. ISHIKAWA, S. KIMURA, M. YANAGISAWA, and T. MASAKI. Endothelin activates the dihydropyridine-sensitive, voltage-dependent Ca^{2+} channel in vascular smooth muscle. *Proc. Natl. Acad. Sci. USA* 86: 3915–3918, 1989.
28. GRIENDLING, K. K., and R. W. ALEXANDER. Angiotensin, other pressors and the transduction of vascular smooth muscle contraction. In: *Hypertension: Pathophysiology, Diagnosis and Management*, edited by J. H. Laragh and B. M. Brenner. New York: Raven, 1990, p. 583–600.
29. GRIENDLING, K. K., B. C. BERK, and R. W. ALEXANDER. Evidence that Na^+H^+ exchange regulates angiotensin II–stimulated diacylglycerol accumulation in vascular smooth muscle cells. *J. Biol. Chem.* 263: 10620–10624, 1988.
30. GRIENDLING, K. K., S. E. RITTENHOUSE, T. A. BROCK, L. S. EKSTEIN, M. A. GIMBRONE, JR., and R. W. ALEXANDER. Sustained diacylglycerol formation from inositol phospholipids in angiotensin II–stimulated vascular smooth muscle cells. *J. Biol. Chem.* 261:5901–5906, 1986.
31. GRIENDLING, K. K., T. TSUDA, and R. W. ALEXANDER. Endothelin stimulates diacylglycerol accumulation and activates protein kinase C in cultured vascular smooth muscle cells. *J. Biol. Chem.* 264: 8237–8240, 1989.
32. GRINSTEIN, S., and A. ROTHSTEIN. Mechanism of regulation of the Na^+/H^+ exchanger. *J. Membr. Biol.* 90: 1–12, 1986.
33. GRYGLEWSKI, R. J., R. M. BOTTING, and J. R. VANE. Mediators produced by the endothelial cell. *Hypertension* 12: 530–548, 1988.
34. HATORI, N., B. P. FINE, A. NAKAMURA, E. J. CRAGOE, JR., and A. AVIV. Angiotensin II effect on cytosolic pH in cultured rat vascular smooth muscle cells. *J. Biol. Chem.* 262: 5073–5078, 1987.
35. HILL, T. D., N. M. DEAN, and A. L. BOYNTON. Inositol 1,3,4,5-tetrakisphosphate induces Ca^{2+} sequestration in rat liver cells. *Nature* 242: 1176–1178, 1988.
36. HIRATA, Y., H. YOSHIMI, T. EMORI, M. SHICIRI, F. MARUMO, T. X. WATANABE, S. KUMAGAYE, K. NAKAJIMA, T. KIMURA, and S. SAKAKIBARA. Receptor binding activity and cytosolic free calcium response by synthetic endothelin analogs in cultured rat vascular smooth muscle cells. *Biochem. Biophys. Res. Commun.* 160: 228–234, 1989.
37. HIRATA, Y., H. YOSHIMI, S. TAKATA, T. X. WATANABE, S. KUMAGAI, K. NAKAJIMA, and S. SAKAKIBRA. Cellular mechanism of action by a novel vasoconstrictor endothelin in cultured rat vascular smooth muscle cells. *Biochem. Biophys. Res. Commun.* 154: 868–875, 1988.
38. ISHIKAWA, T., M. YANAGISAWA, S. KIMURA, K, GOTO, and T. MASAKI, Positive inotropic action of novel vasoconstrictor peptide endothelin on guinea pig atria. *Am. J. Physiol.* 255(*Heart Circ. Physiol.* 24): H970–H973, 1988.
39. ITOH, H., and K. LEDERIS. Contraction of rat thoracic aorta strips induced by phobol 12-myristate 13-acetate. *Am. J. Physiol.* 252 (*Cell Physiol.* 21): C244–C247, 1987.
40. JIANG, M. J., and K. C. MORGAN. Intracellular calcium levels in phorbol ester–induced contractions of vascular smooth muscle. *Am. J. Physiol.* 253 H1365–H1371, 1987.
41. KACZMAREK, L. K. The role of protein kinase C in the regulation of ion channels and neurotransmitter release. *Trends Neurosci.* 10: 30–34, 1987.
42. KAI, H., H. KANAIDE, and M. NAKAMURA. Endothelin-sensitive intracellular Ca^{2+} store overlaps with a caffeine-sensitive one in rat aortic smooth muscle cells in primary culture. *Biochem. Biophys. Res. Commun.* 158: 235–243, 1989.
43. KAYSUYA, Y., T. ISHIKAWA, M. YANAGISAWA, S. KIMURA, K. GOTO, and T. MASAKI. Mechanism of contraction to endothelin in isolated porcine coronary artery. *Am. J. Physiol.* 257(*Heart Circ. Physiol.* 26): H1828–H1835, 1989.
44. KAYSUYA, Y., T. TAKUWA, M. YAAGISAWA, S. KIMURA, K. GOTO, and T. MASAKI. Endothelin-1

induces vasconstriction through two functionally distinct pathways in porcine coronary artery: contribution of phosphoinositide turnover. *Biochem. Biophys. Res. Commun.* 161: 1049–1055, 1989.
45. KHALIL, R. A., N. J. LODGE, K. SAIDA, C. H. GELBAND, and C. VAN BREEMEN. Calcium mobilization in vascular smooth muscle and its relevance to the etiology of hypertension. In: *Hypertension: Pathophysiology, Diagnosis and Management,* edited by J. H. Laragh and B. M. Brenner. New York: Raven, 1990, p. 547–564.
46. KIM, S., S. MORIMOTO, E. KOH, Y. MIYASHITA, and T. OGIHARA. Comparison of effects of a potassium channel opener BRL34915, a specific potassium ionophore valinomycin and calcium channel blockers on endothelin-induced vascular contraction. *Biochem. Biophys. Res. Commun.* 164:1003–1008, 1989.
47. KODAMA, M., H. KANAIDE, S. ABE, K. HIRANO, H. KAI, and M. NAKAMURA. Endothelin-induced Ca-independent contraction of the porcine coronary artery. *Biochem. Biophys. Res. Commun.* 160: 1302–1308, 1989.
48. KOMURA, I., H. KURIHARA, T. SUGIYAMA, F. TAKAKU, and Y. YAZAKI. Endothelin stimulates c-fos and c-myc expression and proliferation of vascular smooth muscle cells. *FEBS Lett.* 238: 249–252, 1988.
49. LEE, T.-S., T. CHAO, K.-Q. HU, and G. L. KING. Endothelin stimulates a sustained 1,2-diacylglycerol increase and protein kinase C activation in bovine aortic smooth muscle cells. *Biochem. Biophys. Res. Commun.* 162: 381–386, 1989.
50. LE MONNIER DE GOUVILLE, A.-C., H. L. LIPPTON, I. CAVERO, W. R. SUMMER, and A. L. HYMAN. Endothelin—a new family of endothelium-derived peptides with widespread biological properties. *Life Sci.* 45: 1499, 15–13, 1989.
51. MARSDEN, P. A., N. R. DANTHULURI, B. M. BRENNER, B. J. BALLERMANN, and T. A. BROCK. Endothelin action on vascular smooth muscle involves inositol trisphosphate and calcium mobilization. *Biochem. Biophys. Res. Commun.* 158: 86–93, 1989.
52. MATSUOKA, M., H. ITOH, T. KOZASA, and Y. KAZIRO. Sequence analysis of cDNA and genomic DNA for a putative pertussis toxin–insensitive guanine nucleotide-binding regulatory protein α subunit. *Proc. Natl. Acad. Sci. USA* 85: 5384–5388, 1988.
53. MIASIRO, N., H. YAMAMOTO, H. KANAIDE, and M. NAKAMURA. Does endothelin mobilize calcium from intracellular store sites in rat aortic vascular smooth muscle cells in primary culture? *Biochem. Biophys. Res. Commun.* 156: 312–317, 1988.
54. MITSUHASI, T., R. C. MORRIS, JR., and H. E. IVES. Endothelin-induced increases in vascular smooth muscle Ca^{2+} do not depend on dihydropyridine-sensitive Ca^{2+} channels. *J. Clin. Invest.* 84: 635–639, 1989.
55. MORGAN, J. P., and K. G. MORGAN. Vascular smooth muscle, the first recorded Ca^{2+} transient. *Pflugers Arch.* 395: 75–77, 1983.
56. MULDOON, L. L., K. D. RODLAND, M. L. FORSYTHE, and B. E. MAGUN. Stimulation of phosphatidylinositol hydrolysis, diacylglycerol release, and gene expression in response to endothelin, a potent new agonist for fibroblasts and smooth muscle cells. *J. Biol. Chem.* 264: 8529–8536, 1989.
57. NEER, E. J., and D. E. CLAPHAM. Structure and function of G-protein β subunit. In: *G proteins,* edited by R. Iyengar and L. Birnbaumer. San Diego, CA: Academic, 1990, p. 41–61.
58. NISHIKAWA, M., J. R. SELLERS, R. S. ADELSTEIN, and H. HIDAKA. Protein kinase C modulates in vitro phosphorylation of the smooth muscle heavy mermyosin by myosin light chain kinase. *J. Biol. Chem.* 259: 8808–8814, 1984.
59. NISHIKAWA, M., S. SHIRAKAWA, and R. S. ADELSTEIN. Phosphorylation of smooth muscle myosin light chain kinase by protein kinase C. *J. Biol. Chem.* 260: 8978–8983, 1985.
60. NISHIZUKA, Y. Studies and perspectives of the protein kinase C family for cellular regulation. *Cancer* 63: 1892–1903, 1989.
61. OHLSTEIN, E. H., S. HOROHONICH, and D. W. P. HAY. Cellular mechanisms of endothelin in rabbit aorta. *J. Pharmacol. Exp. Ther.* 250: 548–555, 1989.
62. OWEN, N. E. Effect of catecholamines on Na/H exchange in vascular smooth muscle cells. *J. Cell Biol.* 103: 2053–2060, 1986.
63. PANG, D. C., A. JOHNS, K. PATTERSON, L. H. P. BOTHELLO, and G. M. RUBANYI. Endothelin-1 stimulates phosphatidylinositol hydrolysis and calcium uptake in isolated canine coronary arteries. *J. Cardiovasc. Sci.* 13(Suppl. 5): S75–S79, 1989.
64. PARIS, S., and J. POUYSSEGUR. Growth factors activate the Na^+/H^+ antiporter in quiescent fibroblasts by increasing its affinity for intracellular H^+. *J. Biol. Chem.* 259: 10989–10994, 1984.
65. RAPOPORT, R. M., K. A. STAUDERMAN, and R. F. HIGHSMITH. Effects of EDCF and endothelin on phosphatidylinositol hydrolysis and contraction in rat aorta. *Am. J. Physiol.* 258(*Cell Physiol.* 27): C122–C131, 1990.

66. RASMUSSEN, H. The calcium messenger system. *N. Engl. J. Med.* 314: 1164–1170, 1986.
67. RASMUSSEN, H., Y. TAKUWA, and S. PARK. Protein kinase C in the regulation of smooth muscle contraction. *FASEB J.* 1: 177–185, 1987.
68. REMBOLD, C. M., and R. A. MURPHY. [Ca^{2+}]-dependent myosin phosphorylation in phorbol diester stimulated smooth muscle contraction. *Am. J. Physiol.* 255(*Cell Physiol.* 24): C719–C723, 1988.
69. RESINK, T. J., T. SCOTT-BURDEN, and F. R. BUHLER. Endothelin stimulates phospholipase C in cultured vascular smooth muscle cells. *Biochem. Biophys. Res. Commun.* 157: 1360–1368, 1988.
69a. RESINK, T. J., T. SCOTT-BURDEN, and F. R. BUHLER. Activation of phospholipase A_2 by endothelin in cultured vascular smooth muscle cells. *Biochem. Biophys. Res. Commun.* 158: 279–286, 1989.
70. RESINK, T. J., T. SCOTT-BURDEN, E. WEBER, and F. R. BUHLER. Phorbol ester promotes a sustained down-regulation of endothelin receptors and cellular responses to endothelin in human vascular smooth muscle cells. *Biochem. Biophys. Res. Commun.* 166: 1213–1219, 1990.
71. REYNOLDS, E. E., J. M. BRUM, E. J. CRAGOE, JR., and C. M. FERRARIO. Effect of Na^+/H^+ exchange inhibitors on agonist-induced contraction of rat aorta. *J. Pharmacol. Exp. Ther.* 247: 1146–1151, 1988.
72. REYNOLDS, E. E., and G. R. DUBYAK. Agonist-induced calcium transients in cultured smooth muscle cells: measurements with Fura-2 loaded monolayers. *Biochem. Biophys. Res. Commun.* 136: 927–934, 1986.
73. REYNOLDS, E. E., L. S. MOK, and S. KUROKAWA. Phorbol ester dissociates endothelin-stimulated phosphoinositide hydrolysis and arachidonic acid release in vascular smooth muscle cells. *Biochem. Biophys. Res. Commun.* 160: 868–873, 1989.
74. RHEE, S. G., P.-G. SUH, S.-H. RYU, and S. Y. LEE. Studies of inositol phospholipid–specific phospholiase C. *Science* 244: 546–550, 1089.
75. ROUBERT, P., V. GILLARD, P. PLAS, J.-M. GUILLON, P.-E. CHABRIER, and P. BRAQUET. Angiotensin II and phorbol esters potently down-regulate endothelin (endothelin-1) binding sites in vascular smooth muscle cells. *Biochem. Biophys. Res. Commun.* 164: 809–815, 1989.
76. SATO, K., H. OZAKI, and H. KARAKI. Changes in cytosolic calcium level in vascular smooth muscle strip measured simultaneously with contraction using fluorescent calcium indicator fura-2. *J. Pharmacol. Exp. Ther.* 246: 294–300, 1988.
77. SECREST, R. J., and M. L. COHEN. Endothelin: differential effects in vascular and nonvascular smooth muscle. *Life Sci.* 45: 1365–1372, 1989.
78. SILBERBERG, S. D., T. C. PODER, and A. E. LACERDA. Endothelin increases single channel calcium currents in coronary arterial smooth muscle cells. *FEBS Lett.* 247: 68–72, 1989.
79. SOCORRO, L., R. W. ALEXANDER, and K. K. GRIENDLING. Cholera toxin modulation of angiotensin II–stimulated inositol phophate production in cultured vascular smooth muscle cells. *Biochem. J.* 265: 799–807.
80. SOMLYO, A. P., and B. HIMPENS. Cell calcium and its regulation in smooth muscle. *FASEB J.* 3: 2266–2276, 1989.
81. SOMLYO, A. V., M. BOND, A. P. SOMLYO, and A. SCARPA. Inositol trisphosphate–induced calcium release and contraction in vascular smooth muscle. *Proc. Natl. Acad. Sci. USA* 82: 52315235, 1985.
82. SPERTI, G., and W. S. COLUCCI. Phorbol ester–stimulated bidirectional transmembrane Ca^{2+} flux in A_7r_5 vascular smooth muscle cells. *Mol. Pharmacol.* 32: 37–42, 1987.
83. ST.-LOUIS, J., D. REGOLI, J. BARABE, and W. K. PARK. Myotropic actions of angiotensin and noradrenaline in strips of rabbit aortae. *Can. J. Physiol. Pharmacol.* 55: 1056–1069, 1977.
84. SUGIURA, M., T. INAGAMI, G. M. T. HARE, and J. A. JOHNS. Endothelin action: inhibition by a protein kinase C inhibitor and involvement of phosphoinositols. *Biochem. Biophys. Res. Commun.* 158: 170–176, 1989.
85. SUNAKO, M., Y. KAWAHARA, K. HIRATA, T. TSUDA, M. YOKOYAMA, H. FUKUZAKI, and Y. TAKAI. Mass analysis of 1,2-diacylglycerol in cultured rabbit vascular smooth muscle cells: comparison of stimulation by angiotensin II and endothelin. *Hypertension* 15: 84–88, 1990.
86. TAMAOKI, T., H. NOMOTO, I. TAKAHASHI, Y. KATO, M. MORIMOTO, and F. TOMITA. Staurosporine, a potent inhibitor of phospholipid/Ca^{++} dependent protein kinase. *Biochem. Biophys. Res. Commun.* 135, 397–402, 1986.
87. TAKUWA, Y., Y. KASUYA, N. TAKUWA, M. KUDO, M. YANAGISAWA, K. GOTO, T. MASAKI, and K. YAMASHITA. Endothelin receptor is coupled to phospholipase C via a pertussis toxin–

insensitive guanine nucleotide binding regulatory protein in vascular smooth muscle cells. *J. Clin. Invest.* 85: 653–658, 1990.
88. TAUBMAN, M. B., B. C. BERK, S. IZUMO, T. TSUA, R. W. ALEXANDER, and B. NADAL-GINARD. Angiotensin II induces c-*fos* mRNA in aortic smooth muscle. *J. Biol. Chem.* 264:526–530, 1989.
89. VAN RENTERGHEM, C., P. VIGNE, J. BARHANIN, A. SCHMID-ALLIANA, C. FRELIN, and M. LAZDUNSKI. Molecular mechanism of action of the vasoconstrictor peptide endothelin. *Biochem. Biophys. Res. Commun.* 157: 977–985, 1988.
90. WALLNOFER, A., S. WEIR, U. RUEGG, and C. CAUVIN. The mechanism of action of endothelin-1 as compared with other agonists in vascular smooth muscle. *J. Cardiovasc. Pharmacol.* 13(Suppl. 5): S23–S31, 1989.
91. YAMAMOTO, H., and C. VAN BREEMEN. Inositol-1,4,5-trisphosphate releases calcium from skinned cultured smooth muscle cells. *Biochem. Biophys. Res. Commun.* 130: 270–274, 1985.
92. YANAGISAWA, M., H. KURIHARA, S. KIMURA, Y. TOMOBE, M. KOBAYASHI, Y. MITSUI, Y. YAZAKI, K. GOTO, and T. MASAKI, A novel potent vasoconstrictor peptide produced by vascular endothelial cells. *Nature* 332: 411–415, 1988.
93. YANAGISAWA, M., and T. MASAKI. Molecular biology and biochemistry of endothelins. *TIPS* 10: 374–378, 1989.

8

Interaction between Endothelin and Endothelium-Derived Relaxing Factor(s)

THOMAS F. LÜSCHER, CHANTAL BOULANGER, ZHIHONG YANG, AND YASUAKI DOHI

In the last decade, a number of vasoactive substances produced and released by endothelial cells have been characterized.[1] Indeed, the cells are not only a source of endothelin, but also of other contracting factors such as angiotensin II, cyclooxygenase-dependent contracting factors (i.e., thromboxane A_2, prostaglandin H_2, and superoxide anions), and a yet unidentified contracting factor released during hypoxia ($EDCF_1$). In addition, the endothelium produces relaxing factors such as endothelium-derived nitric oxide (formerly called "endothelium-derived relaxing factor"), prostacyclin, and an endothelium-derived hyperpolarizing factor of unknown biochemical nature. With the increasing complexity of endothelium-dependent vascular regulatory mechanisms, interactions between vasoactive substances produced by the cells have become more important. This chapter focuses on the interactions of endothelin-1, endothelium-derived nitric oxide, and prostacyclin at the levels of the endothelium and the vascular smooth muscle cell.

INTERACTIONS AT THE LEVEL OF THE ENDOTHELIAL CELL

Endothelin and Production of Relaxing Factors

Prostacyclin
In certain preparations, such as the isolated perfused and pressurized mesenteric resistance arteries of the rat, endothelin-1 can evoke endothelium-dependent relaxations (Fig. 8.1).[2] This response can be observed with intraluminal application of the peptide, but not with extraluminal application. As the relaxations can be prevented by indomethacin, but not by the inhibitor of the nitric oxide formation from L-arginine, N^G-monomethyl-L-arginine (L-NMMA), prostacyclin is the most likely mediator.[2] Indeed, endothelin-1 releases eicosanoids from the rabbit isolated perfused kidney and spleen.[3,4] In the former, the peptide evokes mainly the production of prostacyclin and to a lesser extent that of prostaglandin E_2, whereas in the latter prostaglandin E_2 dominates and less prostacyclin and little thromboxane A_2 are formed.[3] Indomethacin, on the other hand, potentiates the pressor responses obtained during intravenous infusion of endothelin-1, suggesting that the prostanoids also blunt the vasoconstrictor effects of the peptide in vivo.[4]

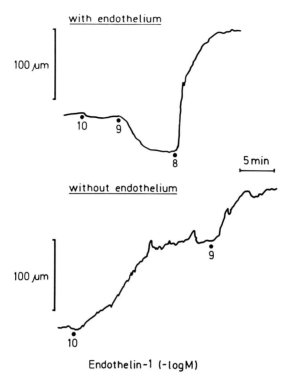

FIGURE 8.1. Effects of intraluminal infusion of endothelin-1 in a perfused and pressurized mesenteric resistance artery of a normotensive Wistar-Kyoto rat (16 weeks of age). In the preparation with endothelium (top), but not in that without endothelium (bottom), the peptide causes a vasodilation.

Endothelium-Derived Relaxing Factor

In the isolated perfused mesentery of the rat and the luminally perfused aorta of the rabbit, endothelin-1 releases endothelium-derived relaxing factor (EDRF).[5] As endothelin-3 seems to be more potent in eliciting these responses than endothelin-1, the endothelin receptor subtypes located on the endothelium may differ from those located on vascular smooth muscle (to mediate contraction),[6-8] or the efficacy and nature of the coupling mechanism may be different.

Endothelium-Derived Nitric Oxide and Endothelin Production

Endothelium-derived nitric oxide is released under basal conditions and after stimulation with certain agonists. Nitric oxide activates soluble guanylyl cyclase not only in vascular smooth muscle but also in endothelial cells themselves (Fig. 8.2).[9] The formation of cyclic 3'-5'-guanosine monophosphate (cGMP) in vascular smooth muscle cells has been recognized as an important pathway to mediate relaxation, but the physiological role of the increased formation of the cyclic nucleotide in endothelial cells is not entirely clear.[1] The increase in cGMP in endothelial cells may act as a negative feedback mechanism on the production of the relaxing factor[1,10]; more recent experimental evi-

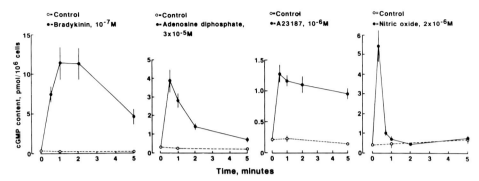

FIGURE 8.2. Effects of bradykinin, adenosine diphosphate, the calcium ionophore A23187, and nitric oxide solution on the production of cyclic 3'-5'-guanosine monophosphate (cGMP) in endothelial cells in culture (passages 1 and 2). (From Boulanger et al.[9])

dence suggests that it is also an important mediator of the interaction between endothelium-derived nitric oxide and endothelin at the level of the endothelial cells.[11]

Thrombin is an agonist that evokes endothelium-dependent relaxations in isolated arteries.[1,12] As the response is prevented by methylene blue or L-NMMA, nitric oxide is the most likely mediator of the response.[12,13] Furthermore, thrombin stimulates the release of endothelin in endothelial cells in culture and in the intact porcine aorta (Fig. 8.3).[11,14] In the intact porcine aorta, L-NMMA augments the thrombin-induced, but not the basal, production of endothelin from the intact porcine aorta (Fig. 8.3).[11] Because the inhibitor of sol-

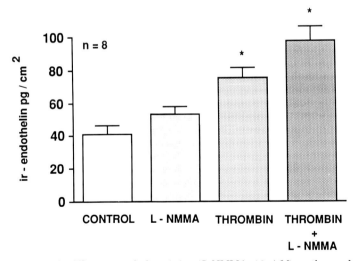

FIGURE 8.3. Effects of L-N^G-monomethyl arginine (L-NMMA; 10^{-4} M) on the production of endothelin under basal conditions and upon stimulation with thrombin (4 U/ml) in porcine aortae with endothelium. The amount of the peptide produced is expressed as picograms of immunoreactive (ir) endothelin released per square centimeter of intimal surface after 4 h incubation (N = 8). Asterisks indicate significant difference compared with control. Note also the significant difference between preparations stimulated with thrombin alone and those with both thrombin and L-NMMA ($P < 0.05$). (From Boulanger and Lüscher.[11])

uble guanylyl cyclase methylene blue exerts similar effects,[11] these results are best explained by an inhibitory effect of endothelium-derived nitric oxide concomitantly released in response to thrombin on the production of the peptide. In line with this interpretation, superoxide dismutase (which inhibits the breakdown of endothelium-derived nitic oxide by superoxide anions[15,16]) markedly inhibits the thrombin-induced endothelin production by the intact porcine aorta (Fig. 8.4). As the nonhydrolyzable analog of cGMP 8-bromo cGMP has comparable inhibitory effects, a cGMP-dependent mechanism must be involved (Fig. 8.4).[11] Thus, during simultaneous stimulation of the production of endothelium-derived nitric oxide and endothelin, the former inhibits the production of the latter via a cGMP-dependent mechanism (Fig. 8.5). This may provide a new mechanism of action of nitrovasodilators in the blood vessel wall. Indeed, activators of soluble guanylate cyclase such as nitroglycerin or the active metabolite of molsidomine SIN-1 also markedly reduce the thrombin-induced endothelin production.[17]

The inhibitory effects of endothelium-derived nitric oxide or nitrovasodilators on the production of endothelin cannot be observed in endothelial cells in culture.[18] Endothelial cells in culture maintain the activity of soluble guanylate cyclase at early passages (i.e., passages 1 and 2) and then progressively lose their capacity to form cGMP in response to nitric oxide donors.[9] In addition (as judged from vascular smooth muscle cells in culture), with increasing passage, cGMP-dependent protein kinases are no longer expressed.[19] Thus these functional alterations occurring under culture conditions may explain different results obtained in intact blood vessels and endothelial cells in culture.

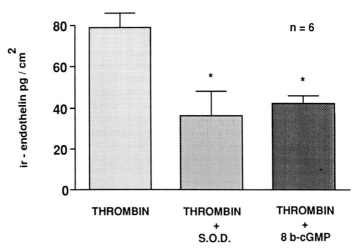

FIGURE 8.4. Effects of superoxide dismutase (SOD; 250 U/ml) and 8-bromo cGMP (10^{-3} M) on the production of endothelin induced by thrombin (4 U/ml) in porcine aortae with endothelium (N = 6). The amount of the peptide produced is expressed as picograms of immunoreactive (ir) endothelin released per square centimeter of intimal surface after 4 h incubation. Asterisks indicate significant effects of SOD and 8-bromo cGMP on the response to thrombin ($P < 0.05$). (From Boulanger and Lüscher.[11])

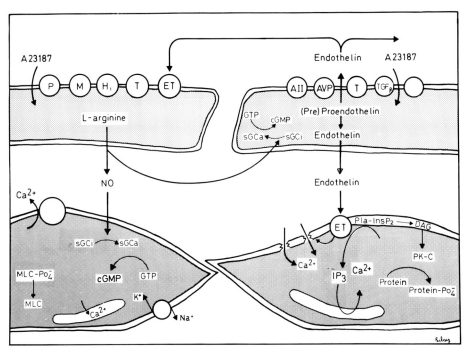

FIGURE 8.5. Schematic representation of the interaction between endothelin (ET) and endothelium-derived nitric oxide (NO) in the blood vessel wall: the release of both substances is stimulated by activation of specific receptors on the endothelial cell membrane (open circles). Nitric oxide activates soluble guanylyl cyclase (sGC) in vascular smooth muscle and endothelial cells. Endothelin activates phospholipase C, which leads to the formation of inositol trisphosphate (IP_3) and diacylglycerol (DAG), which then activates protein kinase C (PK-C). A II, angiotension II; AVP, arginine vasopressin; A23187, a calcium ionophore; H_1, histaminergic receptor; M, muscarinic receptor; MLC, myosin light chain; T, thrombin receptor; TGF_β, transforming growth factor β; sGCi/sGCa, inactive and active soluble guanylyl cyclase, respectively.

INTERACTIONS AT THE LEVEL OF VASCULAR SMOOTH MUSCLE

Large Blood Vessels

EDRF is continuously released under basal conditions.[1] In the human internal mammary artery, endothelial removal does augment the maximal response, but not the sensitivity to endothelin-1.[20] In contrast, in the human saphenous vein, the presence or absence of the endothelium does not affect the contractions to endothelin-1.[20]

In the internal mammary artery (and in canine arteries), endothelium-derived nitric oxide released by either acetylcholine or bradykinin is a very effective inhibitor of endothelin-induced contractions (Fig. 8.6).[20–22] The inhibitory effects of endothelium-derived nitric oxide are of similar potency against contractions induced by the peptide as those evoked by norepinephrine (Fig. 8.7). Similar effects can be obtained with exogenous nitric oxide (Fig. 8.8) or with nitrovasodilators such as sodium nitroprusside.[20,21]

In contrast to arteries, in veins endothelin-1 blunts the potency of EDRF released by bradykinin (Fig. 8.7) as well as that of exogenous nitric oxide (Fig.

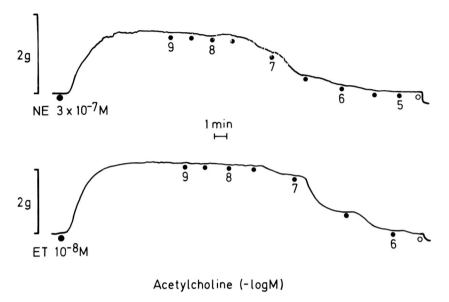

FIGURE 8.6. Effects of acetylcholine in the human internal mammary artery contracted with either norepinephrine (NE) or a comparable concentration of endothelin-1 (ET). Note the similar inhibitory effects of acetylcholine under both conditions. (From Lüscher.[28]).

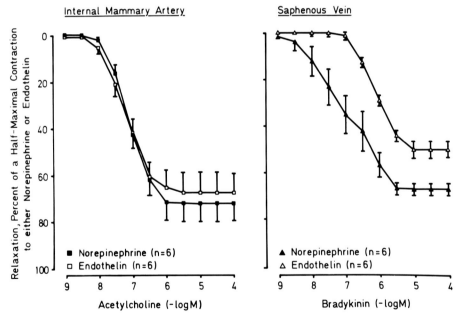

FIGURE 8.7. Effects of endothelium-derived relaxing factor (EDRF) released by acetylcholine in human internal mammary arteries (left) and by bradykinin in saphenous veins (right) on endothelin-1–induced contractions. Bradykinin was used in the veins, as acetylcholine only released small amounts of EDRF. Note that in arteries EDRF potently inhibits contractions to either endothelin-1 (□) or norepinephrine (■). In contrast, in veins endothelin-1 (△) significantly inhibits EDRF-induced relaxations ($P < 0.01$). (From Lüscher et al.[20])

8.8) or nitrovasodilators to reverse endothelin-1–induced contractions.[20,21] Similar effects have been observed in canine veins.[22] As nitrovasodilators and nitric oxide tend to be more effective in reversing norepinephrine-induced contractions in veins than in arteries studied under in vitro conditions, endothelin-1 must specifically interfere with the capacity of venous smooth muscle to relax by a cGMP-dependent mechanism. In canine veins, endothelin-1 evokes more pronounced changes in membrane potential than in arteries.[22] As depolarization reduces the relaxing effects of EDRF and nitrovasodilators,[23] this effect of the peptide may importantly contribute to these differences. Similarly, prostacyclin effectively inhibits endothelin-1–induced contractions in the internal mammary artery, while the prostanoid is much less effective in the saphenous vein.[21]

Resistance Arteries

In isolated perfused and pressurized mesenteric resistance arteries of the rat, endothelial removal markedly augments the contractions evoked by endothelin-1 (Fig. 8.9).[24] EDRF released by acetylcholine can reverse endothelin-1–induced contractions in resistance arteries (Fig. 8.10).[24,25] In perfused blood vessels maximally contracted with the peptide, the acetylcholine-induced reversal of contractions to endothelin-1 tends to be less pronounced than in prep-

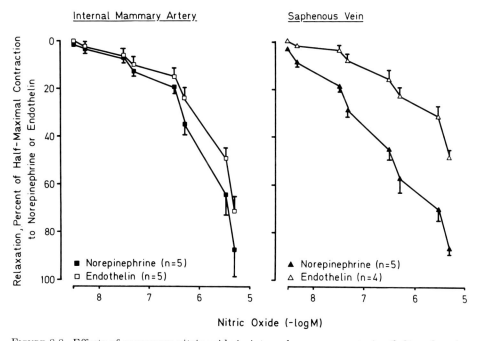

FIGURE 8.8. Effects of exogenous nitric oxide in internal mammary arteries (left) and saphenous veins (right) contracted with a half-maximal concentration of either endothelin-1 (□, △) or norepinephrine (■, ▲). Note the equipotency of nitric oxide irrespective of whether the arteries were contracted with endothelin-1 or norepinephrine (left). In contrast, the vascular effects of nitric oxide are significantly attenuated in veins contracted with endothelin-1 as compared to those contracted with norepinephrine (right; $P < 0.005$). (From Lüscher et al.[20])

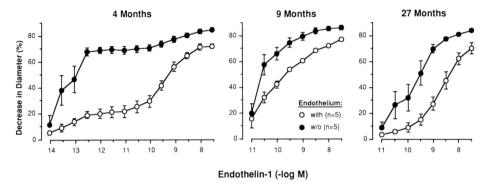

FIGURE 8.9. Contractile effects of extraluminal endothelin-1 in perfused and pressurized mesenteric resistance arteries of Fischer 344 rats of three different ages with (○) and without (●) endothelium. Note the marked inhibition of endothelin-1–induced contractions by the endothelium in young but not in adult and old rats. In addition, the sensitivity to the peptide in preparations without endothelium decreases with age. (From Dohi and Lüscher.[24])

arations studied in a myograph system.[24,25] In perfused mesenteric resistance arteries, particularly high concentrations of endothelin-1 tend to occlude the blood vessels; thus acetylcholine could be applied only extraluminally, although intraluminal activation with the agonist is much more effective than extraluminal activation.[26]

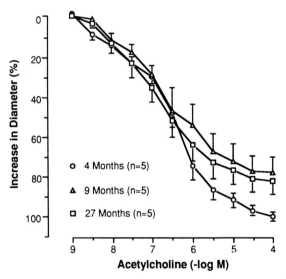

FIGURE 8.10. Effect of aging on acetylcholine-induced contraction in rat mesenteric resistance arteries. Endothelium-dependent relaxations to extraluminal acetylcholine in perfused and pressurized rat mesenteric resistance artieres with endothelium maximally contracted with endothelin-1 (3×10^{-8} M) obtained from 4-month, 9-month, and 27-month-old Fischer 344 rats. The maximal response to the muscarinic agonist is reduced in 9- and 27-month-old rats compared with 4-month-old rats ($P < 0.05$). (From Dohi and Lüscher.[24])

In the human forearm circulation studied in vivo, intraarterial infusion of endothelin-1 causes small increases in forearm blood flow at lower and marked decreases at higher concentrations of the peptide.[27] Concomitant infusion of acetylcholine (to evoke the release of EDRF) or sodium nitroprusside (to activate soluble guanylate cyclase in vascular smooth muscle cells) does not prevent endothelin-1–induced contractions. In contrast, calcium antagonists of all three classes are very effective inhibitors of endothelin-induced contractions. Indeed, under these conditions, the vasodilator effects of the peptide are unmasked. This suggests that in the human forearm microcirculation endothelin-1 dominates the effects of EDRF.

EFFECTS OF AGING AND DISEASE

Aging

In rat mesenteric resistance arteries, aging is associated with a marked reduction of the inhibitory effects of the endothelium against contractions induced by endothelin-1 (Fig. 8.9).[24] In addition, the sensitivity of vascular smooth muscle to the contractile effects of the peptide decreases.

In mesenteric resistance arteries of old rats, the capacity of EDRF to reverse endothelin-1–induced contractions is slightly reduced (Fig. 8–10).[24] However, only the maximal relaxations, but not the sensitivity to acetylcholine, is impaired under these conditions.[24] That the response to the nitric oxide donor SIN-1 is fully maintained in mesenteric resistance arteries of old rats (Fig. 8.11),[24] indicates that aging is associated with a mild reduction in the release and/or production of EDRF.

Hypertension

In mesenteric resistance arteries of the spontaneously hypertensive rat, the endothelium-dependent relaxations induced by intraluminal endothelin-1 are reduced.[2] Similarly, the ability of acetylcholine to reverse endothelin-1–induced contractions is impaired in hypertensive resistance arteries.[2,25] In perfused resistance arteries, this defect can only be demonstrated with intraluminal, and not extraluminal, infusion of acetylcholine, indicating a selective defect of the luminal surface of the hypertensive endothelium in response to acetylcholine.[26] Similar differences can be demonstrated with norepinephrine as the contractile agent, indicating that it is not specific for endothelin-1.[26]

SUMMARY

Endothelin and EDRF can interact at the levels of both vascular smooth muscle and endothelium. At the endothelial level, endothelin can stimulate the release of prostacyclin and EDRF, while nitric oxide can inhibit the production of endothelin. The latter is particularly important, with agonists such as thrombin stimulating the production of both endothelium-derived nitric oxide and endothelin. In large conduit arteries, endothelium-derived nitric oxide and

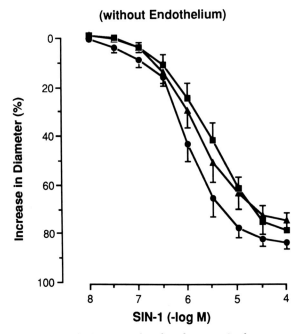

FIGURE 8.11. Relaxations to SIN-1 in perfused and pressurized rat mesenteric resistance arteries without endothelium maximally contracted with endothelin-1 (3×10^{-8} M) obtained from 4-month-old (●), 9-month-old (▲), and 27-month-old (■) Fischer 344 rats. The maximal response to SIN-1 is similar in all age groups. (From Dohi and Lüscher.[24])

nitrovasodilators are potent inhibitors of endothelin-induced contractions. The potency of this inhibitory effect is less pronounced in veins as compared with arteries. In the microcirculation, EDRF and nitrovasodilators are also potent inhibitors of endothelin-induced contractions. Aging and hypertension reduce the inhibitory effects of the endothelium against endothelin-induced contractions.

ACKNOWLEDGMENTS

The authors thank Bernadette Libsig and Sabine Bohnert for technical assistance. Original research reported in this manuscript was made possible by grants from the Swiss National Research Foundation (grant No. 32-25468.88), the Swiss Cardiology Foundation, and the Helmut Horten Foundation and by an educational grant from Hoechst Pharmaceutica, Paris, France. T. F. L. is the recipient of a career development award from the Swiss National Research Foundation (SCORE grant No. 3231-025150).

REFERENCES

1. LÜSCHER, T. F., and P. M. VANHOUTTE. *The endothelium: modulator of cardiovascular function*. Boca Raton, FL: CRC Press, 1990.
2. DOHI, Y., and T. F. LÜSCHER. Endothelin-1 in hypertensive mesenteric resistance arteries. Different intra- and extraluminal dysfunction. *Hypertension* 1991 (in press).
3. RAE, G. A., M. TRYBULEC, G. DE NUCCI, and J. R. VANE. Endothelin-1 releases eicosanoids from rabbit isolated kidney and spleen. *J. Cardiovasc. Pharmacol.* 13(Suppl. 5): 89–92, 1989.

4. DE NUCCI, G., F. THOMAS, P. D'ORLEANS-JUSTE, E. ANTUNES, C. WALDER, T. D. WARNER, and J. R. VANE. Pressor effects of circulating endothelin are limited by its removal in the pulmonary circulation and by the release of prostacyclin and endothelium-derived relaxing factor. *Proc. Natl. Acad. Sci. USA* 85: 9797–9800, 1989.
5. WARNER, T. F., J. A. MITCHELL, G. DE NUCCI, and J. R. VANE. Endothelin-1 and endothelin-3 release EDRF from isolated perfused arterial vessels of the rat and rabbit. *J. Cardiovasc. Pharmacol.* 13(Suppl. 5): 85–88, 1989.
6. WATANABE, K., H. MIYAZAKI, M. KONDOH, Y. MASUDA, S. KIMURA, M. YANAGISAWA, T. MASAKI and K. MURAKAMI. Two distinct types of endothelin-1 receptors are present on chick cardiac membranes. *Biochem. Biophys. Res. Commun.* 161: 1252–1259, 1989.
7. MAGGI, C. A., S. GIULIANI, R. PATACCHINI, P. ROVERO, A. GIACHETTI, and A. MELI. The activity of peptides of the endothelin-1 family in various mammalian smooth muscle preparations. *Eur. J. Pharmacol.* 174: 23–31, 1989.
8. SPOKES, R. A., M. A. GHATEI, and S. R. BLOOM. Studies with endothelin-3 and endothelin-1 on rat blood pressure and isolated tissues: evidence for multiple endothelin-1 receptor subtypes. *J. Cardiovasc. Pharmacol.* 13(Suppl. 5): 191–192, 1989.
9. BOULANGER, C., V. B. SCHINI, S. MONCADA, and P. M. VANHOUTTE. Stimulation of cyclic GMP production in cultured porcine endothelial cells by bradykinin, adenosine diphosphate, calcium ionophore A23187 and nitric oxide. *Br. J. Pharmacol.* 1990 (in press).
10. EVANS, H. G., J. A. SMITH, and M. J. LEWIS. Release of endothelium-derived relaxing factor is inhibited by 8-bromo-cyclic guanosine monophosphate. *J. Cardiovasc. Pharmacol.* 12; 672–677, 1988.
11. BOULANGER, C., and T. F. LÜSCHER. Endothelin is released from the porcine aorta: inhibition by endothelium-derived nitric oxide. *J. Clin. Invest.* 85: 587–590, 1990.
12. LÜSCHER T. F., D. DIEDERICH, R. SIEBENMANN, K. LEHMANN, P. STULZ, L. VON SEGESSER, Z. YANG, M. TURINA, E. GRÄDEL, E. WEBER, and F. R. BÜHLER. Difference between endothelium-dependent relaxations in arterial and in venous coronary bypass grafts. *N. Engl. J. Med.* 319:462–467, 1988.
13. YANG, Z., T, F. LÜSCHER. Unpublished observations, 1990.
14. SCHINI, V. B., H. HENDRICKSON, D. HEUBLEIN, J. BURNETT, JR., and P. VANHOUTTE. Thrombin enhances the release of endothelin from cultured porcine aortic endothelial cells. *Eur. J. Pharmacol.* 165: 333–334, 1989.
15. RUBANYI, G. M., and P. M. VANHOUTTE. Ouabain inhibits endothelium-dependent relaxations to arachidonic acid in canine coronary arteries. *J. Pharmacol. Exp. Ther.* 235: 81–86, 1985.
16. GRYGLEWSKI, R. J., R. M. J. PALMER, and S. MONCADA. Superoxide anion is involved in the breakdown of endothelium-derived vascular relaxing factor. *Nature* 320: 454–456, 1986.
17. BOULANGER, C., and T. F. LÜSCHER. Hirudin nitros inhibit the thrombin-induced release of endothelin from the intact porcine aorta. *Circ. Res.* 68: 91, 1991.
18. BOULANGER, C., and T. F. LÜSCHER. Unpublished observations, 1990.
19. CORNWELL, T. L., and T. M. LINCOLN. Regulation of intracellular Ca^{2+} levels in cultured vascular smooth muscle cells. *J. Biochem. Chem.* 264: 1146–1155, 1989.
20. LÜSCHER, T. F., Z. YANG, M. TSCHUDI, L. VON SEGESSER, P. STULZ, C. BOULANGER, SIEBENMANN, M. TURINA, and F. R. BÜHLER. Interaction between endothelin-1 and endothelium-derived relaxing factor in human arteries and veins. *Circ. Res.* 66: 1088–1094, 1990.
21. YANG, Z., F. R. BÜHLER, D. DIEDERICH, and T. F. LÜSCHER. Different effects of endothelium on cyclic AMP– and cyclic GMP–mediated vascular relaxation in human arteries and veins: comparison with norepinephrine. *J. Cardiovasc. Pharmacol.* 13(Suppl. 5): 129–131, 1989.
22. MILLER, V. M., K. KOMORI, J. C. BURNETT, and P. M. VANHOUTTE. Differential sensitivity to endothelin in canine arteries and veins. *Am. J. Physiol.* 257: 1113–1126, 1989.
23. COLLINS, P., A. H. HENDERSON, D. LANG, and M. J. LEWIS. Endothelium-derived relaxing factor and nitroprusside compared in noradrenaline- and K^+-contracted rabbit and rat aortae. *J. Physiol. (Lond.)* 400: 395–404, 1988.
24. DOHI, Y., and T. F. LÜSCHER. Aging differentially affects direct and indirect actions of endothelin-1 in perfused rat mesenteric resistance arteries. *Br. J. Pharmacol.* 100: 889–893, 1990.
25. LÜSCHER, T. F., D. DIEDERICH, Z. YANG, and F. R. BÜHLER. Endothelin overrides endothelium-derived relaxing factor in hypertensive resistance arteries. *Kidney Int.* 35(1): 331, 1989.
26. DOHI, Y., M. THIEL, F. R. BUHLER, and T. F. LÜSCHER. Activation of endothelial L-arginine pathway in resistance arteries: effect of age and hypertension. *Hypertension* 15: 170–179, 1990.

27. KIOWSKI, W., L. LINDER, T. F. LÜSCHER and F. R. BÜHLER. Endothelin-1 induced vasoconstriction in man: reversal by a calcium channel blockade, but not by nitrovasodilators or endothelium-derived relaxing factor. Circulation (submitted).
28. LÜSCHER, T. F. Endothelial vasoactive factors and regulation of vascular tone in human blood vessels. *Lung* (Suppl.): 27–34, 1990.

9

Endothelin as a Growth Factor in Vascular Remodeling and Vascular Disease

VICTOR J. DZAU, RICHARD E. PRATT, AND JOHN P. COOKE

Vascular remodeling is an adaptive process in response to chronic alterations in flow, pressure, and humoral factor(s). This process involves a complex sequence of events, ranging from the sensing and transduction of physical or humoral signals, to the expression of selective genes within the vessel wall, to the process of cellular growth and extracellular matrix production.[1] Autocrine–paracrine growth factors have been shown to be expressed by vascular endothelial and smooth muscle cells. These substances may play an important role in the remodeling mechanism. A growing body of evidence suggests that the endothelium has the capacity to sense changes in blood flow–shear stress and subsequently to influence vascular smooth muscle contraction by the release of endothelial-derived vasoactive substance(s).[2-10] In addition to its role in regulating vascular tone and blood flow, the endothelium may play a role in long-term vascular remodeling related to changes in flow.[5] Indeed, the endothelium is capable of producing growth-promoting (e.g., platelet-derived growth factors (PDGF) and basic fibroblast growth factor) and growth-inhibiting (e.g., heparin, prostaglandins, and endothelium-derived relaxing factor) substances.[1] The endothelium-derived vasoactive substances can also influence vascular smooth muscle growth. We and others have found that potent vasoconstrictors, such as angiotensin II, induce growth and proliferation of vascular smooth muscle cells.[1,11-15] Therefore, when Yanagisawa and colleagues[3] isolated the potent peptide vasoconstrictor endothelin from cultured endothelial cells, the question arose as to whether endothelin influenced remodeling of the blood vessel. We describe our investigation of the effect of endothelin on vascular smooth muscle cell growth and examine the potential significance of its mitogenic action in vascular remodeling and disease.

EFFECTS OF ENDOTHELIN ON VASCULAR SMOOTH
MUSCLE CELL (VSMC) GROWTH

Endothelin was reported to have growth factor-like properties and to stimulate growth and proliferation of VSMC,[16-18] fibroblasts[19,20] and mesangial cells[21] in culture. A significant increase in the rate of [^{3}H]thymidine incorporation occurs in confluent quiescent VSMC exposed to endothelin in defined serum-free media.[16] We observed a 33%–130% increase in the magnitude of DNA synthesis in comparison to vehicle control. This increase in [^{3}H]thymidine incorporation was dose-dependent, with an EC_{50} of 5×10^{-9} M (Fig. 9.1). Time-course studies consistently showed that the peak of incorporation was between 12 to 18 h after the addition of endothelin. In parallel experiments, 10% fetal calf serum caused a fourfold increase in the rate of [^{3}H]thymidine incorporation by VSMC in the same period of time. Thus, compared with serum, endothelin is a less potent stimulator of DNA synthesis.

Endothelin also stimulates VSMC proliferation. Daily exposure to a single dose of endothelin (10^{-7} M) significantly increased cell number after 2 days in comparison to vehicle control.[16] If the cultured cells were exposed to only one dose of endothelin, no increase in cell number was observed. Using a bioassay for vasoactive substances, we observed that only 40% of the original vasoconstrictor activity of endothelin could be detected in the media after 4 h exposure to the cells, suggesting that endothelin is metabolized under these conditions. This suggests that the mitogenic effect of endothelin requires a sustained or repeated exposure of the VSMC to an effective concentration of the peptide.

Endothelin also stimulated protein synthesis in VSMC as measured by [^{35}S]methionine incorporation into protein.[16] Protein synthesis increased in cells exposed to endothelin (10^{-7} M), with a significant increase observed between 12 and 15 h (49% increase). By comparison, 10% fetal calf serum induced a threefold greater increase in protein synthesis (136% increase) during the same period of time. An analysis of the total protein content of stimulated and unstimulated VSMC cultures confirmed that endothelin had a minimal effect on protein synthesis. After 2 or 4 days of exposure to endothelin (10^{-7} M), we observed no significant increases in the protein content per well despite a doubling of cell numbers. In other words, the protein content per cell decreased by 50%. In contrast, 10% fetal calf serum induced an 81% and an 86% increase in protein content per well after 2 to 4 day stimulations, respectively. After 6 days of endothelin treatment, there was an increase in cellular protein content that resulted in a normalization of protein content per cell. The mechanism by which endothelin induces growth is unknown. Endothelin stimulates inositol

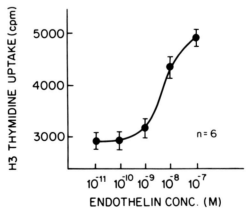

FIGURE 9.1. Dose-dependent relationship between endothelin concentration and DNA synthesis.

trisphosphate (IP_3) production by phospholipase C and increases intracellular calcium by mobilizing intracellular stores[22] and inducing extracellular influx.[3] Several VSMC mitogens, such as PDGF and epidermal growth factor (EGF), also activate phospholipase C and generate IP_3 and diacylglycerol in association with increased intracellular calcium and cellular alkalinization.[23,24] These mitogens stimulate the expression of protooncogenes (such as c-*fos*, c-*jun*, and c-*myc*) that confer competence to the stimulated cells and promote entry into the growth cycle.[25-27] Accordingly, we examined the effects of endothelin on c-*fos* mRNA expression in VSMC. Endothelin stimulated protooncogene expression, with an EC_{50} in the nanomolar range (Fig. 9.2). This cellular effect simulates that of angiotensin II (Ang II), another vasoactive substance with growth-promoting properties. We have shown recently that this effect of Ang II is in part mediated by the autocrine growth factors PDGF and transforming growth factor β (TGF-β).[12,28] Ang II induces the expression of these endogenous growth factors, thereby stimulating VSMC growth by an autocrine mechanism. Whether endothelin exerts its mitogenic effect via these autocrine growth factors is unknown.

POTENTIAL SIGNIFICANCE OF ENDOTHELIN
AS A VASCULAR GROWTH FACTOR

Alteration in the normal growth pattern of VSMC may be involved in the pathophysiology of several vascular diseases. In atherosclerosis, the intimal migration and subsequent proliferation of the VSMC are thought to be early and critical events in the development of the lesions.[29] In hypertension, the growth responses (hypertrophy and hyperplasia) of the VSMC may serve to amplify and perpetuate the blood pressure elevation.[1]

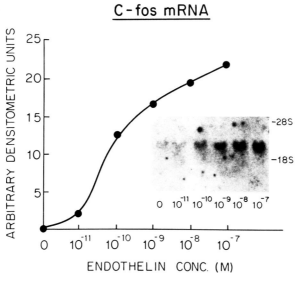

FIGURE 9.2. Effect of endothelin on the expression of c-*fos* mRNA (30 min after exposure). Confluent quiescent VSMC were treated with 10^{-11} to 10^{-7} M endothelin for 30 min.

Alterations in the normal contractile function of VSMC have also been implicated in the pathophysiology of vascular disease. Indeed, experimental and clinical evidence suggests that arteriosclerotic arteries exhibit enhanced reactivity to vasoactive substances.[30–33] The current hypothesis emphasizes the role of endothelial dysfunction in this enhanced response, but the contribution of VSMC proliferation is also important. Of interest is the observation that a number of vasoactive substances can influence VSMC growth. Catecholamine, vasopressin, serotonin, and thromboxane have been reported to induce VSMC proliferationi.[1,12–14,34] Our results show that endothelin also promotes growth in VSMC. Acutely, endothelin stimulated increases in DNA synthesis. The increases observed are modest compared with those induced by serum or purified growth factors. Endothelin induces the expression of c-*fos*, a protooncogene that has been shown to play a key role in growth cycle regulation.[26] Unlike Ang II, which only causes cellular hypertrophy,[11] endothelin, when added to the serum-deprived cells repeatedly over 2 days, results in an increase in cell number. It is likely that endothelin may also act synergistically with other factors to promote hypertrophy or hyperplasia in vivo. The physiological and pathological roles of endothelin in vascular remodeling and disease conditions remain speculative. Endothelin may play a role in disease states in which sustained increases in vascular tone are observed. For example, it may be involved in essential hypertension. Two prominent pathophysiological processes in hypertension are the increase in vascular resistance and the development of vascular hypertrophy. Endothelin can mediate both processes. We postulate that endothelin may be involved with the long-term regulation of vascular structure.[35,36] Indeed, elevated levels of endothelin have been reported in hypertensive patients.[37] However, the pathophysiology of essential hypertension is multifactorial. Whereas endothelin may play a direct role in a subset of hypertensive persons, it is more likely to contribute to hypertension by providing an elevated basal tone to which other processes add and amplify.

Another disorder in which endothelin may also play a role is cerebral vasospasm, a complex process that becomes clinically manifest several days after a subarachnoid hemorrhage. The pathophysiology of cerebral vasospasm is poorly understood, but initially involves increased vascular reactivity, followed later by structural alterations in the vessel wall.[38–41] In animal models, angiographically documented vasoconstriction occurs within minutes of the subarachnoid hemorrhage, following within 48 h by smooth muscle cell and fibroblast proliferation. These phenomena suggest that vasoconstrictive substance(s), mitogenic factors, or substances possessing both properties are involved in this disorder. In the setting of a subarachnoid hemorrhage, prominent changes occur in the vascular milieu, including locally elevated concentrations of thrombin and catecholamines, together with hypoxia and low blood flow.[42–44] These are all known stimulants of endothelium-dependent vasoconstrictor activity, probably caused by endothelin release.[3,45,46] As mentioned above, endothelin can stimulate vasoconstriction as well as vascular remodeling. It is therefore quite plausible that endothelin is released after a subarachnoid hemorrhage and contributes to the sustained vasoconstriction and subsequent vascular smooth muscle proliferation that are characteristic of chronic cerebral vasospasm. Indeed, intracisternal injection of endothelin potently constricts canine cerebral arteries in vivo.[47] The exact role of endothelin in the vascular structural changes of subarachnoid hemorrhage remains to be determined. Endothelin may also be involved in the excessive vasoconstriction seen in other disease states. The clinical presentation of coronary vasospasm can range from an asymptomatic transient vasoconstriction to a sustained episode inducing myonecrosis. If excessive endothelin release were to be involved with coronary vasospasm, it would be in those cases in which sustained vasoconstriction is a prominent feature, as is characteristic of its effect on isolated human coronary arteries.[48] In the setting of active thrombus formation, the associated hypoxia and locally increased thrombin levels could induce endothelin synthesis and provoke prolonged vasospasm. Whether endothelin plays a role in the remodeling of the coronary circulation in the setting of a prolonged alteration in flow caused by coronary stenosis is unknown, but may have important pathophysiological implications.

In summary, we hypothesize that endothelin may be of major pathophysiological importance in vascular disease in which sustained vasoconstriction is accompanied by structural remodeling of the blood vessel. With the development of specific and sensitive radioimmunoassays and the availability of specific antagonists, the role of endothelin in different cardiovascular diseases will be elucidated.

ACKNOWLEDGMENTS

This work was supported by NIH grants HL35610, HL35792, HL19259, HL35352, HL43131, and HL42663 and by NIH Specialized Center of Research in Hypertension HL36568.

REFERENCES

1. Dzau, V. J., and G. H. Gibbons. Cell biology of vascular hypertrophy in systemic hypertension. *Am. J. Cardiol.* 62: 30G–35G, 1988.
2. Furchgott, R. F. Role of endothelium in responses of vascular smooth muscle. *Circ. Res.* 53: 557–573, 1987.
3. Yanagisawa, M., H. Kurihara, S. Kimura, et al. A novel potent vasoconstrictor peptide produced by vascular endothelial cells. *Nature* 332: 411–415, 1988.
4. Cooke, J. P., E. Rossitch Jr., N. Andon, J. Loscalzo, and V. J. Dzau. Flow activates a specific endothelial potassium channel to release an endogenous nitrovasodilator. *J. Am. Coll. Cardiol.* 15: 1A, 1990.
5. Frangos, J. A., S. G. Eskin, L. V. McIntire, and C. L. Ives. Flow effects on prostacyclin production by cultured human endothelial cells. *Science* 227: 1477–1479, 1984.
6. Pohl, V., J. Holtz, R. Busse, and E. Bassenge. Crucial role of the endothelium in the vasodilator response to increased flow in vivo. *Hypertension* 8: 37–44, 1986.
7. Oleson, S. P., D. E. Clapham, and P. F. Davies. Haemodynamic shear stress activates a K$^+$ current in vascular endothelial cells. *Nature* 331: 168–170, 1988.
8. Langille, B. L., and F. O'Donnell. Reductions in arterial diameter produced by chronic diseases in blood flow are endothelium dependent. *Science* 231: 405–407, 1986.
9. Rubanyi, G., J. C. Romero, and P. M. Vanhoutte. Flow-induced release of endothelium-derived relaxing factors. *Am. J. Physiol.* 250 (*Heart Circ. Physiol.* 19): H1145–H1149, 1986.
10. Rubanyi, G. M. Endothelium-dependent pressure-induced contraction of isolated canine carotid arteries. *Am. J. Physiol.* 255 (*Heart Circ. Physiol.* 24): H783–H788, 1988.
11. Geisterfer, A. N., M. J. Peach, and G. K. Owens. Angiotensin II induces hypertrophy, not hyperplasia, of cultured rat aortic smooth muscle cells. *Circ. Res.* 62: 749–757, 1988.
12. Naftilan, A. J., R. E. Pratt, and V. J. Dzau. Induction of PDGF A-chain and c-*myc*-gene expressions by angiotensin II in cultured rat vascular smooth muscle cells. *J. Clin. Invest.* 83: 1419–1424, 1989.

13. BLAES, N., and J. P. BOISSEL. Growth stimulating effects of catecholamines on rat aortic smooth muscle cells in culture. *J. Cell. Physiol.* 116: 167–172, 1983.
14. COUGHLIN, S. R., W. M. F. LEE, P. W. WILLIAMS, G. M. GIELS, and L. T. WILLIAMS. C-*myc* gene expression is stimulated by agents that activate protein kinase C and does not account for the mitogenic effect of PDGF. *Cell* 43: 243–251, 1985.
15. YAMORI, Y., M. MANO, Y. NARA, and R. HORIE. Catecholamine-induced polyploidization in vascular smooth muscle cells. *Circulation* 1987;75(Suppl. I): I-92–I-95.
16. DUBIN D., R. E. PRATT, J. P. COOKE, and V. J. DZAU. Endothelin, a potent vasoconstrictor, is a vascular smooth muscle mitogen. *J. Vasc. Med. Biol.* 1: 150–154, 1989.
17. KOMURO, I., H. KURIHARA, T. SUGIYAMA, F. TAKAKU, and Y. YAZAKI. Endothelin stimulates c-*fos* and c-*myc* expression and proliferation of vascular smooth muscle cells. *FEBS Lett.* 238: 249–252, 1988.
18. NAKAKI, T., M. NAKAYANA, S. YAMAMOTO, and R. KATO. Endothelin-mediated stimulation of DNA synthesis in vascular smooth muscle. *Biochem. Biophys. Res. Commun.* 158: 880–883, 1989.
19. BROWN, K. D., and C. J. LITTLEWOOD. Endothelin stimulates DNA synthesis in Swiss 3T3 cells: synergy with polypeptide growth factors. *Biochem. J.* 263: 977–980, 1989.
20. TAKUWA, M., H. TSUKADA, H. MATSUOKA, and S. YAGI. Inhibitory effect of endothelin on renin release in vitro. *Am. J. Physiol.* 257 (*Endocrinol. Metab.* 20): E833–E838, 1989.
21. SIMONSON, M. S., S. WANN, P. MEN, G. R. DUBYAK, M. KESTER, Y. NAKAZATO, J. R. SEDOR, and M. J. DUNN. Endothelin stimulates phospholipase C, Na^+/H^+ exchange, c-*fos* expression, and mitogenesis in rat mesangial cells. *J. Clin. Invest.* 83: 7708–7712, 1989.
22. MARSDEN, P. A., N. R. DANTHULURI, B. M. BRENNER, B. J. BALLERMAN, and T. A. BROCK. Endothelin action on vascular smooth muscle involves inositol triphosphate and calcium mobilization. *Biochem. Biophys. Res. Commun.* 158: 86–93, 1989.
23. MODENAAR, M. H., R. Y. TSIEN, P. T. VANDER SAAG, and S. W. DE LAAT. Na^+/H^+ exchange and cytoplasmic pH in the action of growth factors in human fibroblasts. *Nature* 305: 645–648, 1983.
24. BERK, B. C., B. A. BROCK, M. A. GIMBRONE, and R. W. ALEXANDER. Early agonist-mediated events in cultured vascular smooth muscle cells. *J. Biol. Chem.* 262: 5065–5072, 1987.
25. MILLER, A. D., T. CURRAN, and F. M. VERMA. C-*fos* protein can induce cellular transformation: a novel mechanism of activation of a cellular oncogene. *Cell* 36: 51–60, 1984.
26. ROZENGURT, E. Early signals in the mitogenic response. *Science* 234: 161–166, 1986.
27. MARX, J. L. The *fos* gene as "master switch." *Science* 237: 854–856, 1987.
28. GIBBONS, G. H., R. E. PRATT, and V. J. DZAU. Vascular myocyte hypertrophy vs hyperplasia: autocrine transforming growth factor-beta determines response to angiotensin II. *Clin. Res.* 287A, 1990.
29. ROSS, R. The pathogenesis of atherosclerosis—An update. *N. Engl. J. Med.* 314: 488–514, 1986.
30. LUDMER, P. L., A. P. SELWYN, T. L. SHOOK, et al. Paradoxical vasoconstriction induced by acetylcholine in atherosclerotic coronary arteries. *N. Engl. J. Med.* 315: 1046–1051, 1986.
31. HENRY, P. D., and M. YOKOYAMA. Supersensitivity of atherosclerotic rabbit aorta to ergonovine: mediation by a serotonergic mechanism. *J. Clin. Invest.* 66: 306–313, 1980.
32. HEISTAD, D. D., M. L. I. ARMSTRONG, M. L. MARCUS, D. J. PIEGORS, and A. L. MARK. Augmented responses to vasoconstrictor stimuli in hypercholesterolemic and atherosclerotic monkeys. *Circ. Res.* 54: 711–718, 1984.
33. ROSSITCH, E. JR., E. ALEXANDER III, and J. P. COOKE. L-arginine normalizes endothelial function in hypercholesterolemic basilar arteries. *FASEB J.* 4: A416, 1990.
34. CAMPBELL-BOSWELL, M., and A. L. ROBERTSON. Effects of angiotensin II and vasopressin on human smooth muscle cells in vitro. *Exp. Mol. Pathol.* 35: 265–276, 1981.
35. DZAU, V. J., J. P. COOKE, and G. M. RUBANYI. Significance of endothelial derived vasoactive substances. *J. Vasc. Med. Biol.* 1: 43–45, 1989.
36. COOKE, J. P., and V. J. DZAU. Possible role for endothelin in vascular disease: an alternative hypothesis. *J. Vasc. Med. Biol.* 1: 316–318, 1989.
37. SAITO, Y., K, NAKAO, T. YAMADA, et al. Plasma endothelin-1-like immunoreactivity level in healthy subjects and patients with hypertension. *Hypertension* 14: 335, 1989.
38. KIM, P., T. M. SUNDT, and P. M. VANHOUTTE. Alterations in endothelium-dependent responsiveness of the canine basilar artery after subarachnoid hemorrhage. *J. Neurosurg.* 69: 239–246, 1988.
39. NAKAGOMI, T., N. F. KASSELL, R. M. LEHMAN, J. A. JANE, T. SASAKI, T. TAKAKURA. Endothelial damage and cerebral vasospasm. In: *Cerebral Vasospasm*, edited by R. H. Wilkinson. New York: Raven, 1988, p. 137–143.
40. MAYBERG, M. R., T. M. LISZCAK, P. M. BLACK, and N. T. ZERVAS. Acute structural changes in

feline cerebral arteries after subarachnoid hemorrhage. In: *Cerebral Vasospasm*, edited by R. H. Wilkinson. New York: Raven, 1988, p. 231–246.
41. SMITH, R. R., B. R. CLOWER, Y. HONMA, and J. M. CRUSE. The constrictive angiopathy of subarachnoid hemorrhage: an immunopathological approach. In: *Cerebral Vasospasm*, edited by R. H. Wilkinson. New York: Raven, 1988, p. 247–252.
42. LISZCAK, T. M., B. G. VASSOS, P. M. BLACK, et al. Cerebral arterial constriction after experimental subarachnoid hemorrhage is associated with blood components within the arterial wall. *J. Neurosurg.* 58: 18–26, 1983.
43. FRASER, R. A., B. M. STEIN, R. E. BARRETT, and J. L. POOL. Noradrenergic mediation of experimental cerebrovascular spasm. *Stroke* 1: 356–362, 1970.
44. FRASER, R. A., J. M. STEIN, W. J. FLOR, S. L. COHAN, and J. PARKHURST. Sequential changes of vascular ultrastructure in experimental cerebral vasospasm: myonecrosis of subarachnoid arteries. *J. Neurosurg.* 41: 49–57, 1974.
45. DEMEY, J. G., and P. M. VANHOUTTE. Anoxia and endothelium-dependent reactivity of the canine femoral artery. *J. Physiol.* 335: 65–74, 1983.
46. RUBANYI, G. M., and P. M. VANHOUTTE. Hypoxia releases a vasoconstrictor substance from the canine vascular endothelium. *J. Physiol.* 364: 45–56, 1985.
47. ASANO, I., I. IKEAKI, Y. SUZUKI, S. SATOH, and M. SHIBUYA. Endothelin and the production of cerebral vasospasm in dogs. *Biochem. Biophys. Res. Commun.* 31: 345–3551, 1989.
48. CHESTER, A. H., M. R. DASHWOOD, J. G. CLARKE, et al. Influence of endothelin on human coronary arteries and localization of its binding sites. *Am. J. Cardiol.* 63: 1395–1398, 1989.

10
Endothelin and the Heart

WINIFRED G. NAYLER AND XIN HUA GU

A little over 3 years have elapsed since Yanagisawa and his colleagues[48] named and described the chemistry and sustained constrictor activity of the polypeptide (endothelin-1) that accumulates in the supernatant of cultured endothelial cells. During these years, interest in this polypeptide has escalated to such a degree that several hundred papers have now been published on the topic. Other endothelins (endothelin-2, endothelin-3, and mouse β-endothelin) have been identified,[49,50] chemically homologous polypeptides have been isolated from the venom of the asp *Atractaspis engaddensis*,[24] and their mode of action has been identified.[49,50] In addition, attention has been directed to their possible involvement in a variety of pathological conditions, including asthma, vasospasm, hypertension, renal failure, and myocardial infarction.[33,34]

Although early interest in the endothelins centered around their ability to evoke a sustained vasoconstrictor response,[48] evidence of a sustained positive inotropic effect soon emerged.[18,19] At the same time, endothelin-specific high-affinity binding sites were identified in cardiac myocytes,[15] cardiac membranes,[11-13] at the atrioventricular node,[47] and in the smooth muscle cells of the coronary vasculature.[38] It also became apparent that at least one of the endothelins—endothelin-1—has a positive chronotropic effect.[20]

POSITIVE INOTROPIC EFFECTS OF ENDOTHELIN-1

The positive inotropic effect of endothelin-1 (Fig. 10.1) lacks species specificity, the effect having been described for human atria,[5] guinea pig and rat atria,[18,19] and ferret[40] and rabbit[44] papillary muscles. The threshold for this effect is in the low nanomolar range, and concentrations of ~10 nM produce a maximum response (Table 10.1). Whereas this inotropic effect of endothelin-1 is dose dependent (Fig. 10.1) and relatively easy to detect in electrically driven isolated atrial and papillary muscle preparations, it is usually concealed in intact animal studies because of the secondary changes associated with the attendant coronary vasoconstriction.[3]

Characteristics of the Positive Inotropic Response to Endothelin-1

The characteristics of the positive inotropic effect of endothelin-1 can be summarized as follows:

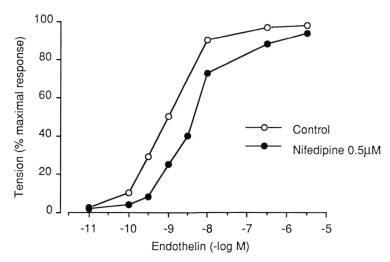

FIGURE 10.1. Dose–response curve for the positive inotropic effects of endothelin-1 on guinea pig left atria in the absence (○) and presence (●) of 0.5 μM nifedipine. Each point is the mean of the percentage maximum response.

1. The response develops slowly and is sustained. Often as much as 15 min is required for the response to reach asymptote, and the inotropic effect may persist for up to 60 or 90 min after the polypeptide has been removed from the coronary perfusate.[32]
2. The response is insensitive to α- and β-adrenergic blockade, to anticholinergic agents, and to histaminergic and serotonergic blockers.[18,19,44]
3. The dose–response curve for the positive inotropic effect of endothelin-1 is shifted to the right by calcium antagonists, including nicardipine[19] and verapamil.[43]
4. The inotropic response is accompanied by a prolongation of the cardiac action potential. In rabbit papillary muscle, for example, 10 nM endothelin-1 lengthens the action potential by 16%, an effect that is associated with an increase of 180% in developed tension.[44]

Comparison with the Positive Inotropic Activity of Other Inotropic Agents

Comparison with Bay K 8644
The positive inotropic effect of the dihydropyridine-based calcium agonist Bay K 8644[39] is also accompanied by a prolongation of the cardiac action potential, because of enhanced Ca^{2+} influx through the voltage-sensitive Ca^{2+} channels. In the case of Bay K 8644, a 16% increase in the duration of the action potential is accompanied by an increase in tension development of <40%.[44] For endothelin-1, a 16% increase in the duration of the action potential is associated with an increase of ~180% in peak tension development! If the positive inotropic effect of endothelin-1 had been due simply to enhanced entry of Ca^{2+} through the voltage-dependent Ca^{2+} channels, the ratio between the lengthening of the cardiac action potential and the magnitude of the inotropic response should be the same for both agents. The failure of this condition to apply, together with the fact that the calcium antagonists move the dose–response curve for the

TABLE 10.1. Evidence of the positive inotropic activity of endothelin-1

Tissue	Threshold (nM)	EC_{50} (nM)	Maximum response (nM)	Reference
Guinea pig atria	0.1	1.08	—	20
Rat atria	1.0	2.6	10	18
Rat atria	1.0	2.0	—	43
Ferret capillary muscle	0.1	—	10	40
Human right atria	—	—	10	5
Rabbit papillary	0.1	—	—	44

EC_{50} is the concentration required to produce a half-maximal response. —, Data not available.

inotropic effect of endothelin-1 to the right rather than suppressing the response, provide support for the hypothesis that the Ca^{2+} component of the positive inotropic effect of endothelin-1 can be divided into at least two components: (1) a voltage-sensitive Ca^{2+} channel component and (2) a voltage-insensitive Ca^{2+} channel component.

Comparison with the Positive Inotropic Effects of Isoproterenol
Again there are differences between the positive inotropic effect of endothelin-1 and that caused by isoproterenol. These differences can be summarized as follows:

1. The positive inotropic effect of isoproterenol, but not of endothelin-1, is antagonized by β-adrenoceptor antagonists.
2. The positive inotropic effect of isoproterenol is accompanied by an increase in cyclic 3'5'-monophosphate, but no such effect occurs with endothelin-1.
3. The positive inotropic effect of isoproterenol is accompanied by a decrease in the time to peak tension and by an accelerated rate of relaxation, whereas these parameters remain unaltered for endothelin-1.[19,40]

Comparison with the Positive Inotropic Effects of Other Vasoactive Peptides
The positive inotropic effects of endothelin-1 differ from those of other vasoactive polypeptides, including angiotension II. Whereas the inotropic effect of angiotension II develops rapidly and is readily reversed on washout, the effects of endothelin-1 develop slowly and are sustained on washout.[19]

Comparison with the Positive Inotropic Effects of Additional Ca^{2+}
There is good agreement between the response with endothelin-1 and that with elevated extracellular Ca^{2+}, both being characterized by an unchanged time to peak tension and an unchanged time for relaxation.[19,40]

The Relative Sensitivity of Atrial and Ventricular Muscle to Endothelin-1

Moravec and colleagues[32] have shown that atrial muscle is more sensitive to endothelin-1 than is ventricular muscle. In human atrial muscle, for example, 50 nM endothelin-1 increases peak developed tension by about 75%, compared with an increase of <20% in human ventricular muscle.[32] The difference appears to be even more pronounced for rat atrial and ventricular preparations, because 50 nM endothelin-1 causes a <40% increase in the peak tension developed by atrial muscle compared with only ~15% for ventricular muscle. This means that atrial muscle is approximately three times more sensitive to endothelin-1 than to ventricular muscle, a difference that probably can be accounted for in terms of the different number of endothelin-specific receptors that are present in the two tissues, atrial muscle containing more receptors than the ventricular muscle (Table 10.2).

The Mechanisms Involved in the Positive Inotropic Effect of Endothelin-1

The positive inotropic effect of endothelin-1 is almost certainly a receptor-mediated event. Because it occurs in electrically paced preparations,[19,43] the possibility of it being dependent on an altered frequency of contraction—as in "the staircase"—can be rejected. However, the following observations probably indicate the major factors that are involved.

1. The response is dependent on the presence of extracellular Ca^{2+}.[19,43]
2. It is accompanied by an increased rate of phosphoinositol metabolism, leading to raised tissue levels of inositol phosphates.[10,43]
3. In isolated, cultured neonatal myocytes endothelin-1 causes a biphasic increase in cytosolic Ca^{2+} (as indicated by Fura-2 fluorescence measurements). This endothelin-induced increase in cytosolic Ca^{2+} is a dose-dependent phenomenon and is characterized by an initial rapidly developing spike followed by a sustained plateau phase[15] reminiscent of that which occurs in smooth muscle cells.[42] The rapidly developing spike-like increase in cytosolic Ca^{2+} is sensitive to agents such as ryanodine and caffeine, which interfere with sarcoplasmic reticulum Ca^{2+} mobilization and storage. The plateau phase, however, is calcium antagonist sensitive.[15]

A logical explanation for these observations is that the positive inotropic effect of endothelin-1 depends on an enhanced influx of Ca^{2+} from the extracellular space and the mobilization of Ca^{2+} from internal stores, the latter event probably being triggered by the raised levels of phosphoinositides.

The stimulant effect of endothelin-1 on phosphoinositol metabolism is calcium antagonist insensitive and is largely, but not totally, independent of extracellular Ca^{2+}.[10] Presumably, therefore, it does not depend on an endothelin-

TABLE 10.2. Density of endothelin-specific binding sites (receptors) in atrial and ventricular muscle

Species	Density (amol/mm^2)	
	Left atria	Left ventricle
Pig	5.7 ± 0.2	2.8 ± 0.4
Rat	5.9 ± 0.4	53.5 ± 0.3
Human	8.0 ± 0.3	6.4 ± 0.4

Data are from Davenport et al.[6]

1–mediated increase in Ca^{2+} influx through the voltage-activated, calcium antagonist–sensitive channels. There are many factors concerning the biochemical mechanisms that are responsible for the positive inotropic effect of endothelin-1, including the possible involvement of enhanced Ca^{2+} entry by way of the Na^+–Ca^{2+} exchanger consequent on the ability of endothelin-1 to stimulate the Na^+–H^+ exchanger,[41] that remain uncertain as far as cardiac muscle is concerned. A possible scheme of events is shown in Figure 10.2.

THE POSITIVE CHRONOTROPIC EFFECT OF ENDOTHELIN-1

Under in vivo conditions, the intravenous injection of endothelin-1 causes a slowed heart rate, an effect that is secondary to the accompanying reduction on coronary flow.[21] When added to isolated, spontaneously beating atrial preparations, however, the reverse response occurs,[20] with a dose-dependent increase in heart rate. This positive chronotropic effect of endothelin-1 occurs at a relatively low concentration. In guinea pig atria, for example, the threshold for the response is <0.1 nM, and the EC_{50} approaches 0.55 nM.[20] This is ~100 times below the EC_{50} value for the positive chronotropic effect of Bay K 8644, which is ~40 nM. It is interesting to note that tachyphylaxis develops to the positive chronotropic effects of endothelin-1, but not to Bay K 8644.[20]

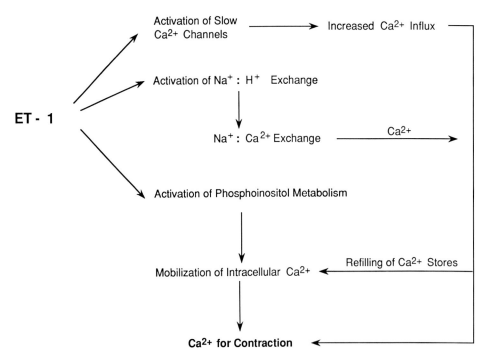

FIGURE 10.2. Schematic representation of the biochemical basis of the positive inotropic effects of endothelin-1 (ET).

THE STIMULANT EFFECT OF ENDOTHELIN-1 ON ATRIAL NATRIURETIC PEPTIDE RELEASE

In addition to its positive inotropic and chronotropic actions, endothelin-1 has a stimulatory effect on the secretion of atrial natriuretic peptide (ANP).[8] Because the response to endothelin-1 has been described for isolated myocytes[8] as well as in isolated atria[17] and intact animal studies, the effects cannot be accounted for in terms of the usual stimuli of stretch or volume expansion.[26,27] The response is dose dependent, has an ED_{50} of ~0.2 nM, and is attenuated by the L-type calcium channel antagonists nicardipine[8] and verapamil.[17] It is unaffected by phentolamine, propranolol, and atropine, indicative of the fact that transmitter release from endogenous nerve endings is not the trigger for the release.

ENDOTHELIN-1 AND THE CORONARY CIRCULATION

The endothelins are potent coronary constrictors. In anesthetized dogs, for example, the bolus intracoronary injection of as little as 30 pmol endothelin-1 per kilogram body weight reduces coronary blood flow by >90% and may even cause total coronary artery occlusion.[25] The effect is not species specific[2,3,22,23] and is insensitive to α-adrenergic and serotonergic blockade, as well as to angiotensin-converting enzyme inhibition and cyclooxygenase inhibition.[3] Coronary veins are more sensitive than are coronary arteries, and the effect is endothelium independent.[4] Nevertheless, it is attenuated by Ca^{2+}-channel blockade.[9,48] Large subepicardial vessels appear to be more sensitive than endocardial vessels, at least in anesthetized dogs.[3]

The coronary constrictor effect of endothelin-1 can have devastating consequences. These consequences (Fig. 10.3) include a decrease in segmental shortening, an increase in end-diastolic pressure, electrocardiographic signs of myocardial ischemia, a net release of lactate (indicative of the switch to anaerobic glycolysis,[3,7,21] and, at high dose levels, ventricular fibrillation and death. To some extent the coronary constrictor effect of this polypeptide is counterbalanced by an outflow of dilator prostanoids.[21] Fortunately several pharmacological interventions, including the administration of nitroglycerine and calcium antagonists, attenuate the constrictor response in large as well as in small resistance vessels.

THE IDENTIFICATION OF ENDOTHELIN-1–BINDING SITES IN INTACT HEARTS, ISOLATED MYOCYTES, AND CARDIAC MEMBRANE FRACTIONS

[^{125}I]endothelin-1 has been widely used to identify the presence of specific high-affinity binding sites for endothelin-1 in a variety of tissues, including the heart[5,6,16] and coronary vasculature.[38] As can be seen from Table 10.3, the density of binding in the atria is more than in the ventricles, which is more than in the coronary arteries, but these levels are below those obtained for the cer-

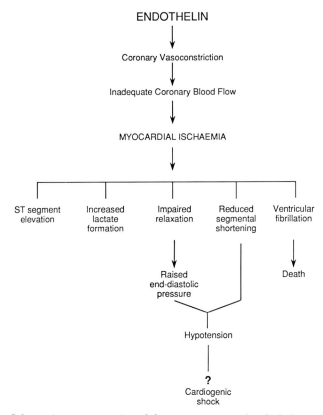

FIGURE 10.3. Schematic representation of the consequences of endothelin-1–induced coronary vasoconstriction.

ebellum, lung alveolar cells, or renal glomeruli. Irrespective of whether autoradiographic or classic binding techniques have been used to study the characteristics of the cardiac endothelin-1–binding sites, the data that have been obtained thus far are remarkably uniform. The sites are of high affinity, with K_d values that are compatible with the idea that these receptors are functional. Thus the K_d values are of the same order of magnitude as the EC_{50} values

TABLE 10.3. Relative density of endothelin-1–binding sites in various tissues

Species	Tissue	Density (amol/mm^2)
Rat	Left atria	5.9 ± 0.4
Rat	Left ventricle	3.5 ± 0.3
Rat	Aorta	2.2 ± 0.3
Pig	Coronary artery	2.5 ± 0.5
Rat	Renal glomeruli	9.2 ± 0.2
Rat	Lung alveolar cells	9.2 ± 0.4
Rat	Cerebellum	10.3 ± 0.5

The technique used was that of radioautography. Results are mean ± of three to six sections from at least three individual experiments. Data are from Davenport et al.[6]

obtained for the stimulant effect of the polypeptide on phosphoinositol metabolism and ANP secretion and its constrictor effect on the coronary vasculature, particularly coronary veins (Table 10.4).

Characterization of Cardiac Endothelin-1–Binding Sites

Binding studies on intact cardiac myocytes and membrane fragments have consistently shown that the specific binding of endothelin-1 is not inhibited by any of the currently available antagonists of the L-type voltage-sensitive Ca^{2+} channels.[11,12,14,15] Other substances that have been shown to have no effect on the specific binding of endothelin-1 (as [^{125}I]labeled endothelin-1) to cardiac membrane fragments include the Na^+ channel inhibitor tetrodotoxin, ω-conotoxin (the inhibitor of *N*-type Ca^{2+} channels), prazosin (an α-adrenoceptor antagonist), and ergonovine,[11,12] which is a potent vasoconstrictor. Other vasoactive peptides, including angiotension II and vasopressin (and their inhibitors), are also without effect, as are inhibitors of the Na^+–H^+ exchanger (amiloride) and the calcium agonist Bay K 8644.[11] One of the chemically homogenous toxins present in the venom of the asp (sarafotoxin S6b), however, is capable of inhibiting the binding (Fig. 10.4).

As far as the various isoforms of endothelin are concerned, endothelin-1 has a higher affinity than endothelin-3 for the cardiac receptor.[10] The same rank order of potency (endothelin-1 > endothelin-3) applies to their ability to stimulate phosphoinositide metabolism. These and other observations[45] have left little doubt as to the probable existence of two distinct types of endothelin-binding sites in cardiac cell membranes, one type having a higher affinity for endothelin-1 and endothelin-2, relative to endothelin-3, and the other preferring endothelin-3. One interesting but as yet untested hypothesis is that the endothelin-1–preferring receptors may modulate the effect of endothelin on phosphoinositol metabolism, whereas the other endothelin-3–preferring receptor may modulate its stimulant effect on prostanoid release.

TABLE 10.4. K_d values for cardiac endothelin-1–binding sites relative to the EC_{50} values for the positive inotropic and chronotropic effects of endothelin-1

Response	K_d (nM)	EC_{50} (nM)	Reference
Isolated myocyte binding	0.6–0.9		15
Ventricular microsomal binding	0.5		11
Cultured vascular smooth muscle cells	0.2		5
Positive inotropic response		0.1	19
Positive chronotropic response		0.4	20
ANP secretion		0.2	8
Coronary vasoconstrion			
Arteries		5.0	4
Veins		0.5	4
Phosphoinositol metabolism		3.5	10

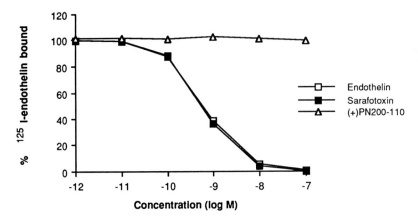

FIGURE 10.4. Inhibitory effect of sarafotoxin S6b on the specific binding of [^{125}I]endothelin-1 to isolated rat cardiac membranes. Note that whereas sarafotoxin S6b inhibited the binding, the calcium antagonist PN200-110 was without effect. Data are from Cernacek and Stewart[1] and from Gu et al.[12]

"Downregulation" of Endothelin-1 Receptors in the Heart

Hirata et al.[15] were the first investigators to draw attention to the fact that cardiac endothelin-1 receptors are capable of "downregulation" in response to the presence of a raised concentration of endothelin-1. Since the publication of that paper other conditions that cause downregulation have been described. One such condition is that of streptozotocin-induced diabetes.[37] In the latter case significant downregulation occurs within 3 days of a single intravenous bolus dose of streptozotocin. Receptor affinity and selectivity is unchanged, and a direct effect of the streptozotocin was excluded.

"Upregulation" of Cardiac Endothelin-1 Receptors

Upregulation of cardiac endothelin-1 receptors is a very real phenomenon. For example, the induction of cyclosporine-induced toxicity in mice is accompanied by a significant increase in endothelin-binding site density without any change in selectivity.[35] Provided that the additional receptors are functional, this cyclosporine-induced increase in the density of the endothelin-1 receptor binding could contribute to the cardiotoxicity of the cyclosporine compound. Thus, by enhancing Ca^{2+} influx and mobilizing internal Ca^{2+}, activation of the "new" endothelin receptors would accelerate the rate at which Ca^{2+} homeostasis is lost, thereby precipitating the conditions that signal cell death and tissue necrosis. Whether such an effort occurs in the peripheral vasculature is currently unknown, but should this happen then it could account for the hypertension that often develops during cyclosporine therapy despite a normal catecholamine excretion and the absence of any major abnormality in the renin–angiotensin system.

Other conditions that cause upregulation of cardiac endothelin-1–binding sites include prolonged episodes of ischemia and postischemic reperfusion.[29] Hypoxia, however, is without effect.[30] The ischemia-induced increase in endothelin-1–binding site density increases as the duration of the ischemic episode is extended, and the increase is exacerbated upon reperfusion.[29] Maneuvers that protect against this ischemia-induced increase include hypothermia[29] and pretreatment with dihydropyridine-based calcium antagonists.[36] Recent studies in our own laboratories have shown that the ischemia-induced increase in endothelin-1–binding site density involves the externalization of the binding sites.[28] Preliminary evidence indicates that the release of free radicals is a possible causative factor (Liu and Nayler, unpublished data). Surprisingly (Table 10.5), systemic hypertension causes a downregulation of the cardiac endothelin-1–binding sites at a time when the binding site density in the brain is increasing.[13]

THE PATHOLOGICAL SIGNIFICANCE OF CHANGES IN THE CIRCULATING LEVELS OF ENDOTHELIN-1

Because endothelin-1 causes sustained coronary vasoconstriction, raises cytosolic Ca^{2+}, and stimulates Ca^{2+} influx, it is logical to consider whether changes in the circulating levels of endothelin-1 that occur in certain pathological conditions are of sufficient magnitude to have a direct deleterious effect on the heart and the coronary vasculature. Some of the conditions that are associated with raised levels of circulating endothelin-1 are myocardial ischemia, cardiogenic shock, endotoxic shock, renal failure, and hypertension. However, although the increased levels are statistically significant, their contribution to the etiology of the disease state is uncertain beause (*1*) the plasma levels remain in the picomolar range and therefore are below the level needed to trigger the events listed in Figure 10.3, and (*2*) the endothelin-induced release of vasodilator substances, including ANP, and the dilator prostanoids compensates, to some extent, for the potent constrictor effect of the polypeptide. To make any assumption relating to the importance of the circulating plasma levels of endothelin-1 is therefore premature, particularly because the endothelin that appears in the plasma may simply be the residual endothelin that has leaked away from the interstitial space. With the increasing evidence of endothelin's ability to act as a local hormone, attention must now be turned

TABLE 10.5. Conditions that cause up- and downregulation of cardiac endothelin-1–binding site density

	Reference
Downregulation	
Streptozotocin-induced diabetes	37
Spontaneous hypertension	13
Upregulation	
Prolonged ischemia	28, 29
Postischemic reperfusion	29
Posthypoxic reoxygenation	30
Cyclosporine-induced toxicity	35
Free radical generation	Unpublished data

toward monitoring its rate of production at its site of origin, which is not always restricted to vascular endothelial cells.

ENDOTHELIN AND MYOCARDIAL INFARCTION

Several observations indicate that endothelin is probably involved in the etiology of events that trigger or exacerbate the damage caused by episodes of inadequate coronary perfusion: (*1*) Endothelin-1 is an extremely potent vasoconstrictor. (*2*) The density of cardiac endothelin-1 receptor increases during prolonged episodes of ischemia.[29,30] (*3*) Plasma levels of endothelin-1 rise during myocardial infarction.[1,31] At the same time the levels of the immediate precursor (big endothelin-1) increase, indicating an increased rate of endothelin-1 production.

Recently Watanabe et al.[46] reported that pretreatment of rats with a monoclonal antibody against endothelin significantly reduced the size of the infarct caused by coronary artery ligation. It is difficult not to speculate, therefore, concerning the involvement of locally generated endothelin in the events that are triggered by inadequate coronary perfusion and that culminate in cardiac cell death and tissue necrosis. What is not clear at the moment, however, is whether the antibody provided protection because of an effect on the vasculature or whether the myocytes were the main target sites.

CONCLUSION

The heart and its vasculature provide a major site of action for endothelin-1 (Fig. 10.5). Certain pathological conditions, including ischemia, postischemic reperfusion, and cyclosporine-induced toxicity, enhance the cardiac endothelin-1–binding site density, an effect that could culminate in Ca^{2+} overload and cell death. Already the use of an endothelin-1 antibody has been shown to be of use in one of these conditions (ischemia). An inhibition of the enzyme that converts pre-endothelin to "mature" endothelin[48] could be of equal benefit.

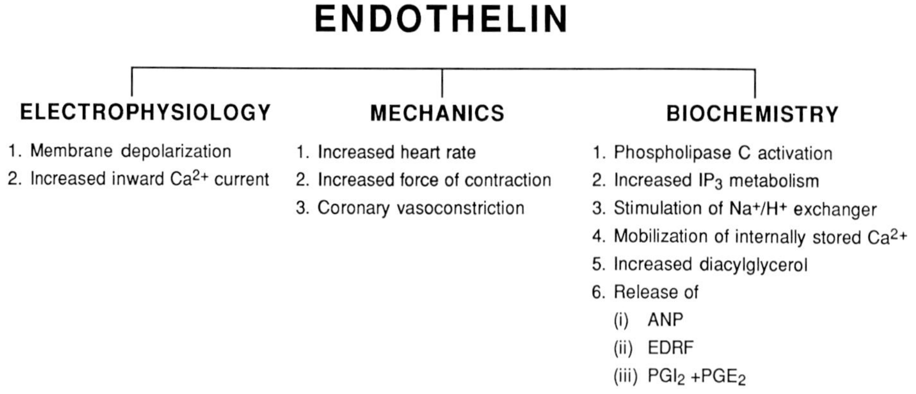

ENDOTHELIN

ELECTROPHYSIOLOGY	MECHANICS	BIOCHEMISTRY
1. Membrane depolarization	1. Increased heart rate	1. Phospholipase C activation
2. Increased inward Ca^{2+} current	2. Increased force of contraction	2. Increased IP_3 metabolism
	3. Coronary vasoconstriction	3. Stimulation of Na^+/H^+ exchanger
		4. Mobilization of internally stored Ca^{2+}
		5. Increased diacylglycerol
		6. Release of
		(i) ANP
		(ii) EDRF
		(iii) $PGI_2 + PGE_2$

FIGURE 10.5. Schematic representation of the overall effects of endothelin-1 on the heart.

REFERENCES

1. CERNACEK, P., and D. J. STEWART. Immunoreactive endothelin in human plasma: marked elevations in patients in cardiogenic shock. *Biochem. Biophys. Res. Commun.* 161: 562–567, 1989.
2. CHESTER, A. H., M. R. DASHWOOD, J. G. CLARKE, S. W. LARKING, G. J. DAVIES, S. TADJKARIMI, A. MASERI, and M. H. YACOUB. Influence of endothelin on human coronary arteries and localization of its binding sites. *Am. J. Cardiol.* 63: 1395–1398, 1989.
3. CLOZEL, J. P., and M. CLOZEL. Effects of endothelin on the coronary vascular bed in open-chest dogs. *Circ. Res.* 65: 1193–1200, 1989.
4. COCKS, T. M., N. L. FAULKNER, K. SUDHIR, and J. ANGUS. Reactivity of endothelin-1 on human and canine large veins compared with large arteries in vitro. *Eur. J. Pharmacol.* 171: 17–24, 1989.
5. DAVENPORT, A. P., A. J. KAUMANN, J. A. HALL, D. J. NUNEZ, and M. J. BROWN. [^{125}I]endothelin binding in mammalian tissue: relation to human atrial inotropic effects and coronary contraction. *Br. J. Pharmacol.* 96: 102, 1989.
6. DAVENPORT, A. P., D. J. NUNEZ, J. A. HALL, A. J. KAUMANN, and M. J. BROWN. Autoradiographic localization of binding sites for porcine [^{125}I]endothelin-1 in humans, pigs, and rats: functional relevance in humans. *J. Cardiovasc. Pharmacol.* 13(Suppl. 5): 166–170, 1989.
7. EZRA, D., R. E. GOLDSTEIN, J. F. CZAJA, and G. Z. FEUERSTEIN. Lethal ischemia due to intracoronary endothelin in pigs. *Am. J. Physiol.* 257 (*Heart Circ. Physiol.* 26): H339–H343, 1989.
8. FUKUDA, Y., Y. HIRATA, H. YOSHIMI, T. KOJIMA, Y. KOBAYASHI, M. YANAGISAWA, and T. MASAKI. Endothelin in a potent secretagogue for atrial natriuretic peptide in cultured rat atrial myocytes. *Biochem. Biophys. Res. Commun.* 155: 167–172, 1988.
9. FUKUDA, K., S. HORI, M. KUSUHARA, T. SATOH, S. KYOTANI, S. HANDA, Y. NAKAMURA, H. OONA, and K. YAMAGUCHI. Effect of endothelin as a coronary vasoconstrictor in the Langendorff-perfused rat heart. *Eur. J. Pharmacol.* 165: 301–304, 1989.
10. GALRON, R., Y. KLOOG, A. BDOLAH, and M. SOKOLOVSKY. Functional endothelin/sarafotoxin receptors in rat heart myocytes: structure–activity relationships and receptor subtypes. *Biochem. Biophys. Res. Commun.* 163: 936–943, 1989.
11. GU X. H., D. J. CASLEY, and W. G. NAYLER. Characterization of [^{125}I]endothelin-1 binding sites in rat cardiac membrane fragments. *J. Cardiovasc. Pharmacol.* 13(Suppl. 5): S171–173, 1989.
12. GU, X. H., D. J. CASLEY, and W. G. NAYLER. Sarafotoxin S6b displaces specifically bound ^{125}I-endothelin. *Eur. J. Pharmacol.* 162: 509–510, 1989.
13. GU, X. H., D. J. CASLEY, M. CINCOTTA, and W. G. NAYLER. [^{125}I]endothelin-1 binding to brain and cardiac membranes from normotensive and spontaneously hypertensive rats. *Eur. J. Pharmacol.* 177: 205–209, 1990.
14. HIRATA, Y., Y. FUKUDA, H. YOSHIMA, T. EMORI, M. SHICHIRA, and F. MARUMA. Specific receptors for endothelin in cultured rat cardiocytes. *Biochem. Biophys. Res. Commun.* 160: 1438–1444, 1989.
15. HIRATA, Y., H. YOSHIMA, T. EMORI, M. SHICHIRA, F. MARUMO, T. X. WATANABE, S. KUMUGAYE, K. NAKAJIMA, T. KIMURA, and S. SAKAKIBARA. Receptor binding activity and cytosolic free calcium response by synthetic endothelin analogues in cultured rat vascular smooth muscle cells. *Biochem. Biophys. Res. Commun.* 160: 228–234, 1989.
16. HOYER, D., C. WAEBER, and J. M. PALACOIS. [^{125}I]endothelin-1 binding sites: autoradiographic studies in the brain and periphery of various species, including humans. *J. Cardiovasc. Pharmacol.* 13(Suppl. 5): S162–S165, 1989.
17. HU, J. R., U. G. BERNINGER, and R. E. LANG. Endothelin stimulates atrial natriuetic peptide (ANP) release from rat atria. *Eur. J. Pharmacol.* 164: 177–178, 1988.
18. HU, J. R., R. VON HARSDORF, and R. E. LANG. Endothelin has potent inotropic effects in rat atria. *Eur. J. Pharmacol.* 158: 275–278, 1988.
19. ISHIKAWA, T., M. YANAGISAWA, S. KIMURA, K. GOTO, and T. MASAKI. Positive inotropic action of novel vasoconstrictor peptide endothelin on guinea pig atria. *Am. J. Physiol.* 255 (*Heart Circ. Physiol.* 24): H970–H973, 1988.
20. ISHIKAWA, T., M. YANAGISAWA, S. KIMURA, K. GOTO, and T. MASAKI. Positive chronotropic effects of endothelin, a novel endothelium-derived vasoconstrictor peptide. *Pflugers Arch.* 413: 108–110, 1989.
21. KARAWATOWSKA-PROLOPCZUK, E., and A. WENNMALM. Effects of endothelin on coronary flow, mechanical performance, oxygen uptake, and formation of purines and on outflow of prostacyclin in the isolated rabbit heart. *Circ. Res.* 66: 46–54, 1990.
22. KASUYA, Y., T. ISHIRAWA, M. YANAGISAWA, S. KIMURA, K. GOTO, and T. MASAKI. Mechanism

of contraction to endothelin in isolated porcine coronary artery. *Am. J. Physiol.* 257(Heart Circ. Physiol. 26): H1828–H1835, 1989.
23. KASUYA, Y., Y. TAKUWAA, M. YANAGISAWA, S. KIMURA, K. GOTO, and T. MASAKI. Endothelin-1 induces vasoconstriction through two functionally distinct pathways in porcine coronary artery: contribution of the phosphoinositide turnover. *Biochem. Biophys. Res. Commun.* 161: 1049–1055, 1989.
24. KLOOG, Y., I. AMBAR, M. SOLOLOVSKY, E. KOCHUA, Z. WOLLBERG, and A. BDHOLAH. Sarafotoxin, a novel vasoconstrictor peptide: phosphoinositide hydrolysis in rat heart and brain. *Science* 242: 268–270, 1988.
25. KURIHARA, H., K. YAMAOKI, R. NAGAI, M. YOSHIZUMI, F. TAKAKU, H. SATOH, and J. INUI. Endothelin: a potent vasoconstrictor associated with coronary vasospasm. *Life Sci.* 44: 1937–1943, 1989.
26. LANG, R. E., H. THOLKEN, D. GANTEN, F. C. LUFT, H. RUSKOAKO, and T. UNGER. Atrial natriuretic factor—a circulating hormone stimulated by volume loading. *Nature* 314: 264–266, 1985.
27. LEDSOME, J. R., N. WILSON, A. COURNEYA, and A. J. RANKIN. Release of atrial natriuretic peptide by atrial distension. *Can. J. Physiol. Pharmacol.* 63: 739–742, 1986.
28. LIU, J. J., D. J. CASLEY, and W. G. NAYLER. Ischaemia causes externalization of endothelin-1 binding sites in rat cardiac membranes. *Biophys. Biochem. Res. Commun.* 164: 1220–1225, 1989.
29. LIU, J. J., R. CHEN, D. J. CASLEY, and W. G. NAYLER. Effect of ischemia and reperfusion on [^{125}I]endothelin-1 binding in rat cardiac membranes. *Am. J. Physiol.* (in press).
30. LIU, J. J., X. H. GU, D. J. CASLEY, and W. G. NAYLER. Reoxygenation but neither hypoxia nor intermittent ischaemia increases [^{125}I]endothelin-1 binding to rat cardiac membranes. *J. Cardiovasc. Pharmacol.* 15: 436–443, 1990.
31. MIYAUCHI, T., M. YANAGISAWA, T. TOMIZAWA, Y. SUGISHITA, N. SUZUKI, M. FUJINO, R. AJISAKA, K. GOTO, and T. MASAKI. Increased plasma concentrations of endothelin-1 and big endothelin-1 in acute myocardial infarction. *Lancet* 1: 53, 1989.
32. MORAVEC, C. S., E. E. REYNOLDS, R. W. STEWART, and M. BOND. Endothelin is a positive inotropic agent in human and rat heart in vitro. *Biochem. Biophys. Res. Commun.* 159: 14–18, 1989.
33. NAYLER, W. G. Endothelin: a mini review. *J. Appl. Cardiol.* 4: 495–504, 1989.
34. NAYLER, W. G. Endothelin: isoforms, binding sites and possible implications in pathology. *Trends Pharmacol. Sci.* 11: 96–99, 1990.
35. NAYLER, W. G., X. H. GU, D. J. CASLEY, S. PANAGIOTOPOULOS, J. LIU, and P. L. MOTTRAM. Cyclosporine increases endothelin-1 binding site density in cardiac cell membranes. *Biochem. Biophys. Res. Commun.* 163: 1270–1274, 1989.
36. NAYLER, W. G., and J. J. LIU. Nifedipine and experimental cardioprotection. *Cardiovasc. Drugs Ther.* (in press).
37. NAYLER, W. G., J. LIU, S. PANAGIOTOPOULOS, and D. J. CASLEY. Streptozotocin-induced diabetes reduces the density of endothelin-binding sites in rat cardiac membranes. *Br. J. Pharmacol.* 97: 993–995, 1989.
38. POWER, R. F., J. WHARTON, S. P. SALAS, S. KANSE, M. GHATEI, S. R. BLOOM, and J. M. POLAK. Autoradiographic localisation of endothelin binding sites in human and porcine coronary arteries. *Eur. J. Pharmacol.* 160: 199–200, 1989.
39. SCHRAMM, M., G. THOMAS, M. TOWART, and G. FRANCKOWIAK. Novel dihydropyridines with positive inotropic action through activation of Ca^{2+} channels. *Nature* 303: 535–537, 1983.
40. SHAH, A. M., M. J. LEWIS, and A. H. HENDERSON. Inotropic effects of endothelin in ferret ventricular myocardium. *Eur. J. Pharmacol.* 163: 365–367, 1989.
41. SIMONSON, M. J., S. WANN, P. MENE, G. R. DUBYAW, M. KESTER, Y. NAKAZATO, and J. R. SEDO. Endothelin stimulates phosphorylase C, Na^+/H^+ exchange, c-*fos* expression, and mitogenesis in rat mesangial cells. *J. Clin. Invest.* 83: 708–712, 1989.
42. SIMPSON, A. W. M., and C. C. ASHLEY. Endothelin-evoked Ca^{2+} transients and oscillations in AIO vascular smooth muscle cells. *Biochem. Biophys. Res. Commun.* 163: 1223–1229, 1989.
43. VIGNE, P., M. LAZDUNSKI, and C. FRELIN. The inotropic effect of endothelin-1 on atria involves hydrolysis of phosphatidylinositol. *Biochem. Biophys. Res. Commun.* 249: 143–146, 1989.
44. WATANABE, T., K. KUSOMOTO, T. KITAYOSHI, and N. SHIMAMOTO. Positive inotropic and vasoconstrictive effects of endothelin-1 in in vivo and in in vitro experiments: characteristics and the role of L-type calcium channels. *J. Cardiovasc. Pharmacol.* 13(Suppl. 5): 108–111, 1989.
45. WATANABE, T., H. MIYAZAKI, M. KONDOH, Y. MASUDA, S. KIMURA, M. YANAGISAWA, T. MASAKI,

and K. MURAKAMI. Two distinct types of endothelin receptors are present on chick cardiac membranes. *Biochem. Biophys. Res. Commun.* 161: 1252–1259, 1989.
46. WATANABE, T., N. SUZUKI, N. SHIMATMOTO, M. JUJINO, and A. IMADA. Endothelin in myocardial infarction. *Nature* 344: 114, 1990.
47. YAMASAKI, H., M. NIWA, K. YAMASHITA, Y. KATAOKA, K. SHIGEMATSU, K. HASHIBA, and M. OZAKI. Specific ^{125}I-endothelin-1 binding sites in the atrioventricular node of the porcine heart. *Eur. J. Pharmacol.* 168: 247–250, 1989.
48. YANAGISAWA, M., H. KURIHARA, S. KIMURA, Y. TOMOBE, M. KOBAYASHI, Y. MITSUI, Y. YAZAKI, K. GOTO, and T. MASAKI. A novel potent constrictor peptide produced by vascular endothelial cells. *Nature* 332: 411–415, 1988.
49. YANAGISAWA, M., and T. MASAKI. Endothelin, a novel endothelium-derived peptide. *Biochem. Pharmacol.* 38: 1877–1883, 1989.
50. YANAGISAWA, M., and T. MASAKI. Molecular biology and biochemistry of the endothelins. *Trends Pharmacol. Sci.* 10: 374–378, 1989.

11

Renal and Systemic Hemodynamic Actions of Endothelin

ANDREW J. KING AND BARRY M. BRENNER

Nearly a century has passed since W. M. Bayliss first described the ability of the vasculature to respond to changes in blood flow and pressure by adjusting local vascular smooth muscle tone.[9] However, the role of the endothelium in the transduction of a variety of mechanical and neurohumoral signals into appropriate modulation of underlying vascular smooth muscle tone has only recently been appreciated.[13,32,95-99,113] Signal transduction appears to be mediated by the release of soluble factors and by direct cell–cell interactions through gap junctions.[24] The endothelial cell produces both vasoconstricting and vasodilating mediators, the balance of which ultimately determines the vascular smooth muscle tone.[31] Several stimuli have been recognized to induce endothelium-dependent vasoconstriction.[96,114] The first reports of peptidase-sensitive endothelium-derived contracting factors appeared nearly 5 years ago.[39,83] The description of the endothelin family of peptides by Yanigisawa and colleagues has led to a burst of investigation of the role of endothelin in the modulation of regional and systemic hemodynamics.[45,121] In addition, it is likely that other soluble endothelial-derived contracting factors also exist.[50,96,98]

In the original report, intravenous endothelin-1 induced a profound pressor response in a chemically denervated rat.[121] Since that time, the in vivo pharmacological effects of the endothelin family of peptides have been studied extensively in a variety of species. However, our understanding of the role endothelin plays in either the maintenance of normal vascular tone[97] or the pathophysiology of vascular disorders[20] remains rudimentary. In this chapter we focus on several aspects of endothelin biology: the systemic and renal hemodynamic effects of exogenous endothelin, the potential role of endothelin in the modulation of glomerular hemodynamics, and preliminary evidence that endothelin is a mediator of vasoconstrictive disorders.

SYSTEMIC RESPONSE TO ENDOTHELIN

Systemic Vasodilatory Response

Bolus intravenous injections of subnanomolar quantities of endothelin-1, endothelin-2, and endothelin-3 induce a transient hypotensive response followed by a prolonged hypertensive phase (Fig. 11.1).[26,43,45,51-53,76,119-121] This pattern of

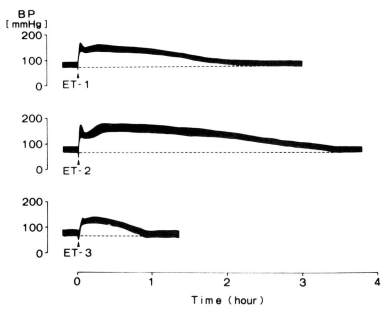

FIGURE 11.1. Representative tracings of pressor response in anesthetized, chemically denervated rats given intravenous endothelin-1 (ET-1), endothelin-2 (ET-2), or endothelin-3 (ET-3). BP, blood pressure. (From Inoue et al.[45])

response occurs in conscious and anesthetized animals of a variety of species.[17,26,33–35,37,40,54,63,76,87,94,101] The hypotensive response to endothelin in the rat is characteristically rapid in onset and short-lived and may occur at nonpressor doses (25 pmol; Fig. 11.2).[52] In vivo, the depressor effects of endothelin-3 are more prominent than for endothelin-1.[45] In the anesthetized rat the early hypotensive phase is due to a decrease in total peripheral resistance, because cardiac index is unchanged during this period (Fig. 11.2).[52] Anesthetized dogs and cats show a similar peripheral vasodilation, although cardiac output rises significantly in this early phase.[53,63] In conscious rats, the depressor response is associated with an appropriate baroreflex response consisting of dose-dependent increases in efferent sympathetic nerve activity, tachycardia, and an increase in cardiac output.[33,54,94] Autonomic blockade with atenolol and methscopolamine obliterates the reflex tachycardia; however, a mild increase in cardiac output persists, suggesting possible direct effects of endothelin on the myocardium.[94] Indeed, endothelin-1 induces a dose-dependent positive inotropic effect in isolated guinea pig atria and to a lesser extent in isolated ventricular trabeculae.[44,46,53,77]

The vasodilatory effects of endothelin vary markedly among different regional vascular beds (Table 11.1). Bolus injection of endothelin uniformly reduces vascular resistance of the hindlimb in anesthetized and conscious animals in the early phase.[18,33,42,54,75,94,119] Carotid and bronchial blood flow are also enhanced in anesthetized pigs and rats, respectively.[33,42,89,119] Compared with the hindquarter, the renal vascular bed is much less sensitive to the vasodilatory effects of endothelin, with immediate vasoconstriction predominating in several studies.[18,32,33,42,94] However, biphasic renal blood flow responses to endothelin infusion have also been reported.[63,111,115] Finally, the mesenteric vas-

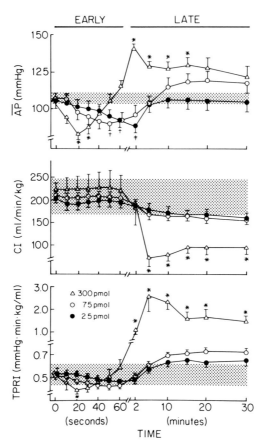

FIGURE 11.2. Average time-course changes in mean arterial pressure (AP), cardiac index (CI), and total peripheral resistance index (TPRI) following single bolus injections of endothelin-1 in the dosages shown. Values are means ± S.E.M. Shaded area represents mean of vehicle ± 2 S.E.M. *$P < 0.05$, 300 pmol vs. vehicle; † $P < 0.05$, 75 pmol vs. vehicle; $P < 0.05$, 25 pmol vs. vehicle. (From King et al.[52])

TABLE 11.1. Regional effects of endothelin-1 and endothelin-3

	Endothelin-1 (porcine; 0.04 nmol)		Endothelin-3 (rat; 0.04 nmol)	
	0.5 min	5.0 min	0.5 min	5.0 min
HR (beats/min)	86 ± 12*	−48 ± 10*	50 ± 11*	−21 ± 6*
MAP (mm Hg)	−16 ± 3*	22 ± 5*	−12 ± 3*	8 ± 3*†
RVR (%)	−1 ± 6	95 ± 14*	−13 ± 6*	30 ± 8*†
MVR (%)	11 ± 13	92 ± 11*	13 ± 8	45 ± 10*†
HVR (%)	−44 ± 5*	46 ± 13*	−39 ± 5*	8 ± 8†

Values are means ± S.E.M.; HR, heart rate; MAP, mean arterial blood pressure; RVR, renal vascular resistance; MVR, mesenteric vascular resistance; HVR, hindlimb vascular assistance. (Adapted from Gardiner et al.[33])

*$p < 0.05$ vs baseline.

†$p < 0.05$ vs endothelin-1.

cular bed uniformly vasoconstricts during the hypertensive phase of endothelin infusion. In aggregate, the vasoconstriction of the mesenteric and renal vascular beds are opposed by dilation of the much larger skeletal muscle vascular bed, leading to a fall in total peripheral vascular resistance.

Several factors may contribute to this vasodilatory response. First, in isolated perfused rat mesentery and luminally perfused rabbit aorta, endothelin induces dose-dependent release of endothelium-derived relaxing factor (EDRF).[25,109] This effect was obliterated by removal of the endothelium or concurrent infusion of oxyhemoglobin. Endothelin-induced vasodilation by EDRF is further supported by studies using N^ω-monomethyl-L-arginine (L-NMMA), an analog of L-arginine (L-Arg) that competitively antagonizes its conversion to the mediator of EDRF, nitric oxide.[1,92] Pretreatment with L-NMMA markedly abrogates the hypotensive effects of endothelin-1 in anesthetized rats.[117] In perfused mesenteries preconstricted with methoxamine, endothelin-1 and endothelin-3 both induce vasodilation at doses 100-fold less than those required for vasoconstriction (0.5 vs. 50 pmol, respectively). In the higher dose range (10–100 pmol), vasodilation persists for endothelin-3, whereas endothelin-1 results in vasoconstriction.[116] The role of endothelin-3 in the modulation of regional vascular tone is even less clear than that for endothelin-1, in that endothelin-1 is the only member of the endothelin family of peptides that is produced by the endothelium, whereas endothelin-3-like immunoreactive material has been identified in homogenates of rat intestine and brain.[72]

Endothelin-1 induces the release of prostaglandin I_2 (PGI_2) much more than thromboxane A_2 (TXA_2) in isolated perfused rat lungs, whereas in guinea pig lungs PGI_2 and TXA_2 were released equally.[26] This response is due, at least in part, to the increase in arachidonate secondary to the activation of phospholipase A_2 by endothelin.[93] Isolated perfused rabbit kidney and spleen also release predominantly vasodilatory prostanoids in response to endothelin.[90] Finally, plasma 6-keto $PGF_{1\alpha}$ levels increase with endothelin-1 in dogs.[38] Pretreatment with indomethacin or piroxicam blunts the hypotensive response in the anesthetized rat[43] and potentiates the vasoconstrictor response.[24,41a] This enhanced pressor response with cyclooxygenase inhibitors has been confirmed by several investigators.[42,76] Similar potentiation was seen in the isolated perfused rabbit spleen and kidney.[90] In contrast, dogs pretreated with acetylsalicylic acid exhibit minimal hypotension following a bolus of endothelin-1 (300 pmol/kg).[38]

Endothelin is a potent secretagogue for the vasorelaxant atrial natriuretic peptide (ANP) in cultured rat atrial myocytes and in isolated perfused hearts.[30,67] Plasma levels of ANP increase with endothelin in vivo.[35,67,73] Infusion of a high dose of ANP can reverse the systemic pressor response to endothelin. However, the role of this peptide in modulating the hemodynamic effects of endothelin is unclear.[41]

Systemic Vasoconstrictor Response

Following the early vasodepressor phase, bolus intravenous injection of endothelin results in a rapid, sustained, and dose-dependent rise in blood pressure (Figs. 11.1, 11.2). Endothelin is generally found to be a slightly less potent pressor agent than angiotensin II.[34,36,52] The relative molar potency of the endothelin subtypes (endothelin-1 > endothelin-2 >> endothelin-3) reflects the vasoconstrictor findings in porcine coronary artery strips, whereas the greatest contractile tensions were seen for endothelin-2.[45] In contrast, the 38/39 amino acid precursor of endothelin, "big" endothelin, has 100-fold less in vitro contractile activity than does mature endothelin and yet is nearly as potent a pressor in vivo as is endothelin-1.[47] The hypertensive effect is sustained for up to 60–90 min following a single bolus injection of endothelin-1, with an LD_{50} of 450 pmol in the anesthetized rat.[45,51,121] The duration of vasoconstriction is markedly less for endothelin-3 than for endothelin-1 or endothelin-2 (Fig. 11.1). Studies of human forearm injections of endothelin-1 demonstrate marked peripheral vasoconstriction.[17] The pressor effect occurs in conscious and anesthetized dogs and rats despite a reduction in cardiac output reflecting extreme peripheral vasoconstriction (Fig. 11.2).[34,50,72] Mild bradycardia is nearly a universal finding in this phase (Table 11.1).[18,35,52,78]

Both the pressor and depressor effects are enhanced by intraarterial vs. intravenous injections.[26] This differential effect may be due to removal of endothelin by the lung. Bioassays of the effluent of isolated perfused lungs estimate a 60% reduction in endothelin in a single pass.[25] Although little is known of the metabolism of endothelin, the circulating half-life of infused endothelin is short (1–7 min), and thus the long-lasting vasoconstrictor effect must reflect persistent binding to and activation of vascular smooth muscle cells.[2,26,97] These findings are in good agreement with the high affinity of endothelin for its receptors.[19,71]

Compared with angiotensin II and norepinephrine, endothelin has similar regional selectivity in its vasoconstrictor effects.[62] The renal, mesenteric, and muscular vascular beds account for the majority of the increase in vascular resistance during the pressor phase (Table 11.1).[115] The pulmonary circulation is less sensitive to the pressor effects of endothelin.[64,115] Intralobar arterial injection of endothelin (40–400 pmol) resulted in slight increases in lobar arterial pressure.[64] In contrast, using labeled microspheres, intravenous infusion of endothelin-1 (1 nmol/kg) led to pulmonary vasodilation concurrent with the systemic vasoconstrictor response.[115] This effect was not blocked by cyclooxygenase and thus may reflect EDRF production and/or the release of other local vasodilators. Endothelin failed to constrict the stomach vascular bed unless indomethacin was present, suggesting a more pronounced buffering effect of local vasodilatory prostanoids.[18,115]

Endothelin is a more potent vasoconstrictor in conscious rats pretreated with chlorisondamine, methscopolamine, captopril, and d($CH_2)_5$Tyr(Me)AVP, indicating a significant neurohumoral response.[40] The effects of ganglionic blockade on the vasoconstrictor phase of endothelin vary among species. The pressor response is not affected by pentolinium in conscious rats,[54] whereas in conscious dogs the endothelin-induced hypertension is blunted by ganglionic blockade.[34,54] Indeed, intracerebroventricular injections of endothelin lead to a dose-related pressor response and to tachycardia in conscious rats, effects blunted by pretreatment with phenoxybenzamine.[86] Furthermore, plasma epinephrine and norepinephrine levels rise in response to intracerebroventricular endothelin injections. In contrast, endothelin inhibits peripheral nerve release of norepinephrine, and thus the interaction of endothelin and the sympathetic nervous system is not fully understood.[118] The bradycardia induced by endothelin in conscious rats is attenuated by simultaneous cholinergic and ganglionic blockade and was similar in magnitude to that induced by equipressor doses of phenylephrine.[54] Renal sympathetic nerve activity in chloralose-anesthetized rats decreases ($-46.4 \pm 12.7\%$) in response to endothelin-induced hypertension ($+53 \pm 5.4$ mm Hg).[54] Thus endothelin does not grossly alter the baroreflex modulation of heart rate.

The contractile response of isolated blood vessels is blunted by depletion of extracellular Ca^{2+} and by Ca^{2+}-channel antagonists.[68,121] However, in the absence of extracellular Ca^{2+} endothelin still induces a rise in intracellular free calcium, indicating mobilization from intracellular stores. In agreement with these observations, pretreatment with dihydropyridine Ca^{2+} channel blockers has minimal effects on the rapid phase of the pressor response, though it markedly attenuates the sustained phase of vasoconstriction.[75,76,87] Verapamil and manganese have similar blunting effects on the systemic pressor response.[15] These studies indicate that the endothelin-induced vasoconstriction in vivo is, at least in part, due to the influx of extracellular calcium through voltage-dependent dihydropyridine-sensitive Ca^{2+} channels.

In summary, bolus endothelin causes a biphasic pressor response in conscious and anesthetized animals of a variety of species. Endothelin-induced vasoactive effects show marked regional selectivity, and endothelin does not seem to interfere with baroreflexes. All isoforms of endothelin are potent vasoconstrictors. The biphasic response and the sustained nature of the vasoconstriction distinguish endothelin from other endothelial-derived vasoactive factors.

RENAL ACTIONS OF ENDOTHELIN

Role of Vasoactive Mediators in Renal Hemodynamics

The glomerular capillary bed is interposed between two high-resistance vessels, the afferent and efferent arterioles. The relative resistances of these two vessels control the glomerular capillary hydraulic pressure and plasma flow rate.[12] Indeed, modulation of the pre- and postglomerular resistances allows the kidney to maintain filtration over a wide range of renal perfusion pressures. The glomerular tuft is a target organ for a variety of vasoactive hormones that exert considerable influence on the filtration process, in part by modulation of the relative resistances of the afferent and efferent arterioles.[27] This vascular bed is capable of synthesizing a number of these mediators and thus may profoundly affect renal function without significantly altering systemic hemodynamics.[27] Endothelin possesses many of the characteristics associated with other known endogenous modulators of renal function. This section will provide evidence that supports a role for endothelin in renal physiology.

Renal Endothelin Receptors and Synthesis

Autoradiographic studies of [^{125}I]endothelin-1 binding to the renal parenchyma indicate a pattern of binding similar to that of angiotensin II and ANP (Fig.

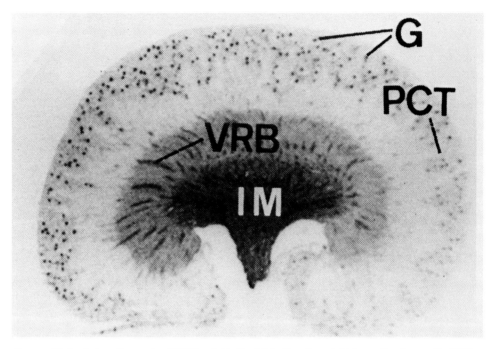

FIGURE 11.3. Autoradiograph of [^{125}I]endothelin-1 binding to rat kidney. G, glomeruli; PCT, proximal convoluted tubules; VRB, vasa recta bundles; IM, inner medulla. (From Kohzuki et al.[55])

11.3).[55,60] In particular, glomeruli, vasa recta bundles, and the inner medulla bind labeled endothelin with high affinity.[55] In addition, high-affinity binding sites have been identified on mesangial cells.[7,100] Scatchard analysis studies of endothelin binding to mesangial cell membranes were not uniformly linear, suggesting the possibility of multiple receptor subtypes.[7] Endothelin induces a brisk increase in cytosolic free calcium, suggesting that these binding sites represent functional cell surface receptors.[8,68,105]

Expression of a 2.3 kb preproendothelin-1 mRNA in cultured bovine glomerular endothelial cells has been shown by Northern analysis using a full length preproendothelin-1 cDNA.[70] Several agonists known to cause an increase in glomerular endothelial cell cytosolic free calcium, such as bradykinin, thrombin, ATP, and platelet-activating factor, upregulate expression of preproendothelin-1 mRNA in these cells.[69] Furthermore, glomerular endothelial cells release endothelin under basal conditions and in response to stimulation by bradykinin (100 nM).[69] Other renal epithelial cell lines (MDCK, LLCPK$_1$) have also been shown to release endothelin-1 in a time-dependent fashion.[59,103]

Thus, in addition to possessing potential receptors, the glomerular microvasculature is capable of producing endothelin-1. Furthermore, this production of endothelin is responsive to several vasoactive mediators, raising the possibility of a local role for endothelin in the modulation of renal hemodynamics.

Whole-Kidney Response

As indicated above, the renal vasculature is highly sensitive to the vasoconstrictive effects of infused endothelin.[7,14,28,41,49,66,74,88,122] Endothelin is 30 times more potent as a renal vasoconstrictor than is angiotensin II.[14] Although several investigators have noted a renal vasodilatory response, the magnitude and duration of this effect is less than the systemic vasodilatory response.[15,63,82,111,119] In the isolated rabbit kidney, endothelin-induced renal vasoconstriction is only slightly enhanced by coinfusion with methylene blue or oxyhemoglobin, suggesting minimal stimulation of EDRF.[14]

Similar to the systemic response, bolus intravenous injection of endothelin-1 induces a dose-dependent, long lasting fall in renal plasma flow rate (RPF) in anesthetized rats (Fig. 11–4).[49] Effects on glomerular filtration rate (GFR) are dose dependent as well; 75 pmol leads to a modest (23%) fall in RPF with no change in GFR and thus a rise in filtration fraction (Fig. 11.4),[51] whereas higher doses result in parallel decreases in RPF and GFR. Others have noted a parallel reduction of GFR and RPF when endothelin was infused into the renal artery of rats or after intravenous infusion in dogs.[41,49] Similar

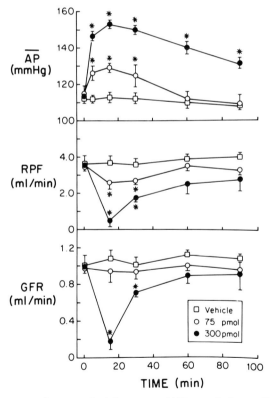

FIGURE 11.4. Time-course of mean arterial pressure (AP), renal plasma flow rate (RPF), and glomerular filtration rate (GFR) following a single bolus of endothelin. Note that the maximal effect occurs within 15–20 min and that the duration of response is sustained. Values are means ± S.E.M. *$P < 0.05$ vs. baseline, N = 4–5 per dose. (From ref. King et al.[51])

parallel reductions of RPF and GFR are seen in rat and rabbit isolated perfused kidneys, indicating that the renal response is independent of the systemic effects.[14,28] Indeed, the endothelin-induced increase in perfusion pressure partially offsets the vasoconstrictor effects on RPF, as evidenced by a more marked reduction in RPF and GFR in kidneys not exposed to the rise in systemic pressure.[51]

Coinfusion of endothelin with nicardipine into rat renal arteries blunts the fall in RPF and GFR.[49] Similar attenuation of endothelin-induced renal vasoconstriction is observed in isolated perfused hydronephrotic and normal rat kidneys.[66] Thus the mechanism of renal vasoconstriction is, at least in part, due to the activation of renal vascular dihydropyridine-sensitive Ca^{2+} channels. In contrast, intrarenal arterial coinfusion of verapamil or manganese with endothelin in dogs fails to blunt the reduction in RPF or GFR.[15] The reason for this apparent discrepancy is unclear.

The renal and systemic hemodynamic effects of endothelin are largely overcome by the coinfusion of pharmacological doses of ANP.[41,49] As noted earlier, endothelin is a potent secretagogue of ANP in cultured rat atrial myocytes and in the perfused rat heart.[30,67] Although plasma ANP levels increase with infusion of endothelin, the extent to which ANP represents an important counterregulatory response in vivo is unknown.[35,49,74,82]

Endothelin induces a modest natriuresis when administered in pressor doses that do not severely compromise GFR[51] as well as with subpressor doses.[80] Indeed, ouabain-sensitive oxygen consumption by inner medullary collecting duct cells (IMCD) is reduced by endothelin-1 (50 pM half-maximal effect), an effect not seen in cortical proximal tubule cells.[125] A reduction in the initial rate of ouabain-sensitive ^{86}Rb uptake by endothelin suggests that endothelin decreases Na^+,K^+-ATPase activity. Similar effects on IMCD cells have been reported previously for PGE_2, raising the possibility that this endothelin response is mediated by prostanoids.[47] Indeed, IMCD cells release PGE_2 in response to endothelin,[125] and the cyclooxygenase inhibitor ibuprofen effectively blocks the endothelin-induced reduction in ^{86}Rb uptake in IMCD cells, suggesting that the inhibition of Na^+,K^+-ATPase is mediated by prostaglandins. Finally, pressor doses of endothelin in conscious dogs increases free water clearance without changes in plasma arginine vasopressin levels.[35] The mechanism of disruption of urinary concentrating ability by endothelin remains to be defined.

Autoradiographic labeling of renal parenchyma by [^{125}I]endothelin reveals intense staining of the inner medulla (Fig. 11.3).[35] In addition, as noted earlier, several renal epithelial cell lines synthesize and release endothelin.[59,103] These findings in conjunction with the tubule effects of endothelin raise the possibility that locally generated endothelin modulates renal medullary transport functions independent of its effects on glomerular hemodynamics.

Effects of Endothelin on the Glomerular Microcirculation

In addition to large-vessel vasoconstriction, endothelin induces profound vasoconstriction in the microcirculation of the hamster cheek pouch,[10,11] the piglet pial arterioles,[3] and in the glomerular afferent and efferent arterioles.[8,41,51,66,121] Studies of the glomerular microcirculation reveal striking effects of endothelin on the determinants of glomerular ultrafiltration. Infusion of mildly pressor doses of endothelin-1 into anesthetized rats (0.63 pmol/min), lead to higher values of transglomerular capillary hydraulic pressure (ΔP) when compared with vehicle-infused controls (Fig. 11.5).[51] This glomerular capillary hypertension was due to a proportionately greater increase in efferent (R_E) versus afferent (R_A) arteriolar resistances. This profiltration force is balanced by a markedly lower glomerular capillary ultrafiltration coefficient (K_f) and a numerically lower rate of glomerular capillary plasma flow (Q_A).[51] Because of the offsetting nature of these effects, the single nephron GFR is kept relatively constant under these conditions of low-dose endothelin infusion.[51] In a separate study, higher doses of endothelin-1 (10 pmol/min) also lead to a proportionately greater increase in efferent versus afferent arteriolar tone.[8] This dose caused more intense renal vasoconstriction, leading to a fall in Q_A that, in conjunction with a marked reduction in K_f, led to a fall in single nephron GFR.[8] In keeping with these findings, studies of isolated perfused glomerular arterioles demonstrate half-maximal concentrations of endothelin needed to induce contraction of efferent arterioles ($3.2 \pm 4.0 \times 10^{-12}M$) to be three orders of magnitude less than those for afferent arterioles ($1.4 \pm 2.5 \times 10^{-9}M$).[124] This vasoconstriction is unaffected by coinfusion of saralasin, a competitive antagonist of angiotensin II.[124] In contrast, infusion of 0.16 pmol/min of endothelin-1 into a branch of the renal artery led mainly to afferent arteriolar vasoconstriction, a fall in glomerular capillary pressure (P_{GC}) and no significant change in K_f.[58] A predominance of afferent constriction was also noted in the rat isolated perfused hydronephrotic kidney.[66] The different glomerular effects of endothelin-1 noted in these studies is likely due to differences in the dosage and route of delivery of the peptide. Overall, these studies indicate that endothelin-1 causes vasoconstriction of the pre- and postglomerular arterioles, reductions in RPF and K_f, and variable effects on GFR. These findings suggest that the glomerular tuft is highly sensitive to endothelin-1 and thus is a potential target tissue for endothelin. The doses used in these studies are likely to represent pharmacological amounts. However, the ability of glomerular capillary endothelial cells to produce endothelin raises the possibility that locally generated endothelin may have important local effects on adjacent renal vascular and epithelial cells.

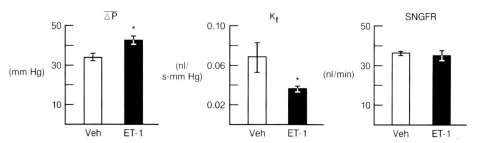

FIGURE 11.5. Effects of low-dose endothelin-1 infusion (0.63 pmol/min) on determinants of glomerular ultrafiltration. $\overline{\Delta P}$, mean transglomerular capillary hydraulic pressure gradient; K_f, glomerular capillary ultrafiltration coefficient; SNGFR, single nephron glomerular filtration rate. Values are means $\pm$ S.E.M. *$P < 0.05$ vs. vehicle. (Adapted from King et al.[51])

Effects of Endothelin on Mesangial Cells

Whether the fall in K_f after endothelin infusion is due to a decrease in the hydraulic permeability or to a decrease in the capillary surface area of glomerular capillaries is unknown. However, several vasoconstrictors, such as angiotensin II, norepinephrine, and arginine vasopressin have been shown to trigger a contractile response in cultured mesangial cells, suggesting that a fall in the surface area may account for the changes in K_f.[4] Indeed, endothelin-1 induces a fall in mesangial cell planar surface area at subnanomolar concentrations, which is identical to that of equimolar angiotensin II (Fig. 11.6).[22] Several other laboratories have confirmed this mesangial contractile effect of endothelin, noting rearrangements of F-actin filaments consistent with a motile response.[8,126] This contractile response may be attenuated by endothelin-induced mesangial cell prostanoid production. Endothelin-stimulated phospholipase A_2 activity leads to increased tritiated arachidonate release from mesangial cells and facilitates release of $PGE_2 > PGI_2 > TXA_2$.[104,126] Overall, these findings suggest that reduction in filtration surface area may be central to the fall in K_f.

Mesangial cells respond to low-dose endothelin-1 (10 pM) infusion with a slow rise in cytosolic free calcium ([Ca^{2+}]$_i$), which is blocked by EGTA but not by nifedipine (Fig. 11–7).[105] Higher doses lead to a rapid increase in [Ca^{2+}]$_i$,

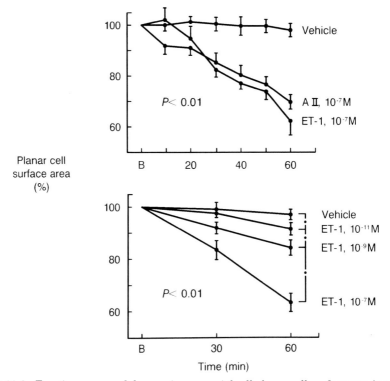

FIGURE 11.6. *Top:* time-course of changes in mesangial cell planar cell surface area in response to endothelin-1 (ET-1), angiotensin II (A II), or vehicle. *Bottom:* dose–response studies of endothelin-1–induced mesangial contraction. (Adapted from Culebras et al.[22])

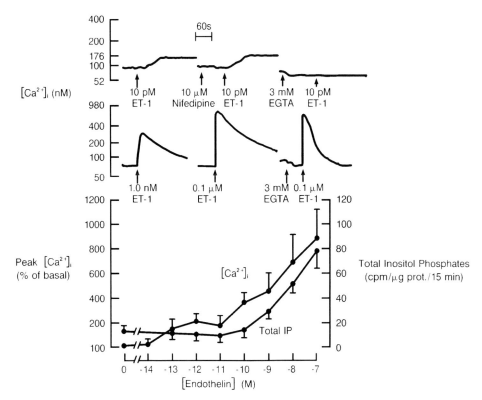

FIGURE 11.7. Effects of endothelin-1 on cytosolic free calcium and inositol phospholipid turnover in mesangial cells. *Top:* slow but sustained increase in $[Ca^{2+}]_i$ at 10 pM endothelin-1 (ET-1), and effect of 10 μM nifedipine and 3 mM EGTA pretreatment in $[Ca^{2+}]_i$ observed at ≥ 0.1 nM endothelin ($[Ca^{2+}]_i$ returned to baseline after 10–12 min), and effect of 3.0 mM EGTA pretreatment at 0.1 μM endothelin ($[Ca^{2+}]_i$ returned to baseline after 3.5–4.0 min). *Bottom:* dose dependence of peak $[Ca^{2+}]_i$ and turnover of total inositol phosphates. Data are mean ± S.E.M. (From Simonson et al.[105])

followed by a slow return toward baseline. EGTA has no effect on the early rise in $[Ca^{2+}]_i$, although this chelator abrogated the later rise in $[Ca^{2+}]_i$. These results suggest that, as with vascular smooth muscle cells, the early rapid rise in $[Ca^{2+}]_i$ is associated with mobilization of intracellular stores, whereas influx of extracellular calcium contributes to the delayed phase.[8,105] The rise in $[Ca^{2+}]_i$ is paralleled by an increase in inositol phospholipid turnover, indicating activation of phospholipase C (Fig. 11.7). Finally, endothelin is a potent comitogen in quiescent mesangial cells, an effect not seen with Swiss 3T3 fibroblasts.[8,105] This response may relate to endothelin-induced activation of Na^+–H^+ exchange.[105] In addition, endothelin significantly stimulates expression of the protooncogenes c-*fos* in mesangial cells and c-*myc*/c-*fos* in vascular smooth muscle cells.[56,105]

Thus the mesangial cell, with its high-affinity endothelin-binding sites, may represent an important target for either locally generated or circulating endothelin. Furthermore, by eliciting a contractile effect on the mesangial cells, endothelin may exert profound effects on glomerular hemodynamics. Taken together, the ability of endothelin to increase glomerular capillary hydraulic pressure and to induce mitogenesis in mesangial cells may indicate a role for endothelin in the pathogenesis of progressive renal disease.

POTENTIAL ROLE OF ENDOTHELIN IN VOLUME HOMEOSTASIS

Pressor doses of endothelin lead to dose-dependent increases in hematocrit and plasma protein concentration, suggesting loss of plasma water into the interstitium.[40,49,50] A reduction in plasma volume and thus venous return may contribute to the fall in cardiac output seen with high doses of endothelin.[41,52] Leukotriene C_4, another potent endogenous vasoconstrictor, has been shown to increase vascular permeability in postcapillary venules, leading to a similar loss of plasma volume, an effect that contributes to a significant reduction in cardiac output.[5,23] Whether endothelin-induced hemoconcentration reflects a direct effect on vascular permeability or is mediated by peptide-induced capillary hypertension is unclear. A reduction of plasma volume may also be mediated by the aforementioned natriuresis induced by nonpressor and pressor doses of endothelin.[51,80] In addition to hemodynamic and direct renal Na^+ transport effects, endothelin-induced release of ANP may contribute to this natriuresis. As discussed above, endothelin is a potent stimulus for secretion of basal and stretch-induced ANP release.[30,67] However, in preliminary studies, coinfusion of endothelin and anti-ANP antibody does not prevent the natriuresis.[80]

Endothelin may also significantly modulate the renin–angiotensin–aldosterone system.[85,108,109] The peptide inhibits release of renin from dispersed juxtaglomerular cells in a superfusion system at concentrations as low as $10^{-11}M$, an effect that is largely, though not exclusively, dependent on the presence of extracellular calcium.[108,109] Inhibition of basal and isoproterenol-stimulated renin release has also been demonstrated in isolated rat glomeruli (IC_{50}, $1.0 \times 10^{-9}M$) and in rat kidney cortical slices.[71,89] Indeed, subpressor doses of endothelin tend to decrease plasma renin activity (PRA) ($P < 0.1$) and produce a significant fall in plasma aldosterone levels in anesthetized dogs,[85] whereas PRA and plasma aldosterone levels rose significantly in rats and dogs given pressor doses of endothelin.[35,74,82,85] This stimulation of PRA may be secondary to a reduction of sodium delivery to the macula densa because of a fall in GFR. Thus endothelin may play a role in local renal autoregulatory responses and in the release of systemic angiotensin II.

Adrenal zona glomerulosa cells possess more binding sites specific for endothelin-1 than for endothelin-3.[20] In addition, both endothelin-1 and, to a lesser extent, endothelin-3 induce dose-dependent secretion of aldosterone (10^{-8}–$10^{-10}M$) from adrenal glomerulosa cells, though not as potently as angiotensin II.[21] Finally, endothelin interferes with peripheral nerve norepinephrine release.[111] Plasma levels of both norepinephrine and epinephrine are reduced by pressor doses of endothelin, although this is likely due to activation of compensatory baroreflexes.[54]

Taken together, endothelin interacts with all of the major volume regulatory systems. At this time, the extent to which endothelin modulates extracellular fluid volume homeostasis in vivo is not yet clear. Studies of this nature

POTENTIAL ROLE OF ENDOTHELIN IN DISORDERS OF RENAL VASOCONSTRICTION

Acute and chronic cyclosporine (CyA) therapy in humans and animals induces marked renal vasoconstriction with a reduction of GFR.[81] This nephrotoxicity presents a major problem in renal, heart, and other organ transplantation. Several preliminary studies have implicated endothelin in this renal vasoconstrictor response.[25,57] CyA-induced vasoconstriction in isolated perfused rat kidneys was largely overcome by coinfusion of antiendothelin antibody, but not by nonimmune serum.[25] In vivo studies of intravenous CyA (20 mg/kg) in rats, with simultaneous intrarenal artery antiendothelin antibody, corroborate these in vitro findings.[57] Rats treated with CyA had plasma levels of endothelin (58 ± 17 pg/ml) markedly higher than normals (<2 pg/ml),[57] further suggesting a role of endothelin in acute CyA-induced renal vasoconstriction. Whether endothelin is an important mediator in chronic cyclosporine toxicity awaits further investigation.

Acute ischemic renal failure in the rat is also characterized by renal vasoconstriction and reduced GFR, the duration of which depends on the duration of ischemia.[29] Coinfusion of antiendothelin antibody into a branch of a rat renal artery significantly ameliorated renal vasoconstriction induced by 25 min of ischemia performed 48 h prior to study.[58] Whether this reflects an increase in local production of endothelin and/or a change in receptor number or binding is as yet unclear. However, endothelin-binding-site density of rat cardiac membranes is markedly increased by global ischemia (20–90 min). Small changes in affinity occurred with ischemia of >30 min duration.[65] Increased binding-site density was further enhanced by reperfusion and was prevented by hypothermic perfusion. Whether similar changes take place in the kidney is not yet known. Taken together, the effects of ischemia on endothelin receptor density and the stimulation of endothelin production by cyclosporine may have important implications in the pathogenesis of early renal allograft dysfunction.

Acute infusion of endotoxin in rats is well known to induce renal vasoconstriction.[6] Rats and sheep infused with endotoxin have marked increases in plasma endothelin concentrations that may contribute to this vasoconstrictions.[78,106] Plasma levels of endothelin are elevated in patients with acute renal failure from a variety of causes and fall with subsequent improvement of renal function. Potentially, intervention in endothelin production and/or binding may be beneficial in the treatment of acute renal failure.

POTENTIAL ROLE OF ENDOTHELIN IN SYSTEMIC HYPERTENSION

Endothelial cells possess several characteristics that make them suitable for the transduction of local changes in pressure and flow into appropriate adjustments of vascular smooth muscle tone. First, they form a continuous monolayer of cells lining the entire vasculature. In addition, they possess specific receptors for a variety of circulating vasoactive mediators that require a functional response from the endothelium in order to elicit an appropriate change in vascular smooth muscle tone.[31,114]

The endothelial cell membrane responds to mechanical stimuli, such as membrane stretch or shear stress, by activating ion channels.[61,84] The effector branch of the endothelial cell response to these mechanical as well as neurohumoral signals is mediated by (1) direct cell–cell contact through gap junctions and (2) the ability to produce soluble vasodilating and vasoconstricting factors.[24,95–97] Disruption of either the sensor or effector capabilities of the endothelial cell may lead to systemic hypertension.[112] The role of endothelin in systemic hypertension is unclear at present. However, the extreme potency and sustained vasoconstrictor responses associated with endothelin make this peptide a potentially important mediator of hypertension. Mesenteric and renal artery rings from 12-week-old spontaneous hypertensive rats are more sensitive to the vasoconstrictor effects of endothelin than are age-matched Wistar-Kyoto rats.[76,110] However, the systemic pressor response to intravenous endothelin was not different between the two groups.[76] Compared with normals, isolated aortic rings from Goldblatt rats and deoxycorticosterone acetate (DOCA)–salt hypertensive rats did not differ in median effective concentrations or maximal responses to endothelin-1.[16] However, in the mesenteric microcirculation, the magnitude of response to endothelin was greater for the two hypertensive groups than in normals.[16] Thus intact vessels from several models of systemic hypertension differ from normals in their response to endothelin. Whether these differences are pertinent to the pathophysiology of systemic hypertension remains to be determined.

Endothelin production has been shown to be sensitive to shear stress and to a variety of endogenous substances.[121,123] Whether acute or chronic damage to endothelial cells alters endothelin production is unknown. In addition to its vasoconstrictive effects, endothelin has potent mitogenic effects that may be involved in vascular wall remodeling associated with chronic hypertension.[8,56,105] Untreated patients with essential hypertension have increased plasma levels of endothelin-1-like immunoreactivity compared with age-matched controls.[100] However, the physiological significance of circulating endothelin in normals or in hypertensives is not yet known. Potentially, pharmacological maneuvers directed at inhibiting the conversion of "big" endothelin to the active 21 amino acid endothelin-1 moiety, or antagonists of the endothelin receptor, may define a population of endothelin-dependent hypertensives. Further investigation into the biology of endothelin holds great promise toward understanding the pathogenesis of vasospastic disorders.

CONCLUSIONS

Endothelin is distinguished from other endogenous vasoactive peptides by its biphasic pressure response and its sustained duration of action. There is marked regional variability in the sensitivity to either the vasodilatory or the vasoconstrictor effects of endothelin. The renal vascular bed is exquisitely sensitive to the vasoactive and mitogenic effects of the peptide and may be an important target organ for endothelin. The extent to which endothelin functions as an autocrine, paracrine, or circulating mediator is as yet unclear. Evidence suggests that endothelin may play a role as a mediator of autoregulatory responses and in chronic volume and blood pressure regulation. Eventual development of effective blockers of endothelin action will aid in understanding the in vivo effects of endothelin.

ACKNOWLEDGMENTS

A. J. K. is the recipient of Individual National Research Service Award (1F32DK08003) from the National Institute of Health. Studies were supported in part by NIH grant DK35930.

REFERENCES

1. AISAKA, K., S. S. GROSS, O. W. GRIFFITH, and R. LEVI. N^ω-methylarginine, an inhibitor of endothelium-derived nitric oxide synthesis, is a potent pressor agent in the guinea pig: does nitric oxide regulate blood pressure in vivo? *Biochem. Biophys. Res. Commun.* 160: 881–886, 1989.
2. ANGGARD, E., S. GALTON, G. RAE, R. THOMAS, L. MCLOUGHLIN, G. DE NUCCI, and J. R. VANE. The fate of radioiodinated endothelin-1 and endothelin-3 in the rat. *J. Cardiovasc. Pharmacol.* 6(Suppl. 4): S46–S49, 1989.
3. ARMSTEAD, W. M., R. MIRRO, C. W. LEFFLER, and D. W. BUSIJA. Influence of endothelin on piglet cerebral microcirculation. *Am. J. Physiol.* 257 (*Heart Circ. Physiol.* 26): H707–H710, 1989.
4. AUSIELLO, D. A., J. L. KREISBERG, C. ROY, and J. M. KARNOVSKY. Contraction of cultured rat glomerular cells of apparent mesangial origin after stimulation with angiotensin and arginine-vasopressin. *J. Clin. Invest.* 65: 754–760, 1980.
5. BADR, K. F., C. BAYLIS, J. M. PFEFFER, M. A. PFEFFER, R. J. SOBERMAN, R. A. LEWIS, K. F. AUSTEN, E. J. COREY, and B. M. BRENNER. Renal and systemic hemodynamic response to intravenous infusion of leukotriene C_4 in the rat. *Circ. Res.* 54: 492–499, 1984.
6. BADR, K. F., V. E. KELLEY, H. G. RENNKE, and B. M. BRENNER. Roles for thromboxane A_2 and leukotrienes in endotoxin-induced acute renal failure. *Kidney Int.* 30: 474–480, 1986.
7. BADR, K. F., K. A. MUNGER, M. SUGIURA, R. M. SNAJDAR, M. SCHWARTZBERG, and T. INAGAMI. High and low affinity binding sites for endothelin on cultured rat glomerular mesangial cells. *Biochem. Biophys. Res. Commun.* 161: 776–781, 1989.
8. BADR, K. F., J. J. MURRAY, M. D. BREYER, K. TAKAHASHI, T. INAGAMI, and R. C. HARRIS. Mesangial cell, glomerular and renal vascular responses to endothelin in the rat kidney. *J. Clin. Invest.* 83: 336–342, 1989.
9. BAYLISS, W. M. On the local reaction of the arterial wall to changes in internal pressure. *J. Physiol. (Lond.)* 28: 220–231, 1902.
10. BRAIN, S. D., D. C. CROSSMAN, T. L. BUCKLEY, and T. J. WILLIAMS. Endothelin-1: demonstration of potent effects on the microcirculation of humans and other species. *J. Cardiovasc. Pharmacol.* 6(Suppl. 4): S147–S149, 1989.
11. BRAIN, S. D., J. R. TIPPINS, and T. J. WILLIAMS. Endothelin induces potent microvascular constriction. *Br. J. Pharmacol.* 95: 1005–1007, 1988.
12. BRENNER, B. M., L. D. DWORKIN, and I. ICHIKAWA. Glomerular ultrafiltration. In: *The Kidney*, edited by B. M. Brenner and F. C. Rector. Philadelphia: W. B. Saunders, 1986, p. 124–144.
13. BRENNER, B. M., J. L. TROY, and B. J. BALLERMANN. Endothelium-dependent vascular responses. *J. Clin. Invest.* 84: 1373–1378, 1989.
14. CAIRNS, H. S., M. E. ROGERSON, L. D. FAIRBANKS, G. H. NEILD, and J. WESTWICK. Endothelin induces an increase in renal vascular resistance and a fall in glomerular filtration rate in the rabbit isolated perfused kidney. *Br. J. Pharmacol.* 98: 155–160, 1989.
15. CAO, L., and R. O. BANKS. Cardiovascular and renal actions of endothelin: effects of calcium-channel blockers. *Am. J. Physiol.* 258 (*Renal Fluid Electrolyte Physiol.* 27): F254–F258, 1990.
16. CATELLI DE CARVALHO, M. H., D. NIGRO, R. SCIVOLETTO, H. V. BARBEIRO, M. APARDECIA DE

OLIVEIRA, G. DE NUCCI, and Z. B. FORTES. Comparison of the effect of endothelin on microvessels and macrovessels in Goldblatt II and deoxycorticosterone acetate–salt hypertensive rats. *Hypertension* 15(Suppl. I): I-68–I-71, 1989.
17. CLARKE, J. G., N. BENJAMIN, S. W. LARKIN, D. J. WEBB, G. J. DAVIES, and A. MASERI. Endothelin is a potent long-lasting vasoconstrictor in men. *Am. J. Physiol.* 257 (*Heart Circ. Physiol.* 26): H2033–H2035, 1989.
18. CLOZEL, M., and J. P. CLOZEL. Effects of endothelin on regional blood flows in squirrel monkeys. *J. Pharmacol. Exp. Ther.* 250: 1125–1131, 1989.
19. CLOZEL, M., W. FISCHLI, and C. GUILLY. Specific binding of endothelin on human vascular smooth muscle cells in culture. *J. Clin. Invest.* 83: 1758–1761, 1990.
20. COOK, J. P., and V. J. DZAU. Possible role of endothelin in vascular diseases: an alternative hypothesis. *J. Vasc. Med. Biol.* 1: 316–318, 1989.
21. COZZA, E. N., C. E. GOMEZ-SANCHEZ, M. F. FOECKING, and S. CHIOU. Endothelin binding to cultured calf adrenal zona glomerulosa cells and stimulation of aldosterone secretion. *J. Clin. Invest.* 84: 1032–1035, 1989.
22. CULEBRAS, M., I. MONTANES, A. LOPEZ-FARRE, I. MILLAS, and J. M. LOPEZ-NOVOA. Effect of endothelin on renal function and on the contraction of cultured rat mesangial cells. *Med. Sci. Res.* 17: 245–246, 1989.
23. DAHLEN, S. E., P. BJORK, P. HEDQVIST, K. E. ARFORS, S. HAMMARSTROM, J. A. LINDGREN, and B. SAMUELSON. Leukotrienes promote plasma leakage and leukocyte adhesion in postcapillary venules: in vivo effects with relevance to the acute inflammatory response. *Proc. Natl. Acad. Sci. USA* 78: 3887–3891, 1981.
24. DAVIES, P. F., S. P. OLESON, D. E. CLAPHAM, E. M. MORREL, and J. SCHOEN. Endothelial communication: state of the art lecture. *Hypertension* 11: 563–572, 1988.
25. DEDAN, J., N. PERICO, and G. REMUZZI. Role of endothelin in cyclosporine-induced renal vasoconstriction, abstracted. *Kidney Int.* 37: 479, 1990.
26. DE NUCCI, G. D., R. THOMAS, P. D'ORLEANS-JUSTE, E. ANTUNES, C. WALDER, T. D. WARNER, and J. R. VANE. Pressor effects of circulating endothelin are limited by its removal in the pulmonary circulation and by the release of prostacyclin and endothelium-derived relaxing factor. *Proc. Natl. Acad. Sci. U.S.A.* 85: 9797–9800, 1988.
27. DWORKIN, L. D., I. ICHIKAWA, and B. M. BRENNER. Hormonal modulation of glomerular function. *Am. J. Physiol.* 244 (*Renal Fluid Electrolyte Physiol.* 13): F95–F104, 1983.
28. FIRTH, J. D., P. J. RATCLIFFE, A. E. G. RAINE, and J. G. G. LEDINGHAM. Endothelin: an important factor in acute renal failure. *Lancet* 2: 1179–1181, 1988.
29. FREGA, N. S., D. R. DiBONA, B. GUERTLER, and A. LEAF. Ischemic renal injury. *Kidney Int.* 10(Suppl.): S17–S25, 1976.
30. FUKADA, Y., Y. HIRATA, H. YOSHIMI, T. KOJIMA, Y. KOBAYASHI, M. YANAGISAWA, and T. MASAKI. Endothelin is a potent secretagogue for atrial natriuretic peptide in cultured rat atrial myocytes. *Biochem. Biophys. Res. Commun.* 155: 167–172, 1988.
31. FURCHGOTT, R. F. The role of endothelium in the responses of vascular smooth muscle to drugs. *Annu. Rev. Pharmacol. Toxicol.* 24: 175–197, 1984.
32. FURCHGOTT, R. F., and J. V. ZAWADZKI. The obligatory role of endothelial cells in the relaxation of arterial smooth muscle by acetylcholine. *Nature* 288: 373–376, 1980.
33. GARDINER, S. M., A. M. COMPTON, and T. BENNETT. Regional hemodynamic effects of endothelin-1 in conscious, unrestrained, Wistar rats. *J. Cardiovasc. Pharmacol.* 6(Suppl. 4): S202–S204, 1989.
34. GIVEN, M. B., R. F. LOWE, H. LIPPTON, A. L. HYMAN, G. E. SANDER, and T. D. GILES. Hemodynamic actions of endothelin in conscious and anesthetized dogs. *Peptides* 10:41–44, 1989.
35. GOETZ, K. L., B. C. WANG, J. B. MADWED, J. L. ZHU, and R. J. LEADLEY. Cardiovascular, renal, and endocrine responses to intravenous endothelin in conscious dogs. *Am. J. Physiol.* 255 (*Regulatory Integrative Comp. Physiol.* 24): R1064–R1068, 1988.
36. GOTO, K., Y. KASUYA, N. MATSUKI, Y. TAKUWA, H. KURIHARA, T. ISHIKAWA, S. KIMURA, M. YANAGISAWA, and T. MASAKI. Endothelin activates the dihydropyridine-sensitive, voltage-dependent Ca^{++} channel in vascular smooth muscle. *Proc. Natl. Acad. Sci. USA* 86: 3915–3918, 1989.
37. HAN, S. P., A. J. TRAPANI, K. F. FOK, T. C. WESTFALL, and M. M. KNEUPFER. Effects of endothelin on regional hemodynamics in conscious rats. *Eur. J. Pharmacol.* 159: 303–305, 1989.
38. HERMAN, F., K. MAGYAR, P. E. CHABRIER, P. BRAQUET, and J. FILEP. Prostacyclin mediates antiaggregatory and hypotensive actions of endothelin in anesthetized beagle dogs. *Br. J. Pharmacol.* 98: 38–40, 1989.
39. HICKEY, K. A., G. RUBANYI, R. J. PAUL, and R. F. HIGHSMITH. Characterization of a coronary vasoconstrictor produced by cultured endothelial cells. *Am. J. Physiol.* 248 (*Cell Physiol.* 17): C550–C556, 1985.

40. HINOJOSA-LABORDE, C., J. W. OSBORN, and A. W. COWLEY. Hemodynamic effects of endothelin in conscious rats. *Am. J. Physiol.* 256 (*Heart Circ. Physiol.* 25): H1742–H1746, 1989.
41. HIRATA, Y., H. MATSUOKA, K. KIMURA, K. FUKUI, H. HAYAKAWA, E. SUZUKI, T. SUGIMOTO, T. SUGIMOTO, M. YANAGISAWA, and T. MASAKI. Renal vasoconstriction by the endothelial cell–derived peptide endothelin in spontaneously hypertensive rats. *Circ. Res.* 65: 1370–1379, 1989.
42. HOFFMAN, A., E. GROSSMAN, K. P. OHMAN, E. MARKS, and H. R. KEISER. Endothelin induces an initial increase in cardiac output associated with selective in rats. *Life Sci.* 45: 249–255, 1989.
43. HOM, G. J., B. TOUHEY, and G. M. RUBANYI. Potential mechanisms of endothelin-induced transient decreases in arterial pressure in spontaneously hypertensive rats. In: *Endothelium-Derived Contracting Factors,* edited by G. M. Rubanyi and P. M. Vanhoutte. Basel: Varger, 1990, p. 98–103.
44. HU, J. R., R. VON HARSDORF, and R. E. LANG. Endothelin has potent inotropic effects in rat atria. *Eur. J. Pharmacol.* 158: 275–278, 1988.
45. INOUE, A., M. YANAGISAWA, S. KIMURA, Y. KASUYA, T. MIYAUCHI, K. GOTO, and T. MASAKI. The human endothelin family: three structurally and pharmacologically distinct isopeptides predicted by three separate genes. *Proc. Natl. Acad. Sci. USA* 86: 2863–2867, 1989.
46. ISHIKAWA, T., M. YANAGISAWA, S. KIMURA, K. GOTO, and T. MASAKI. Positive inotropic action of novel vasoconstrictor peptide endothelin on guinea pig atria. *Am. J. Physiol.* 255 (*Heart Circ. Physiol.* 24): H970–H973, 1988.
47. JABS, K., M. L. ZEIDEL, and P. SILVA. Prostaglandin E_2 inhibits Na^+/K^+ ATPase activity in the inner medullary collecting duct. *Am. J. Physiol.* 257 (*Renal Fluid Electrolyte Physiol.* 26): F424–F430, 1989.
48. KASHIWABARA, T., Y. INAGAKI, H. OHTA, A. IWAMATSU, M. NOMIZU, A. MORITA, and K. NISHIKORI. Putative precursors of endothelin have less vasoconstrictor activity in vitro but a potent effect in vivo. *FEBS Lett.* 247: 73–76, 1989.
49. KATOH, T., H. CHANG, S. UCHIDA, T. OKUDA, and K. KURAKAWA. Direct effects of endothelin in the rat kidney. *Am. J. Physiol.* 258 (*Renal Fluid Electrolyte Physiol.* 27): F397–F402, 1990.
50. KATUSIC, Z. S., J. T. SHEPHERD, and P. M. VANHOUTTE. Endothelium-dependent contraction to stretch in canine basilar arteries. *Am. J. Physiol.* 252 (*Heart Circ. Physiol.* 21): H671–H673, 1987.
51. KING, A. J., B. M. BRENNER, and S. ANDERSON. Endothelin: a potent renal and systemic vasoconstrictor peptide. *Am. J. Physiol.* 256 (*Renal Fluid Electrolyte Physiol.* 25): F1051–F1058, 1989.
52. KING, A. J., J. M. PFEFFER, M. A. PFEFFER, and B. M. BRENNER. Systemic hemodynamic effects of endothelin in the rat. *Am. J. Physiol.* 258 (*Heart Circ. Physiol.* 23): H787–H792, 1990.
53. KITAYOSHI, T., T. WATANABE, and N. SHIMAMOTO. Cardiovascular effects of endothelin in dogs: positive inotropic action in vivo. *Eur. J. Pharmacol.* 166: 519–522, 1989.
54. KNEUPFER, M. M., S. P. HAN, A. J. TRAPANI, K. F. FOK, and T. C. WESTFALL. Regional hemodynamic and baroreflex effects of endothelin in rats. *Am. J. Physiol.* 257 (*Heart Circ. Physiol.* 26): H918–H926, 1989.
55. KOHZUKI, M., C. I. JOHNSTON, S. Y. CHAI, D. J. CASLEY, and F. A. O. MENDELSOHN. Localization of endothelin receptors in rat kidney. *Eur. J. Pharmacol.* 160: 193–194, 1989.
56. KOMURO, I., H. KURIHARA, T. SUGIYAMA, M. YOSHIZUMI, F. TAKAKU, and Y. YAZAKI. Endothelin stimulates c-*fos* and c-*myc* expression and proliferation of vascular smooth muscle cells. *FEBS Lett.* 238: 249–252, 1988.
57. KON, V., M. SUGIURA, T. INAGAMI, R. L. HOOVER, A. FOGO, B. R. HARVIE, and I. ICHIKAWA. Cyclosporine causes endothelin-dependent renal failure, abstracted. *Kidney Int.* 37: 486, 1990.
58. KON, V., T. YOSHIOKA, A. FOGO, and I. ICHIKAWA. Glomerular actions of endothelin in vivo. *J. Clin. Invest.* 83: 1762–1767, 1989.
59. KOSAKA, T., N. SUZUKI, H. MATSUMOTO, Y. ITOH, T. YASUHARA, H. ONDA, and M. FUJINO. Synthesis of the vasoconstrictor peptide endothelin in kidney cells. *FEBS Lett.* 249:42–46, 1989.
60. KOSEKI, C., M. IMAI, Y. HIRATA, M. YANAGISAWA, and T. MASAKI. Autoradiographic distribution in rat tissues of binding sites for endothelin: a neuropeptide? *Am. J. Physiol.* 256 (*Regulatory Integrative Comp. Physiol.* 25): R858–R866, 1989.
61. LANSMAN, J. B., T. J. HALLAM, and T. J. RINK. Single stretch-activated ion channels in vascular endothelial cells as mechanotransducers. *Nature* 325: 811–813, 1987.
62. LAPPE, R. W., K. W. BARRON, J. E. FAVER, and M. J. BRODY. Selective antagonism of hu-

moral versus neural vasoconstrictor responses by nisoldipine. *Hypertension* 7: 216–222, 1985.
63. LIPPTON, H., J. GOFF, and A. HYMAN. Effects of endothelin in the systemic and renal vascular beds in vivo. *Eur. J. Pharmacol.* 155: 197–199, 1988.
64. LIPPTON, H. L., T. A. HAUTH, W. R. SUMMER, and A. L. HYMAN. Endothelin produces pulmonary vasoconstriction and systemic vasodilation. *J. Appl. Physiol.* 66(2): 1008–1012, 1989.
65. LIU, J., R. CHEN, D. J. CASLEY, and W. G. NAYLER. Ischemia and reperfusion increase ^{125}I-labeled endothelin-1 binding in rat cardiac membranes. *Am. J. Physiol.* 258 (*Heart Circ. Physiol.* 27): H829–H835, 1990.
66. LOUTZENHISER, R., M. EPSTEIN, K. HAYASHI, and C. HORTON. Direct visualization of effects of endothelin on the renal microvasculature. *Am. J. Physiol.* 258 (*Renal Fluid Electrolyte Physiol.* 27): F61–F68, 1990.
67. MANTYMAA, P., J. LEPPALUOTO, and H. RUSKOAHO. Endothelin stimulates basal and stretch-induced atrial natriuretic peptide secretion from the perfused rat heart. *Endocrinology* 126: 587–595, 1990.
68. MARSDEN, P. A., N. R. DANTHULURI, B. M. BRENNER, B. J. BALLERMANN, and T. A. BROCK. Endothelin action on vascular smooth muscle involves inositol trisphosphate and calcium mobilization. *Biochem. Biophys. Res. Commun.* 158: 86–93, 1989.
69. MARSDEN, P. A., D. M. DORFMAN, B. M. BRENNER, S. H. ORKIN, and B. J. BALLERMANN. Regulation of endothelin gene expression in glomerular endothelial cells in culture, abstracted. *Kidney Int.* 37: 373, 1990.
70. MARSDEN, P. A., E. R. MARTIN, D. M. DORFMAN, T. A. BROCK, B. M. BRENNER, T. COLLINS, and B. J. BALLERMANN. Endothelin: gene expression, release and action in cultured cells of the renal glomerulus, abstracted. *Am. J. Hypertens.* 2: 49A, 1989.
71. MARTIN, E. R., P. A. MARSDEN, B. M. BRENNER, and B. J. BALLERMANN. Identification and characterization of endothelin binding sites in rat renal papillary and glomerular membranes. *Biochem. Biophys. Res. Commun.* 162: 130–137, 1989.
72. MATSUMOTO, H., N. SUZUKI, H. ONDA, and M. FUJINO. Abundance of endothelin-3 in rat intestine, pituitary gland and brain. *Biochem. Biophys. Res. Commun.* 164: 74–80, 1989.
73. MATSUMURA, Y., K. HISAKI, T. OKYAMA, K. HAYASHI, and S. MORIMOTO. Effects of endothelin on renal function and renin secretion in anesthetized rats. *Eur. J. Pharmacol.* 166: 577–580, 1989.
74. MILLER, W. L., M. M. REDFIELD, and J. C. BURNETT. Integrated cardiac, renal and endocrine actions of endothelin. *J. Clin. Invest.* 83: 317–320, 1989.
75. MINKES, R. K., L. A. MACMILLAN, J. A. BELLAN, M. D. KERSTEIN, D. B. MCNAMARA, and P. J. KADOWITZ. Analysis of regional responses to endothelin in hindquarter vascular bed of cats. *Am. J. Physiol.* 256 (*Heart Circ. Physiol.* 25): H598–H602, 1989.
76. MIYAUCHI, T., T. ISHIKAWA, Y. TOMOBE, M. YANAGISAWA, S. KIMURA, Y. SUGISHITA, I. ITO, K. GOTO, and T. MASAKI. Characteristics of pressor response to endothelin in spontaneously hypertensive and Wistar-Kyoto rats. *Hypertension* 14: 427–434, 1989.
77. MORAVEC, S. S., E. E. REYNOLDS, R. W. STEWART, and M. BOND. Endothelium is a positive inotropic agent in human and rat heart in vitro. *Biochem. Biophys. Res. Commun.* 159: 14–18, 1989.
78. MOREL, D. R., J. S. LACROIX, A. HEMSEN, D. A. STEINIG, J. F. PITTET, and J. M. LUNDBERG. Increased plasma and pulmonary lymph levels of endothelin during endotoxin shock. *Eur. J. Pharmacol.* 167: 427–428, 1989.
79. MORTENSEN, L. H., and G. D. FINK. Hemodynamic effect of human and rat endothelin administration into conscious rats. *Am. J. Physiol.* 258 (*Heart Circ. Physiol.* 27): H362–H368, 1990.
80. MUNGER, K. A., M. SUGIURA, T. INAGAMI, K. TAKAKASHI, and K. F. BADR. Mechanisms of endothelin-induced natriuresis in the rat, abstracted. *Clin. Res.* 37: 497A, 1989.
81. MYERS, B. Cyclosporine nephrotoxicity. *Kidney Int.* 30: 964–974, 1986.
82. NAKAMOTO, H., H. SUZUKI, M. MURAKAMI, Y. KAGEYAMA, A. OHISHI, K. FUKUDA, S. HORI, and T. SARUTA. Effects of endothelin on systemic and renal haemodynamics and neuroendocrine hormones in conscious dogs. *Clin. Sci.* 77: 567–572, 1989.
83. O'BRIEN, R. F., R. J. ROBBINS, and I. F. MCMURTRY. Endothelial cells in culture produce a vasoconstrictor substance. *J. Cell. Physiol.* 132: 263–270, 1987.
84. OLESON, S. P., D. E. CLAPHAM, and P. F. DAVIES. Haemodynamic shear stress activates a K^+ current in vascular endothelial cells. *Nature* 331: 168–170, 1988.
85. OTSUKA, A., H. MIKAMI, K. KATAHIRA, T. TSUNETOSHI, K. MINAMITANI, and T. OGIHARA. Changes in plasma renin activity and aldosterone concentration in response to endothelin injection in dogs. *Acta Endocrinol.* 121: 361–364, 1989.

86. OUCHI, Y., S. KIM, A. C. SOUZA, S. IIJIMA, A. HATTORI, H. ORIMO, M. YOSHIZUMI, H. KURIHARA, and Y. YAZAKI. Central effect of endothelin on blood pressure in conscious rats. *Am. J. Physiol.* 256 (*Heart Circ. Physiol.* 25): H1747–H1751, 1989.
87. PERNOW, J. Characterization of the cardiovascular actions of endothelin in vivo: comparisons with neuropeptide Y and angiotensin II. *Acta Physiol. Scand.* 137: 421–426, 1989.
88. PERNOW, J., J. F. BOUTTIER, A. FRANCO-CERECEDA, J. S. LACROIX, R. NATRAN, and J. M. LUNDBERG. Potent selective vasoconstrictor effects of endothelin in the pig kidney in vivo. *Acta Physiol. Scand.* 134: 573–574, 1988.
89. PERNOW, J., A. FRANCO-CERECEDA, R. MATRAN, and J. M. LUNDBERG. Effect of endothelin-1 on regional vascular resistances in the pig. *J. Cardiovasc. Pharmacol.* 6(Suppl. 4): S205–S206, 1989.
90. RAE, G. A., M. TRYBULEC, G. DE NUCCI, and J. R. VANE. Endothelin-1 releases eicosanoids from rabbit isolated perfused kidney and spleen. *J. Cardiovasc. Pharmacol.* 6(Suppl. 4): S89–S92, 1989.
91. RAKUGI, H., M. NAKAMARU, H. SAITO, J. HIGAKI, and T. OGIHARA. Endothelin inhibits renin release from isolated rat glomeruli. *Biochem. Biophys. Res. Commun.* 155: 1244–1247, 1988.
92. REES, D. D., R. M. J. PALMER, H. F. HODSON, and S. MONCADA. A specific inhibitor of nitric oxide formation from L-arginine attenuates endothelium-dependent relaxation. *Br. J. Pharmacol.* 96: 418–424, 1989.
93. RESINK, T. J., T. SCOTT-BURDEN, and F. R. BUHLER. Activation of phospholipase A_2 by endothelin in cultured vascular smooth muscle cells. *Biochem. Biophys. Res. Commun.* 158: 279–286, 1989.
94. ROHMEISS, P., J. PHOTIADIS, S. ROHMEISS, and T. UNGER. Hemodynamic actions of intravenous endothelin in the rat: comparison with sodium nitroprusside and methoxamine. *Am. J. Physiol.* 258 (*Heart Circ. Physiol.* 27): H337–H346, 1990.
95. RUBANYI, G. M. Endothelium-dependent pressure-induced contraction of isolated canine carotid arteries. *Am. J. Physiol.* 255 (*Heart Circ. Physiol.* 24): H783–H788, 1988.
96. RUBANYI, G. M. Endothelium derived vasoconstrictor factors. In: *Endothelial Cell,* edited by U.S. Ryan. Boca Raton, FL: CRC Press, vol. III, p. 61–74, 1988.
97. RUBANYI, G. M. Maintenance of "basal" vascular tone may represent a physiological role for endothelin. *J. Vasc. Med. Biol.* 1: 315–316, 1989.
98. RUBANYI, G., and P. M. VANHOUTTE. Hypoxia releases vasoconstrictor mediator(s) from the vascular endothelium. *J. Physiol. (Lond.)* 364: 45–56, 1985.
99. RUBANYI, G. M., and P. M. VANHOUTTE. Flow-induced release of endothelium-derived relaxing factor. *Am. J. Physiol.* 250 (*Heart Circ. Physiol.* 19): H1145–H1149, 1986.
100. SAITO, Y., K. NAKAO, M. MUKOYAMA, and H. IMURA. Increased plasma endothelin level in patients with essential hypertension. *N. Engl. J. Med.* 322(3): 205, 1990.
101. SCOGGINS, B. A., C. D. SPENCE, D. G. PARKES, M. MCDONALD, J. D. WADE, and J. P. COGHLAN. Cardiovascular actions of human endothelin in conscious sheep. *Clin. Exp. Pharmacol. Physiol.* 16: 235–238, 1989.
102. SHIBA, R., M. YANAGISAWA, T. MIYAUCHI, Y. ISHII, S. KIMURA, Y. UCHIYAMA, T. MASAKI, and K. GOTO. Elimination of intravenously injected endothelin-1 from the circulation of the rat. *J. Cardiovasc. Pharmacol.* 6(Suppl. 4): S98–S101, 1989.
103. SHICHIRI, M., Y. HIRATA, T. EMORI, K. OHTA, T. NAKAJIMA, K. SATO, A. SATO, and F. MARUMO. Secretion of endothelin and related peptides from renal epithelial cell lines. *FEBS Lett.* 253: 203–206, 1989.
104. SIMONSON, M. S., and M. J. DUNN. Endothelin-1 stimulates contraction of rat glomerular mesangial cells and potentiates β-adrenergic–mediated cyclic adenosine monophosphate accumulation. *J. Clin. Invest.* 85: 790–797, 1990.
105. SIMONSON, M. S., S. WANN, P. MENE, G. R. DUBYAK, M. KESTER, Y. NAKAZATO, J. R. SEDOR, and M. J. DUNN. Endothelin stimulates phospholipase C, Na^+/H^+ exchange, c-*fos* expression, and mitogenesis in rat mesangial cells. *J. Clin. Invest.* 83: 708–712, 1989.
106. SUGIURA, M., T. INAGAMI, and V. KON. Endotoxin stimulates endothelin release in vivo and in vitro as determined by radioimmunoassay. *Biochem. Biophys. Res. Commun.* 161: 1220–1227, 1989.
107. SUGIURA, M., R. M. SNAJDAR, M. SCHWARTZBERG, K. F. BADR, and T. INAGAMI. Identification of two types of specific endothelin receptors in rat mesangial cells. *Biochem. Biophys. Res. Commun.* 162: 1396–1401, 1989.
108. TAKAGI, M., H. MATSUOKA, K. ATARASHI, and S. YAGI. Endothelin: a new inhibitor of renin release. *Biochem. Biophys. Res. Commun.* 157: 1164–1168, 1988.
109. TAKAGI, M., H. TSUKADA, H. MATSUOKA, and S. YAGI. Inhibitory effect of endothelin on renin release in vitro. *Am. J. Physiol.* 257 (*Endocrinol. Metab.* 20): E833–E838, 1989.

110. TOMOBE, Y., T. MIYAUCHI, A. SAITO, M. YANAGISAWA, S. KIMURA, K. GOTO, and T. MASAKI. Effects of endothelin on the renal artery from spontaneously hypertensive and Wistar-Kyoto rats. *Eur. J. Pharmacol.* 152: 373–374, 1988.
111. TSUCHIYA, K., M. NARUSE, T. SANAKA, K. NARUSE, K. NITTA, H. DEMURA, and N. SUGINO. Effects of endothelin on renal regional blood flow in dogs. *Eur. J. Pharmacol.* 166: 541–543, 1989.
112. VANHOUTTE, P. M. The endothelium—modulator of vascular smooth muscle tone. *N. Engl. J. Med. 318: 512–513, 1988.*
113. VANHOUTTE, P. M. Endothelium and control of vascular function. *Hypertension* 13: 658–667, 1989.
114. VANHOUTTE, P. M., G. RUBANYI, V. M. MILLER, and D. S. HOUSTON. Modulation of vascular smooth muscle contraction by the endothelium. *Annu. Rev. Physiol.* 48: 307–320, 1986.
115. WALDER, C. E., G. R. THOMAS, C. THIEMERMANN, and J. R. VANE. The hemodynamic effects of endothelin-1 in the rat. *J. Cardiovasc. Pharmacol.* 6(Suppl. 4): S93–S97, 1989.
116. WARNER, T. D., G. D. DE NUCCI, and J. R. VANE. Rat endothelin is a vasodilator in the perfused mesentery of the rat. *Eur. J. Pharmacol.* 159: 325–326, 1989.
117. WHITTLE, B. J. R., J. LOPEZ-BELMONTE, and D. D. REES. Modulation of the vasodepressor actions of acetylcholine, bradykinin, substance P and endothelin in the rat by a specific inhibitor of nitric oxide formation. *Br. J. Pharmacol.* 98: 646–652, 1989.
118. WIKLUND, N. P., A. OHLEN, and B. CEDERQUVIST. Inhibition of adrenergic neuroeffector transmission by endothelin in the guinea pig femoral artery. *Acta Physiol. Scand.* 134:311–312, 1988.
119. WRIGHT, C. E., and J. R. FOZARD. Regional vasodilator is a prominent feature of the haemodynamic response to endothelin in anesthetized, spontaneously hypertensive rats. *Eur. J. Pharmacol.* 155: 201–203, 1988.
120. YANAGISAWA, M., A. INOUE, T. ISHIKAWA, Y. KASUYA, S. KIMURA, S. I. KUMAGAYE, K. NAKAJIMA, T. X. WATANABE, S. SAKAKIBARA, K. GOTO, and T. MASAKI. Primary structure, synthesis, and biological activity of rat endothelin, an endothelium-derived vasoconstrictor peptide. *Proc. Natl. Acad. Sci. USA* 85: 6964–6967, 1988.
121. YANAGISAWA, M., H. KURIHARA, S. KIMURA, Y. TOMOBE, M. KOBAYASHI, Y. MITSUI, Y. YAZAKI, K. GOTO, and T. MASAKI. A novel potent vasoconstrictor peptide produced by vascular endothelial cells. *Nature* 332: 411–415, 1988.
122. YOKOKAWA, K., M. KOHNO, K. MURAKAWA, K. YASUNARI, T. HORIO, T. INOUE, and T. TAKEDA. Acute effects of endothelin on renal hemodynamics and blood pressure in anesthetized rats. *Am. J. Hypertens.* 2: 715–717, 1989.
123. YOSHIZUMI, M., H. KURIHARA, T. FUMIMARO, M. YANAGISAWA, T. MASAKI, and Y. YAZAKE. Hemodynamic shear stress regulates endothelin gene expression in cultured endothelial cells, abstracted. *Circ. Res.* 78: II–182, 1988.
124. YUAN, B. H., I. F. MCMURTRY, and J. D. CONGER. Effect of endothelin on isolated perfused rat afferent and efferent arterioles. *Clin. Res.* 37: 586A, 1989.
125. ZEIDEL, M. L., H. R. BRADY, B. C. KONE, S. R. GULLANS, and B. M. BRENNER. Endothelin, a peptide inhibitor of Na^+/K^+ ATPase in intact renal tubular epithelial cells. *Am. J. Physiol.* 257 (*Cell Physiol.* 26): C1101–C1107, 1989.
126. ZOJA, C., A. BENIGNI, D. RENZI, A. PICCINELLI, N. PERICO, and G. REMUZZI. Endothelin and eicosanoid synthesis in cultured mesangial cells. *Kidney Int.* 37:927–933, 1990.

12

Endothelin, a Ubiquitous Peptide: Morphological Demonstration of Immunoreactive and Synthetic Sites and Receptors in the Respiratory Tract and Central Nervous System

S. J. GIBSON, D. R. SPRINGALL, AND JULIA M. POLAK

After the original description of endothelin by Yanagisawa and colleagues[68] in 1988 and the subsequent demonstration of the endothelin family,[29] much attention has been devoted to the actions and effects of the peptides. To define their role in normal function and disease, however, much useful information can be gained by a knowledge of the precise anatomical sites of synthesis, storage, and action. There are many techniques in the province of the morphologist that permit such analysis and allow the detailed description of cellular and subcellular localization of peptides, their synthetic machinery, and receptor and binding sites.

TECHNIQUES

Synthesis

Using molecular biological techniques it is possible to determine whether a cell has the synthetic machinery to produce endothelin. The methods involve the detection, by hybridization with labeled probes, of specific mRNA(s) encoding the peptide. The studies can be performed at the whole-tissue level using extracts (Northern blotting), the nucleic acids being separated by electrophoresis followed by transfer to a membrane for hybridization to detect the mRNA of interest.[3] Alternatively, the mRNA may be hybridized in situ in sections of appropriately fixed tissue, albeit at a lower stringency than is useable for Northern blots.

The in situ technique of hybridization has the great advantage of permitting localization of the mRNA at the cellular level in a heterogeneous tissue. The original method used specific cDNA sequences[15] as probes, but now many workers prefer to use "riboprobes." These are RNA fragments complementary (cRNA) to the mRNA being investigated, and are produced by inserting the specific cDNA sequences into a vector capable of generating sense (noncomplementary) and antisense (complementary) RNA probes.[19] This is illustrated in Figure 12.1, together with the endothelin probes used in our studies. Although the presence of mRNA does not prove that synthesis is taking place (the mRNA

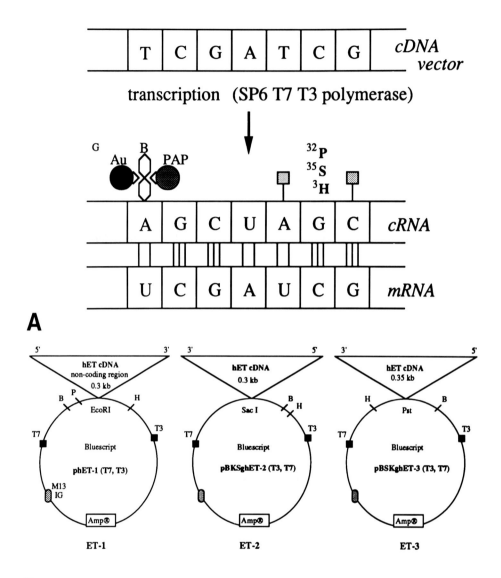

FIGURE 12.1. *A:* schematic representation of the method for obtaining a complementary RNA (cRNA) probe from a cDNA insert in a plasmid vector (top) and its use for hybridization. The vector is cleaved by specific endonucleases to yield a linearized DNA template containing the cDNA and a transcription promoter site. Using a specific RNA polymerase enzyme (SP6, T7, T3 as appropriate), the template is transcribed in vitro to generate a cRNA probe. The cRNA hybridizes to the mRNA in the tissue section by base pairing as shown (bottom). To permit subsequent detection of the probe, one of the indicated labels is used. These are usually radioactive compounds, but nonradioactive labels such as peroxidase, gold, or biotin can be employed. *B:* plasmid vectors from which endothelin cRNA probes employed in the present study were generated.

may not be translated), much stronger evidence is afforded if the peptide can be localized to the same cell.

Localization of Peptide

As with Northern blotting of whole-tissue extracts to detect specific mRNA, whole tissues can be analyzed for their peptide content by radioimmunoassay of extracts.[6] This technique also permits characterization of the peptides present by various chromatographic procedures. However, the same disadvantages also accrue, in that the cellular localization of the peptide is not determined. The principal method for investigating cellular localization is immunocytochemistry, either at the light or electron microscopic level.[48,49] Such techniques now have a high sensitivity, because of novel enhancing procedures,[63] and specificity, and they are also amenable to quantification of antigen levels in at least a comparative fashion.[43] By using antisera to specific fragments of the peptide or its precursor, some clues to the post-translational processing can be obtained. The endothelin antisera used in these studies are shown diagrammatically in Figure 12.2.

In addition to application of antisera to specific peptides, a variety of antisera are available with which to define certain cell populations or intracellular features. Thus, for example, antibodies to protein gene product 9.5 (PGP 9.5),[61] synaptophysin,[39] and chromogranins[38] can be used as markers of endocrine cells and nerves and those to Factor VIII–related antigen (von Willebrand's factor) as markers of endothelial cells. Endothelium can also be labeled by a variety of other procedures such as binding of lectins, including *Ulex europaeus* agglutinin, either directly labeled or subsequently detected by immunocyto-

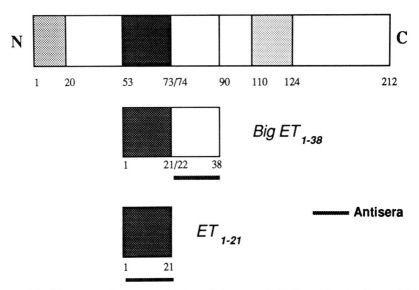

FIGURE 12.2. Diagrammatic representation of the proendothelin molecule, from which big endothelin$_{22-38}$ and endothelin$_{1-21}$ are generated. Antisera employed in this study recognize the regions indicated by the dark underline. N, N-terminal; C, C-terminal; ET, endothelin.

chemistry.[58] The markers available are summarized in Figure 12.3. Thus, by using double immunostaining of a single section or by staining pairs of serial sections, the cell type producing endothelin can be characterized. With serial sections the so-called reverse face section method, orienting a pair of serial sections such that the adjacent cut face is uppermost in both, yields the best results.

Having localized the peptide and its mRNA, strong evidence is provided for synthesis in cells where both are found together. To define the sites of action and to relate to pharmacology findings, it is necessary to localize the binding sites of the peptide.

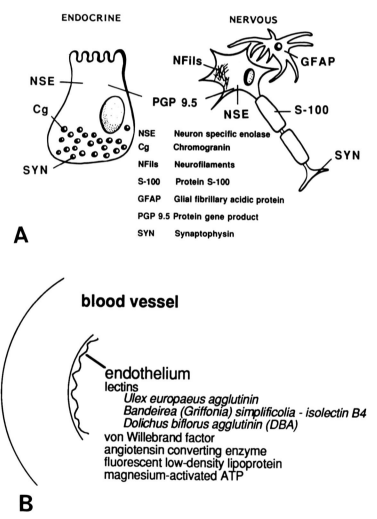

FIGURE 12.3. *A:* diagrammatic representation showing the distribution of some general markers of endocrine and neural cells. *B:* markers that are available for the identification of endothelial cells.

Binding Site and Receptor Analysis

Much information about receptors can be obtained from classic isolated membrane studies investigating the kinetics and amounts of radiolabeled ligand binding to tissue homogenates. However, variations of these techniques can be applied at the cellular level to quantify and determine the kinetics of the binding in various cell types in tissue sections (Figure 12.4). The technique, called "in vitro autoradiography," was first introduced by Young and Kuhar[71] in 1979. It can be applied at two levels of resolution, both of which require an initial incubation of a section of unfixed tissue with the radiolabeled ligand in the presence or absence of excess unlabeled ligand or receptor antagonist. For analysis at the gross level the slide is apposed to autoradiographic film; anal-

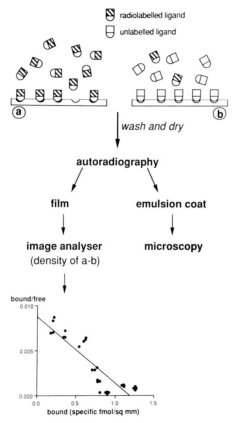

FIGURE 12.4. The in vitro autoradiographic technique for receptor localization. Serial tissue sections are incubated with radiolabeled (^{125}I or ^{3}H) ligand without (*a*) or with (*b*) an excess of unlabeled ligand (competitive displacement, to detect nonspecific binding of the ligand). Autoradiography to visualize the radiolabel is performed either by apposing the slides to autoradiography film or by dipping the slides in liquid emulsion. After exposure and development, optical densities of specific areas on the films may be quantified by image analysis in comparison with coexposed standards, the specific binding derived, and kinetic data produced (e.g., by Scatchard plot, as shown). The emulsion-coated slides may be used for microscopic localization of binding sites (e.g., small blood vessels), which is difficult to achieve with film autoradiograms.

ysis at the microscopic level entails either gluing (to ensure registration of the resultant autoradiogram and the section) coverslips coated with emulsion to the slide or coating the section itself by dipping the slide into liquid emulsion. The first two methods are amenable to quantification, but the emulsion-coated preparations may not be because of dissociation of ligand during the fixation and the dipping of the section into liquified (45°C) emulsion. Despite this, they offer simplicity and excellent resolution.

For quantification, the amount of ligand bound to different cells or tissues within the section can be determined by measuring the optical density of relevant areas of the autoradiograms. Comparison with the optical density of coexposed radioactive standards[11] permits calculation of radioactivity per unit area, which can be transformed into the amount of peptide bound per unit area if the specific activity of the ligand is known. It is thus possible to perform kinetic analysis and thereby determine the saturability and competitive displacement of binding and the binding constants; this is necessary to establish the receptor-like characteristics of binding.

ENDOTHELIN IN THE LUNG

Following the original description of endothelin[68] and its family,[29], reports of radioimmunoassays appeared suggesting the presence of high levels of endothelin-like immunoreactivity in extracts of a variety of tissues, including the lung of pig[32] and rat.[33] However, there was no information on the localization and synthetic sites of the peptide; we therefore investigated these as part of our program to study regulatory peptides in the lung and their involvement in development and disease. Lung tissues from mice of 3–12 weeks old and from adult rats were studied, in addition to fetal, neonatal, and adult humans.

Rat and Mouse

Immunostaining with antiserum to endothelin-1 (Cambridge Research Biochemicals [CRB], Cambridge, England) was seen in the epithelium of the airways of both rats and mice. In the larger bronchioles, immunostaining was found mainly in the serous and mucous secretory cells; many of the ciliated and basal cells were unstained. In the smaller bronchioles the predominant cell type is the Clara cell, and these were consistently strongly stained, as were the cuboidal cells in the respiratory bronchioles (Fig. 12.5). There was no noticeable difference between the two species or the ages regarding the pattern of staining.

This finding is relevant in light of the report that endothelin is formed by cultured airway epithelial cells of dog and pig.[5] Whether the airway epithelium of humans also produces endothelin is less certain. We have preliminary results suggesting that it is possible, that the predominant site of synthesis in human lung is endocrine cells of airway epithelium and the endothelium of blood vessels.

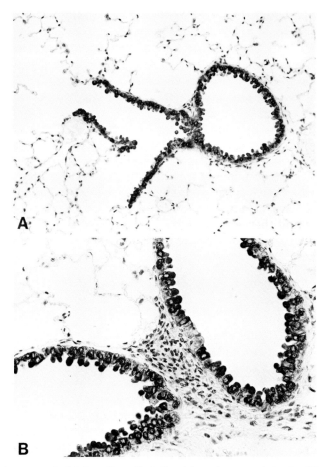

FIGURE 12.5. *A:* wax-embedded section of adult rat lung immunostained with antisera to endothelin-1 using the peroxidase–antiperoxidase method and counterstained with hematoxylin. Immunostained cells (Clara cells) alternate with the unstained ciliated cells. × 100. *B:* a high-power photomicrograph depicting a similar finding. × 250.

Humans

Fetal lung samples (gestational ages 10–39 weeks; N = 32), postnatal lung samples (aged 10 days to 1 year; N = 12), and adult lung samples (aged 42–68 years; N = 16) were studied. Tissues were divided and fixed for immunocytochemistry in either Bouin's solution or a buffered picric acid–formalin (Zamboni) solution[72] or in 10% formalin, for 6–12 h. For the localization of mRNA, tissues were fixed in 4% paraformaldehyde in 0.1 M borate buffer (pH 9.6) for 6 h.

Endocrine cells in the human lung are characterized by their immunoreactivity for the general endocrine markers synaptophysin, chromogranin, and PGP 9.5 and for various regions of the precursor of gastrin-releasing peptide (GRP), the mammalian form of bombesin.[4,38,39,61] Accordingly, antisera to these markers were used to visualize the endocrine cells in our preparations.

Sections were immunostained using the avidin–biotin complex method. To characterize the immunostained cells, one of a pair of serial sections was immunostained with endothelin antiserum, and the other was stained with antiserum to GRP, chromogranin, PGP 9.5, or synaptophysin. The specificity of endothelin immunostaining was determined by preabsorption of the primary antisera with synthetic endothelin-1 (CRB), with synthetic endothelin-2 or endothelin-3 (Scientific Marketing, UK), or with synthetic C-terminal peptide of big endothelin-1 (CRB, UK) at a concentration of 10 nmol/ml of the diluted antiserum.

In situ hybridization was performed with RNA probes prepared from complementary DNA fragments corresponding to the 3′ noncoding region of endothelin-1 (pT7hET91),[30] endothelin-2 (pghET20SG1), and endothelin-3 (pghET3E1PP1) derived from the human chromosomal gene.[29] Cryostat sections (10 μm) from paraformaldehyde-fixed tissues of all ages were hybridized with the [^{35}S]- or [^{32}P]-labeled endothelin-1, endothelin-2, and endothelin-3 cRNA probes (1×10^6 cpm/section). Hybridization of lung sections with sense probes or treatment of sections with RNase solution prior to hybridization was used as a negative control for the specificity of the hybridization signals.

Endothelin-like immunoreactivity was found mainly in endocrine cells of pulmonary airways of developing and adult human lung fixed in formalin or Zamboni's solution. Immunostaining was present in both large and small airways and was more apparent in the proximal airways. The cells were shown to be endocrine cells by their immunoreactivity for chromogranin, synaptophysin, PGP 9.5, and GRP (Figs. 12.6 and 12.7). The fetal immunoreactive cells were first seen at a gestational age of 12 weeks, and they were seen as single cells or clusters. The cells often had processes moving a long distance laterally along the basement membrane or toward the airway lumen. Endothelin-like immunoreactivity was observed in fetal lungs at gestational ages of 12–39 weeks, but the frequency of endothelin-immunoreactive cells was highest at gestational ages 20–22 weeks and declined during postnatal life. Immunoreactive endocrine cells were rarely seen in adult lung (Fig. 12.8). Immunostaining was also seen in the endothelium of pulmonary blood vessels (Fig. 11.8).

Endocrine cell immunoreactivity was seen with both endothelin$_{1-21}$ (amino acids) and big endothelin$_{22-38}$ (amino acids) antisera, although the relative proportions of stained cells varied with gestational age in fetal tissues. In third-trimester fetal lung, more cells were stained with the antiserum to the C-terminal peptide of big endothelin-1 than with the antiserum to endothelin-1; before this, endothelin-1–immunoreactive cells predominated. In the adult cases, strong immunoreactivity with both endothelin antisera was seen in other cell types of the airway epithelium in 50% of the cases. No immunostaining with either of the endothelin antisera was seen in nerve fibers.

Following in situ hybridization with endothelin cRNA probes, endothelin RNA transcripts were expressed in blood vessels and in a subset of cells in the mucosa of small and large airways, having a similar distribution as those immunostained with the endothelin antisera (Fig. 12.9). However, endothelin RNA transcripts were seen in fewer specimens (in almost one-third of the total number of cases studied) than were immunostained for endothelin mature peptide (more than two-thirds of the total number of cases studied), and there

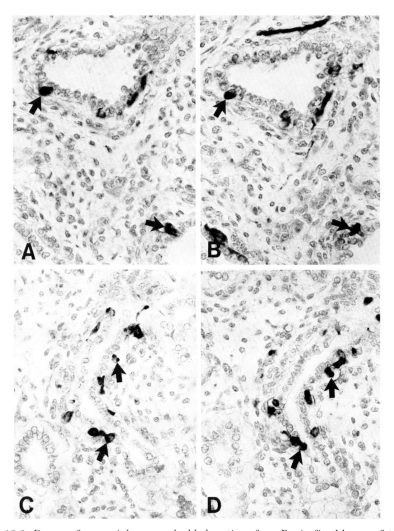

FIGURE 12.6. Reverse face, serial, wax-embedded sections from Bouin-fixed human fetal lung at 17 weeks of gestation. Sections immunostained with antisera to endothelin-1 (A,C), chromogranin (B), and synaptophysin (D) using the avidin–biotin–peroxidase complex method with nickel enhancement and counterstained with hematoxylin. Arrows indicate some cells in which endothelin immunoreactivity coexists with one of the well-established neuroendocrine cell markers. × 250.

were more endothelin-immunoreactive cells than endothelin mRNA-positive cells in each specimen. Immunocytochemistry and in situ hybridization performed on pairs of reverse face serial sections showed that expression of endothelin mRNA and immunoreactivity were colocalized in the same cell. Cells expressing either endothelin mRNA or endothelin immunoreactivity alone were also seen in the serial sections. Cells that were immunoreactive-only formed the greatest proportion of the population, followed by those with mRNA-only and a small proportion that were both.

Binding sites for endothelin were seen on airways and blood vessel smooth muscle and in the lung parenchyma (Fig. 12.10). The binding was saturable,

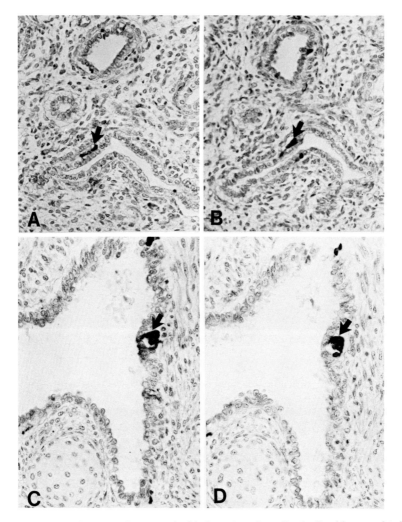

FIGURE 12.7. Reverse face, serial, wax-embedded sections from Bouin-fixed human fetal lung at 17 weeks of gestation. The sections were immunostained with antisera to endothelin-1 (A,C), PGP 9.5 (B), and pro-GRP (D) using the avidin–biotin–peroxidase complex method with nickel enhancement and counterstained with hematoxylin. Arrows indicate colocalization of endothelin immunoreactivity with one of the neuroendocrine cell markers in the same cell. × 250.

competitively inhibited by endothelin-1 and endothelin-2, and, to a lesser extent, by endothelin-3, with a K_d of 0.1 nM in rat, pig, and humans.

Our data have shown the occurrence of endothelin-like immunoreactivity in a high proportion of endocrine cells producing synaptophysin, chromogranin, and PGP 9.5 and in a greater number than those containing pro-GRP. Therefore, the localization of endothelin in human pulmonary endocrine cells seems assured and its synthesis strongly suggested by the presence of mRNA encoding the peptide. Nonendocrine epithelial cell staining is less certain because it was absent from a large number of cases, although endothelin mRNA has been demonstrated[40] and we have shown the peptide to be present in rats and mice at this site.[52] Endothelin-like immunoreactivity seen in the airway

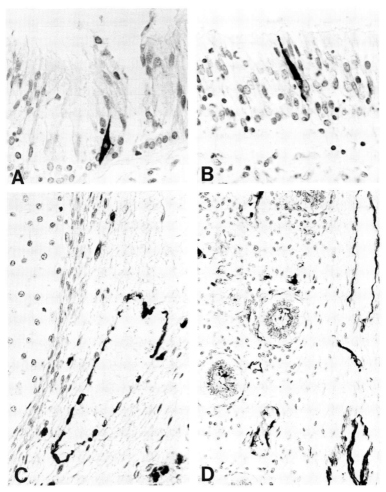

FIGURE 12.8. *A,B:* human adult bronchial pseudostratified columnar epithelium immunostained with antisera to human big endothelin (see Fig. 12–2) using the avidin–biotin–peroxidase complex method with nickel enhancement. × 300. *C,D:* strong immunoreactivity is associated with the endothelial cells of blood vessels in the hilar region (*C*) and at the periphery of the lung (*D*). × 200.

epithelium of adult lung in some cases could be disease or function related, and this merits further study.

Apparently, there were fewer cells expressing endothelin mRNA that were immunoreactive for the mature peptide. This is in agreement with the previously proposed theory that the mRNA encoding pulmonary endocrine cell peptides appears earlier in fetal life than the peak of peptide immunoreactivity.[59] Alternatively, it could be due to the fact that the antiserum to endothelin-1 also recognizes endothelin-2 and endothelin-3. If this were true, it suggests that at least two isoforms are present in endocrine cells of lung. Endothelin is an endothelium-derived peptide, and its RNA transcripts have been reported to be expressed in vascular endothelial cells in tissues[40] and cell cultures.[68] Binding sites for endothelin have been demonstrated in the vasculature of

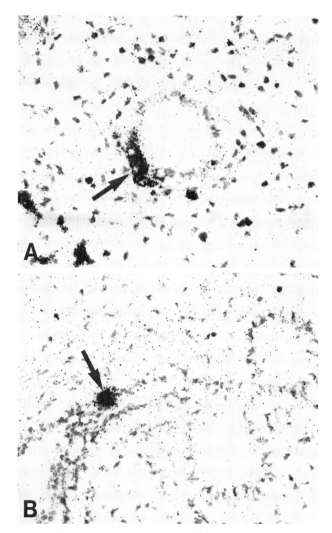

FIGURE 12.9. Sections of human adult (A) and fetal (B) lung fixed by immersion in 4% paraformaldehyde in borate buffer (0.1 M, pH 9.5), hybridized with a [^{35}S] labeled human endothelin-1 cRNA probe, and counterstained with hematoxylin. Cells showing a positive hybridization signal are indicated by arrows. × 300.

many systems, including the intrapulmonary vessels.[50] Therefore, it was not unexpected to see the expression of endothelin immunoreactivity and mRNA in some endothelial cells of pulmonary vessels in the human lung.

It is important to mention that, because of the common sequences between the endothelin-2 and endothelin-3 probes used here, it is very difficult to be certain about the specificity of these probes in detecting the relative mRNA under the normal hybridization stringency. Immunostaining was prevented by absorption of the endothelin-1 antiserum with synthetic endothelin-2 or endothelin-3, indicating that endothelin-1 antiserum cross-reacts with endothelin-2 and endothelin-3. The unchanged immunoreaction following absorption of the C-terminal peptide of big endothelin-1 antiserum with synthetic

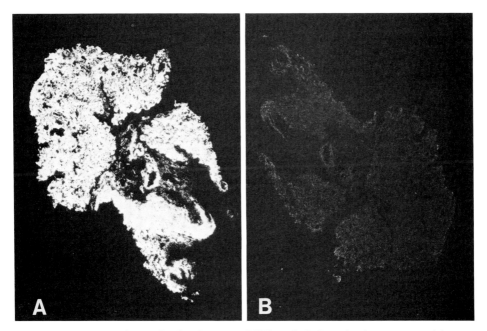

FIGURE 12.10. Autoradiographic localization of [^{125}I] endothelin-1 binding in normal human lung. A: total binding to thick cryostat sections incubated with 150 pmol of [^{125}I] human/porcine endothelin-1 for 30 min at 22°C. Following washing and drying, sections were apposed to Hyperfilm at 4°C for 2 days. The binding was localized to the bronchial smooth muscles, blood vessels, and lung parenchyma. B: A section adjacent to A incubated with unlabeled human/porcine endothelin-1 (1 μM) to displace competitively the specific binding. The autoradiogram shows the location of the nonspecific binding.

endothelin-1, endothelin-2, or endothelin-3 indicates that the antiserum is specific. However, this does not exclude the possibility that the big endothelin antibody recognizes proendothelin or intermediate fragments; presumably the endothelin-1 antibody also binds to big endothelin and to proendothelin. Therefore, the finding of a switch in the predominant immunoreactivity in the third trimester of gestation from endothelin-1 to big endothelin can only be thought to suggest that there is a post-translational processing change at that stage. The precise nature of that change or alterations in other factors such as secretion of the peptide(s) awaits elucidation of the electron microscopical localization of the different forms of endothelin and a more detailed analysis of additional cases.

ENDOTHELIN IN THE NERVOUS SYSTEM

The observation that a peptide, originally isolated from the culture supernatant of endothelial cells, was also present in neurons is perhaps somewhat surprising. However, as discussed in the preceding section, even in the lung endothelin is not merely restricted to endothelial cells but is localized to several other cell types. Furthermore, if we consider that a large number of peptides, first isolated from tissues other than the nervous system, have subsequently been found to occur in several cell types including neurons, e.g., vasoactive intestinal polypeptide,[47,53] the neuronal localization of endothelin is not quite so unexpected.

The search for the precise localization of endothelin in the nervous system was stimulated by several observations. First, early receptor studies showed the presence of high-affinity binding sites for endothelin in peripheral and central nervous tissues. Second, it was shown that endothelin had a direct effect on the neural excitability in newborn rat spinal cord.[69] Of particular significance was the finding that endothelin produced a dose-dependent ventral root depolarization, which was attenuated by nicardipine (a dihydropyridine [DHP]-sensitive Ca^{2+}-channel blocker) or by spantide, the substance P antagonist. Although Yoshizawa and coworkers[69] were unable to deduce whether the effect of endothelin on motoneurons was direct or indirect, they summized that endothelin probably activates DHP-sensitive Ca^{2+} channels on the terminals of substance P–containing nerves, evokes release of substance P, which in turn depolarizes motoneurons, and produces a ventral root depolarization. It is known that dorsal root stimulation induces ventral root depolarization, which is thought to be mediated via release of substance P from primary sensory neurons.[1] However, endothelin could also elicit a release of substance P from neurons intrinsic to the spinal cord. Therefore, to determine whether endothelin was actually present in either primary sensory neurons of the dorsal root ganglia or in the spinal cord, the occurrence of endothelin in these tissues in human postmortem cases was examined in our laboratory. In addition, neonatal piglets (2–3 days after birth) were also studied. These studies, which showed that endothelin peptide and mRNA were indeed present in neurons of the spinal cord and dorsal root ganglia, were later followed by a more extensive investigation of endothelin immunoreactivity and gene expression in the human brain. A separate section is devoted to the localization of endothelin binding sites within the brain (see below under Brain).

To date, few morphological studies have investigated the relative distributions of different members of the endothelin family of related peptides in nervous tissues. Immunocytochemical studies have employed antisera that mostly recognize the common COOH terminus of endothelin (Fig. 12.2)[16,17,70] However, radioimmunoassay and chromatographic analyses in pig spinal cord[56] and rat brain[41] have demonstrated the presence of endothelin-1 and endothelin-3 (13:1-fold, respectively, in pig spinal cord). These data correspond with the reports of endothelin-3 mRNA and binding sites in addition to endothelin-1 in neuronal localizations in rat brain.[13,40] Furthermore, in pig spinal cord, radioimmunoassay detects relatively little "big" endothelin-1_{22-38} (<1% of endothelin-1 immunoreactivity), which suggests that the physiologically most active form of endothelin, endothelin-1_{1-21}, is derived by cleavage at Trp^{21}-Val.[22,56]

SPINAL CORD AND DORSAL ROOT GANGLIA

Immunocytochemistry

Neurons immunoreactive for endothelin have been demonstrated in the spinal cord of humans and pig[16,56,69] and in the dorsal root ganglia of man[16] and pig (unpublished data) (Fig. 12.11; see also 12.14A). In neurologically normal human spinal cord, the distribution of endothelin immunoreactivity was similar in each segmental level studied (cervical, thoracic, lumbar, and sacral) as revealed by means of antisera raised to endothelin-I (CRB, ICI, Cambridge, England; for further details, see Giaid et al.[16]). Immunostaining was restricted to the spinal grey matter, where few immunoreactive fibers were noted. When present, these were localized mainly to the upper region of the dorsal horn (laminae I–III), where the terminals of primary sensory neurons end (Fig. 12.11A).[18] Immunoreactive neuronal cell bodies were, however, seen throughout the spinal cord. Although there were some immunoreactive neurons in the dorsal horn, the majority were restricted to the ventral spinal cord, where a

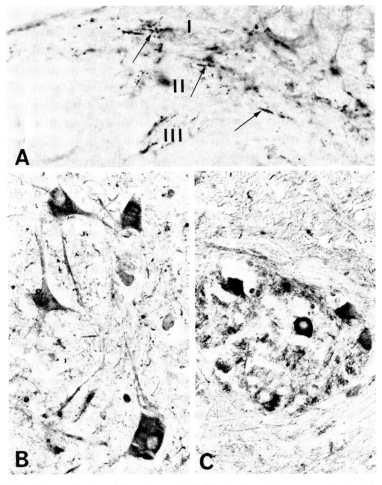

FIGURE 12.11. Human lumbar (A) and sacral (B,C) spinal cord immunostained with antisera to endothelin-1 using the avidin–biotin–peroxidase complex method. A: in the dorsal horn, a few fibers and punctate structures are present in laminae I–III (some arrowed) × 500. B: many motoneurons in the lateral motor columns are endothelin immunoreactive. × 350. C: in the pudendal motor nucleus (Onuf's nucleus) several strongly immunoreactive neurons (arrowed) are apparent. × 350. Bouin-fixed spinal cord tissues, obtained 4 h postmortem. Cryostat sections (20 μm).

number were identified as motoneurons (Fig. 12.11,B,C). Several immunoreactive neurons (presumptive preganglionic neurons) were noted in centers associated with autonomic function in the intermediolateral cell columns of thoracolumbar and lumbosacral segments of the spinal cord. In pig tissues, fibers and structures resembling fibers (punctate, dot-like structures) and cell bodies immunoreactive for endothelin were demonstrated in the dorsal[56,69] and ventral[56] spinal cord. To date, however, we have noted sparse fiber-like structures in the porcine dorsal horn but only occasional immunoreactive cell bodies. This apparent discrepancy between the reported distribution of endothelin[56,69] and our own observations in the porcine spinal cord would appear to be related to different sources of antisera to endothelin-1, fixation methods, ages of animals, or a combination of all three. With respect to the distribution of endothelin-immunoreactive neurons in the ventral horn, the pattern of endothelin immunoreactivity was similar to that in humans.

In the dorsal root ganglia at cervical, lumbar, thoracic, and sacral levels of the spinal cord, endothelin-immunoreactive neurons were abundant in both humans[16] and pig (unpublished data) (see Fig. 12.14A). Immunostained neurons were of both large and small sizes, but immunoreactive fibers were not seen.

In both spinal cord and ganglia of humans, fewer immunoreactive neurons were seen in postmortem cases that exceeded 10 h, suggesting that the stored peptide is relatively labile. In pig tissues, however, which were fixed by intracardial perfusion, there was little variation in the abundance of immunostaining.

In Situ Hybridization

The use of cRNA probes for endothelin-1 and probes sharing coding sequences for endothelin-1, endothelin-2, and endothelin-3 (Fig. 12.1) revealed numerous labeled neurons in the spinal cord and dorsal root ganglia of humans[16] and pig (reported here). In both species, the distribution of endothelin transcripts was similar, with labeled neurons occurring mostly in regions in which endothelin-immunoreactive neurons were noted.

In the dorsal spinal cord, few labeled neurons were present in laminae I–II. Silver grains overlying the cytoplasm of neurons in laminae IV–V and neurons of the autonomic intermediolateral cell columns were frequently noted. The greatest numbers of labeled neurons were seen in the ventral horn, particularly in lamina IX, where the motoneurons are located (Fig. 12.12).

In the dorsal root ganglia, the majority of sensory neuronal cell bodies were labeled following in situ hybridization with endothelin cRNA probes (Fig. 12.12C). Silver grains were aggregated over both small and large neurons, with little difference in the density of grains on cells of different diameters (Fig. 12.13).

Relative Abundance and Colocalization of Endothelin with Other Peptides in Sensory Neurons of the Dorsal Root Ganglia and in Spinal Motoneurons

Substance P is one of many peptides found in the sensory neurons of the dorsal root ganglia.[10,18] Although a number of roles have been attributed to this pep-

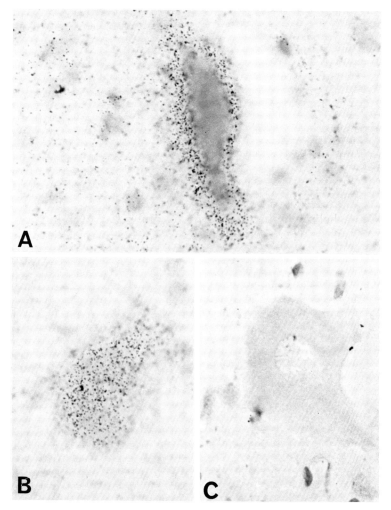

FIGURE 12.12. *A,B:* high-power photomicrographs of motoneurons in the human lumbar spinal cord following hybrization using [^{32}P] labeled human endothelin-1 cRNA probe. Silver grains depict the sites of mRNA overlying the cytoplasm of the motoneurons. *C:* motoneuron from a control tissue section hybridized with the sense (mRNA) endothelin-1 probe. Few silver grains are apparent. Spinal cord was fixed by immersion in 4% paraformaldehyde in borate buffer (0.1 M, pH 9.5). Cryostat (20 μm), counterstained with hematoxylin. × 670.

tide, substance P is most well known for its involvement in the regulation of the processing of sensory information.[25,54,64] The terminals of sensory neurons of the dorsal root ganglia represent a major source of substance P in the spinal cord. Because endothelin may evoke release of substance P via activation of DHP-sensitive Ca^{2+} channels on the terminals of substance P nerves in the spinal cord, it was of particular interest to determine whether endothelin was colocalized with substance P in the dorsal root ganglia. Calcitonin gene–related peptide (CGRP), a peptide produced by alternative processing of the primary transcript of the calcitonin gene,[51] like substance P, is also expressed widely in the sensory nervous system, but is more ubiquitous than substance P in the dorsal root ganglia.[20] CGRP has also been identified as a peptide in-

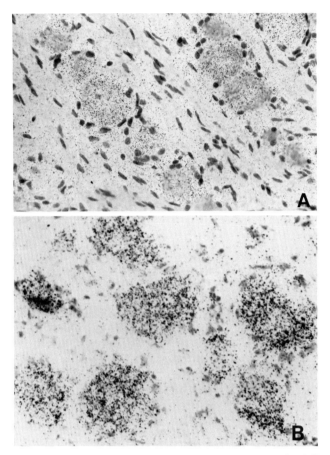

FIGURE 12.13. Sections of porcine (A) and human (B) dorsal root ganglia following *in situ* hybridization with [^{32}P] (A)- and [^{35}S] (B)-labeled human endothelin-1 cRNA probes. Small and large neurons are labeled. Porcine tissue was fixed by intracardial perfusion with 4% paraformaldehyde in borate buffer (0.1 M, pH 9.5), and human tissue was fixed by immersion in the same. Cryostat sections (10 μm) counterstained with hematoxylin. A, × 340; B, × 500.

volved in nociception.[36,67] In addition, CGRP is present in the soma of ventral horn motoneurons and in terminals of motor nerves of a number of mammalian species, including humans.[20,42,45,51]

In motor nerves, CGRP has been identified as a regulatory factor for motor end-plate synthesis of acetylcholine receptors[37] and for the sprouting capacity of motor nerve terminals.[62] Thus investigation of the relationship between expression of endothelin and its mRNA with substance P and CGRP in the dorsal root ganglia and with CGRP in motoneurons may shed more light on the potential role of endothelin as a sensory neurotransmitter–modulator or as a growth factor associated with motor nerves.

Immunocytochemistry and in situ hybridization studies were performed on pairs of reverse face serial sections of dorsal root ganglia and spinal cord of humans[16] and pig. One section of each pair was incubated with antiserum to endothelin-1 and the other with antiserum to either substance P or CGRP. In a similar manner pairs of sections were processed for in situ hybridization

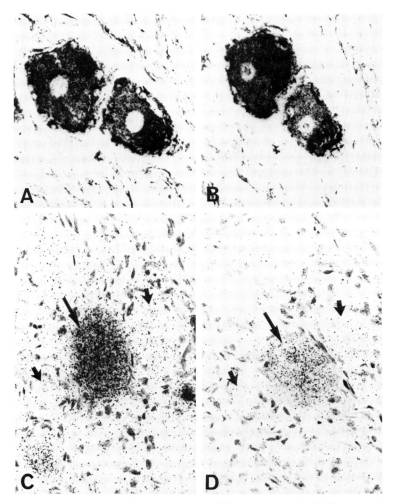

FIGURE 12.14. Reverse face serial sections through human cervical (*A,B*) and lumbar (*C,D*) dorsal root ganglia. Some sections were immunostained with antisera to endothelin-1 (*A*) and α-calcitonin gene–related peptide (α-CGRP) (*B*). Both immunoreactivities are colocalized to the same neurons. Some sections were hybridized with cRNA probes to endothelin-1 (*C*) and β preprotachykinin (*D*). One cell (large arrows) contains both mRNAs. Two cells that are not labeled are shown on either side of the positive cell by the smaller arrows. × 500. (From Giaid et al.[16])

employing cRNA probes for endothelin-1, CGRP, or the β-preprotachykinin gene that encodes substance P.[12] A semiquantitative analysis of the frequency of cells expressing peptide immunoreactivities or mRNAs was made from three cases of adult human tissues.[16]

In the dorsal root ganglia, substance P immunoreactivity (5% of total neuronal population) was localized predominately to a population of small cells, whereas CGRP-immunoreactive neurons were more abundant (20%), with immunostaining observed in small- and medium-sized neurons. All substance P– and most CGRP-immunoreactive neurons were colocalized to a subset of neurons that were endothelin positive (Figs. 12.14A and 12.15A). In situ hy-

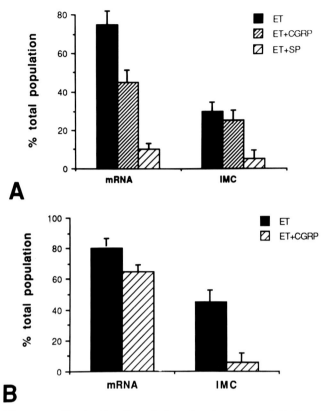

FIGURE 12.15. Histograms illustrating the relative abundance of endothelin mRNA and immunoreactivity with other peptides in the dorsal root ganglia (A) and motoneurons (B) in the lumbar spinal cord of humans. The columns represent the average cell counts taken from three adult cases of neurologically normal spinal cord ± S.E.M. ET, endothelin; CGRP, calcitonin gene–related peptide; and SP, substance P.

bridization revealed a higher percentage of positively labeled cells for each of the three peptide than was revealed by immunocytochemistry. β-Preprotachykinin transcripts were evident in ~10% of the population and CGRP transcripts in 70%. All cells expressing β-preprotachykinin mRNA and the majority of those expressing CGRP mRNA coexpressed endothelin-1 transcripts (Figs. 12.14B, 12.15A).

In contrast to the ganglia, human spinal cord showed relatively few motoneurons that were CGRP immunoreactive (although in pig the majority of motoneurons express CGRP immunoreactivity).[20] In both species, however, most also display endothelin-like immunoreactivity (Fig. 12–15A). However, in man and pig nearly all motoneurons expressed CGRP mRNA, and in man 75%–80% of these were also labeled with the endothelin-1 cRNA probe (Fig. 12.15B).

Endothelin in Neurologically Diseased Spinal Cord

The fact that considerable numbers of motoneurons express endothelin-1 transcripts in the spinal cord implies that endothelin biosynthesis is linked to normal functioning of motor pathways, including the neuromuscular junction. Preliminary study of spinal cords obtained at autopsy (postmortem time 4 h) from two patients with motor neuron disease, a disease in which there is severe muscle wasting and loss of spinal motoneurons,[23,26] showed fewer motoneurons with markedly reduced labeling following hybridization with endothelin-1 cRNA probe. Demonstration of endothelin immunoreactivity and binding sites in the motor end-plate region would be an important step in assessing whether endothelin-like CGRP could act as a motoneuron-derived nerve-muscle growth factor.[37] However, to date endothelin has shown no effect on the activity of skeletal muscle,[66] and thus the action of endothelin in motor nerves remains conjectural.

BRAIN

Immunocytochemistry

By means of antisera raised to the COOH-terminal region of endothelin (which is homologous in the endothelin family of peptides), immunoreactive neurons have been demonstrated in both pig and rat hypothalami.[70] The hypothalamopituitary axis was a focal point because pharmacological studies have shown that endothelin modulates the release of substance P (Calvo and Bloom, personal communication) and vasopressin[55] from the hypothalamus. In both species, endothelin-immunoreactive neurons occur in regions where vasopressin- and oxytocin-immunoreactive neurons are found, namely, the paraventricular and supraoptic hypothalamic nuclei. Immunoreactive fibers and punctate structures (reminiscent of fibers) were noted in the median eminence and posterior lobe of the pituitary, where some structures resembled the swollen profiles that are characteristic of Herring bodies. This implies that endothelin-immunoreactive hypothalamic neurons project to the posterior pituitary from where the peptide is released and may act as a circulating hormone. In addition, further support for the role of endothelin in neurosecretory function comes from studies on water-deprived rats.[70] After 4 days of water deprivation, endothelin immunoreactivity in the posterior pituitary had largely disappeared, suggesting endothelin is released under certain physiological stimuli.

A more comprehensive mapping study of the distribution of endothelin immunoreactivity in the brain has been conducted in man[17] with antisera raised to endothelin-1 (Fig. 12.16).[16] As in pig and rat, numerous endothelin-immunoreactive neurons were seen in the hypothalamus (peri- and paraventricular, supraoptic, and lateral nuclei; Fig. 12.16A). Many immunoreactive neurons were also found in other brain areas. In the cortex (frontal, parietal, and temporal), layers III–VI, a subset of pyramidal and nonpyramidal cells was immunoreactive. Endothelin immunoreactivity was also apparent in the hippocampus (mainly pyramidal cells; Fig. 12.16C) and cerebellum (Purkinje and some Golgi cells; Fig. 12.16F). Scattered neurons were noted in other regions, including the caudate, amygdala, raphe nucleus, substantia nigra, and medulla oblongata (Fig. 12.16D, E). Immunoreactive fibers and punctate structures predominated in the cortex and hypothalamus, with fewer in brain stem and medulla oblongata.

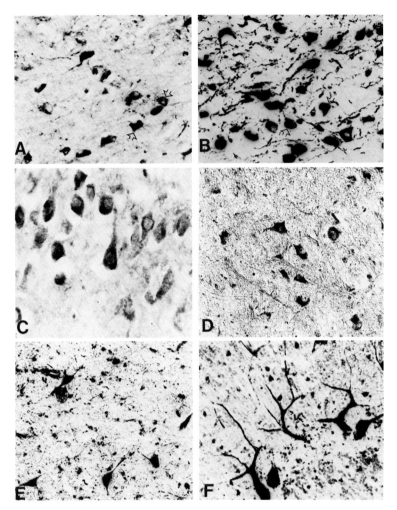

FIGURE 12.16. A,C–F: Neurons immunoreactive for endothelin in various regions of the human brain. Paraventricular nucleus of the hypothalamus (A), hippocampus (C), caudate (D), raphe nucleus (E) and Purkinje cells (F) cells of the cerebellum. Many neurons are endothelin immunoreactive. (A,B): reverse face serial sections through the paraventricular nucleus. Some neurons that are endothelin immunoreactive (A) appear also to contain neurophysin immunoreactivity (B arrows). Tissues were fixed by immersion in Bouin's solution. Cryostat sections (20 μm) immunostained using the avidin–biotin–peroxidase complex method. A,B, × 250; C, × 210; D–F, × 220.

In Situ Hybridization

Endothelin mRNA has been demonstrated at low levels in the developing cerebellum of 19-day-old fetal rats, and a comparison of probes specific for rat endothelin-1 and endothelin-3 showed that rat brain tissue expressed predominantly endothelin-3 mRNA.[40] In pig, human endothelin-1 cRNA probes revealed labeled neurons in the paraventricular nucleus of the hypothalamus. (Because the 3' noncoding sequences of human and porcine preproendothelin-1 mRNAs retain significant nucleotide identity, pig endothelin-1 mRNA can readily cross-hybridize with the human endothelin-1 probe under the stringency conditions employed for in situ hybridization.)

In human brain,[17] numerous neurons were labeled throughout following hybridization with human endothelin-1 cRNA probes and were seen in regions where endothelin-immunoreactive cells had been noted (Fig. 12.17). Positively labeled neurons were present in layers III–VI of the cortex (Fig. 12.17A). A large population of neurons in hippocampal regions (CA 1–4, the hilus of the dentate gyrus) expressed endothelin mRNA. Some labeled neurons were present in the hypothalamic nuclei and caudate, with scattered labeled neurons in the amygdala, raphe nucleus, and dorsal vagal motor nuclei of the medulla oblongata (Fig. 12.17B,D). In the substantia nigra, the majority of neurons of

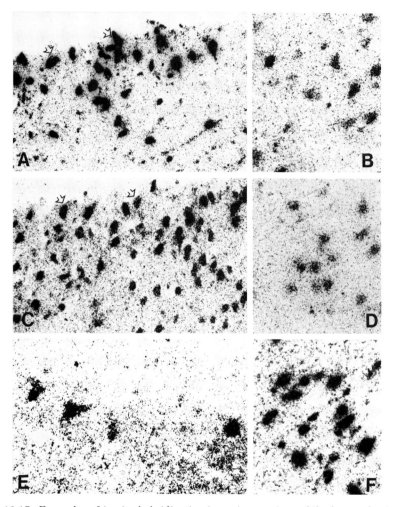

FIGURE 12.17. Examples of *in situ* hybridization in various regions of the human brain as revealed by [^{35}S] labeled cRNA probes for endothelin-1 (*A,B,D–F*) and human neuropeptide Y (*C*). Many neurons express endothelin-1 transcripts in the cortex (*A*), caudate (*B*), raphe nucleus (*D*), cerebellum (*E*), and substantia nigra (*F*). A,C: Reverse face serial sections through the cortex incubated with endothelin-1 (*A*) and neuropeptide Y (*C*) cRNA probes. Some neurons (arrows) appear to contain both mRNAs. A–D, × 250; E, × 310; F, × 270.

the pars reticulata and a few cells in the pars compacta expressed endothelin-1 transcripts (Fig. 12.17F). In the cerebellum most Purkinje cells were heavily labeled with silver grains (Fig. 12.17E).

Combined studies using immunocytochemistry and in situ hybridization (for methodological details, see Chan-Palay et al.[9] and Steel et al.[57]) on sections of human brain (cortex) showed that endothelin and its mRNA were present in the same cell. However, some cells displayed either endothelin immunoreactivity or its mRNA alone. In general, the numbers of neurons labeled after hybridization were greater than those that were immunoreactive, an observation that was also noted in the spinal cord not only for endothelin but also for other peptides.[22]

Colocalization of Endothelin with Other Peptides

Because the distribution of endothelin-immunoreactive neurons in the hypothalamus is similar to that of vasopressin and oxytocin and because all three peptides show a similar response to water deprivation experiments,[70] colocalization was probable. In human paraventricular nucleus, a number of neurons immunoreactive for neurophysin (carrier protein for vasopressin and oxytocin[21,65]) were also endothelin immunoreactive (Fig. 12.16A,B). However, the number of neurophysin-immunoreactive neurons was greater.[17] In pig hypothalamus also, preliminary reports indicate that most endothelin-immunoreactive neurons are colocalized with vasopressin and oxytocin in the paraventricular nucleus.[70]

In human cortex, neurons expressing endothelin and its mRNA were abundant. Neuropeptide Y is a peptide that is also present in high concentrations in the cortex.[7,60] The role of cortical interneurons containing neuropeptide Y is uncertain, but the interneurons have been shown to be abnormal in specific neurological diseases, and a significant reduction has been found in several cortical areas in tissues from patients with Alzheimer's disease.[8] Colocalization studies on reverse face serial sections of human cortex revealed the presence of endothelin and neuropeptide Y immunoreactivities or their respective immunoreactivities in a number of cortical neurons (Fig. 12.17A,C).[17] Whether this population of neuropeptide Y–endothelin neurons shows selective sparing or loss in Alzheimer's disease remains to be elucidated, as does the significance of endothelin expression in neurodegenerative disease.

ENDOTHELIAL CELLS IN THE NERVOUS SYSTEM

In human and pig spinal cord and brain, endothelin immunoreactivity was seen in some cells lining the endothelium of blood vessels. The intimal layer of larger blood vessels was rarely stained, but generally the microvasculature was well represented (Fig. 12.18). Demonstration of endothelin immunoreactivity and endothelial cells was dependent on the antiserum employed.[16] Although all antisera recognized the endothelin family of peptides (staining abolished by endothelin-1, endothelin-2, endothelin-3, and sarafotoxin), some antisera stained both neurons and endothelial cells, whereas others stained

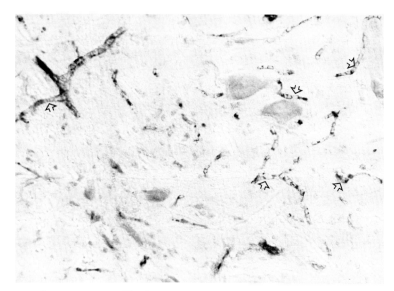

FIGURE 12.18. Immunoreactive blood vessels in the porcine spinal lumbar spinal cord (arrows). Note that with this antiserum to endothelin-1 the motoneurons are nonimmunoreactive. Bouin's solution, perfusion-fixed spinal cord. Cryostat section (20 μm) immunostained using the avidin–biotin-peroxidase complex method. × 190.

either neurons or endothelial cells. Immunoreactive cells were characterized as endothelial by immunostaining with the specific lectin marker *Ulex europaeus* agglutinin 1[58] in the same or adjacent section. In situ hybridization also revealed that endothelin transcripts were expressed in the intimal layer of some blood vessels.[16,17]

LOCALIZATION OF RECEPTOR BINDING SITES FOR ENDOTHELIN IN THE NERVOUS SYSTEM

Most in vitro autoradiography studies have employed radiolabeled [^{125}I]endothelin-1 to demonstrate receptor binding sites in rat and human nervous tissues.[13,28,31,40,50] In adult rat, displaceable binding sites for endothelin-1 are widely distributed in the nervous system, including peripheral[50] and central[13,28,31,35,40] nervous tissues. In the central nervous system, endothelin-1–binding sites are associated with specific brain regions and nuclei. The highest densities of binding sites are localized to the hypothalamus, thalamic areas, lateral ventricular structures, e.g., amygdaloid nucleus, subfornical regions, the basal ganglia (globus pallidus and caudate putamen), and brain stem structures, the substantia nigra, olive, and cerebellum. In the fetal rat, in contrast to other organs (lung, heart, kidney, and so forth) the cerebellum also contains the highest density of binding sites (although it has the lowest levels of endothelin-1 mRNA expression).[40] In the adult rat the lowest densities are found in the cortex and posterior pituitary.

In humans, the distribution of endothelin-1–binding sites follows a similar pattern.[28,31] The highest labeling occurs in the hippocampus and the granular layer of the cerebellum, and the lowest densities of binding occur in the cortex, caudate, and spinal cord. Because there is evidence to suggest that endothelin-1 and endothelin-3 isoforms occur in the nervous system (endothelin-1 being the predominant form), Fuxe et al.[13] compared the localization of binding sites for these isoforms in various regions of the rat brain. There was a significant codistribution of both. In hypothalamic, thalamic, amygdaloid, and habenular nuclei, endothelin-3 binding represented 50% of endothelin-1, but in cortical and hippocampal areas, endothelin-3 represented only 10% of endothelin-1 binding.

Several observations suggest that endothelin may act as a neuropeptide. The distribution of endothelin binding sites thus far described do not follow that of the vasculature. Administration of intraaortic endothelin-1 does not cross the blood–brain barrier,[35] whereas intracerebroventricular injection of endothelin induced behavioral changes associated with cerebellar function,[46] indicating possible release via a neuronal source. Further evidence for a role as a neuromodulator derives from receptor-binding studies in rats with ibotenic acid–induced lesions of the hippocampus and neostriatum,[13] where there was a significant loss of endothelin-1 binding and of neurons in the lesioned area, which demonstrates indirectly that endothelin-binding sites are localized to neurons.

The presence of high densities of binding sites in cerebellum and hippocampus corresponds closely with the distribution of endothelin-producing cells in these areas in rat, pig, and humans (see preceding section). However, there is a clear mismatch between the distribution of binding sites and cells expressing endothelin peptide and/or mRNA in a number of regions in man.[17,28,31] For example, low densities of binding sites are reported in the cortex, caudate, and spinal cord, where many endothelin-producing cells have been identified. It is possible, therefore, that endothelin is synthesized in these regions but acts on receptors that are located in distant structures in the nervous system.[40]

Finally, studies in rat brain have shown that there is a competitive interaction between endothelin and the homologous snake peptide sarafotoxin.[2] It is thought that endothelins and sarafotoxins share common binding sites and mechanisms of action. Both peptides have been implicated in the production of the intracellular second messenger inositol trisphosphate,[44] which releases intracellular stores of calcium for several neurotransmitters, hormones, and growth factors.[2,34] However, despite the proposed role of endothelin on voltage-sensitive Ca^{2+} channels[68,69] and the existence of binding sites on neurons, endothelin has no detectable effects on Ca^{2+} metabolism in the rat brain.[24]

CONCLUSIONS

It is clear from the morphological data reviewed here, which relate to only two systems in the body, the respiratory and nervous systems, that endothelin is indeed a peptide with a ubiquitous distribution. In the lung, the expression of endothelin in pulmonary airways and endothelium and the presence of endothelin receptors support the idea that endothelin is a local hormone (see also MacCumber[40]) in the respiratory tract with diverse actions, including growth promotion in the fetal stages.

The presence of endothelin immunoreactivity and its mRNA in neurons of the nervous system and the localization of high-affinity specific binding sites to many discrete brain regions suggests that endothelin exists as a neuropeptide. The precise role of endothelin in neurons is, however, unknown. Clearly, it occurs in neurons that regulate sensory (dorsal root ganglion), motor (motoneuron, Purkinje cell), and autonomic (preganglionic spinal neuron, dorsal vagal motor nucleus, paraventricular hypothalamic neuron) functions and in the hypothalamopituitary axis. Several roles are thus inferred either as a neurotransmitter or neuromodulator in the spinal cord and brain or as a neurosecretory hormone, circulating via the hypothalamopituitary pathways. Evidence for these actions is fast accumulating, with neurotransmitter–modulator actions suggested from the involvement of endothelin with the second messenger inositol[2] and a neurosecretory function suggested from physiological experimentation in rats.[70] Finally, but of no lesser importance, neurally derived endothelin may play an important role as a tropic factor and therefore should be considered of potential importance in neurological diseases, such as motor neuron disease and Alzheimer's disease.

Although emphasis has been placed on the neural localization of endothelin immunoreactivity and its mRNA and binding sites, the effects of endothelin, which could presumably derive from endothelial cells in injured central nervous system, should not be overlooked. In both the brain and spinal cord, central administration of pharmacological doses of endothelin has been shown to induce profound effects, resulting in ischemic lesions in the rat brain[14] and in profound damage to motoneurons in the spinal cord,[27] most likely because of the vasoconstrictor properties of the peptide.

ACKNOWLEDGMENTS

This work was supported by The Grand Charity of Freemasons and by the Motor Neurone Disease Association of Great Britain.

REFERENCES

1. AKAGI, H., S. KONISHI, M. OTSUKA, and M. YANAGISAWA. The role of substance P as a neurotransmitter in the reflexes of slow time courses in the neonatal rat spinal cord. *Br. J. Phamacol.* 84: 663–673, 1985.
2. AMBAR, I., Y. KLOOG, I. SCHAVARTZ, E. HAZUM, and N. SOKOLOVSKY. Competitive interaction between endothelin and sarafotoxin binding and phosphoinositide hydrolysis in rat atria and brain. *Biochem. Biophys. Res. Commun.* 158: 195–201, 1989.
3. ANDREWS, P. C., K. A. BRAYTON, and J. E. DIXON. Post-translational proteolytic processing of precursors to regulatory peptides. In: *Regulatory Peptides,* edited by J. M. Polak. Basel: Birkhauser Verlag, 1989, p. 192–209.
4. BHATNAGAR, M., D. R. SPRINGALL, M. A. GHATEI, P. W. J. BURNET, Q. HAMID, A. GIAID, N. B. N. IBRAHIM, F. CUTTITTA, E. R. SPINDEL, R. PENKETH, C. RODE, S. R. BLOOM, and J. M. POLAK. Localisation of mRNA and co-expression and molecular forms of GRP gene products in endocrine cells of fetal human lung. *Histochemistry* 90: 299–307, 1988.
5. BLACK, P. N., M. A. GHATEI, K. TAKASHI, D. BRETHERTON-WATT, T. KRAUSZ, C. T. COLLERY, and S. R. BLOOM. Formation of endothelin by cultured airway epithelial cells. *FEBS Lett.* 255: 129–132, 1989.
6. BLOOM, S. R., and R. G. LONG. *Radioimmunoassay of Gut Regulatory Peptides.* London: W. B. Saunders, 1982.
7. CHAN-PALAY, V., Y. S. ALLEN, W. LANG, U. HAESLER, and J. M. POLAK. I. Cytology and distribution in normal human cerebral cortex of neurons immunoreactive with antisera against neuropeptide Y. *J. Comp. Neurol.* 238: 382–389, 1985.

8. CHAN-PALAY, V., W. LANG, Y. S. ALLEN, U. HAESLER, and J. M. POLAK. II. Cortical neurons immunoreactive with antisera against neuropeptide Y are altered in Alzheimer's type dementia. *J. Comp. Neurol.* 238: 390–400, 1985.
9. CHAN-PALAY, V., G. YASARGIL, Q. HAMID, J. M. POLAK, and S. L. PALAY. Simultaneous demonstrations of neuropeptide Y gene expression and peptide storage in single neurons of the human brain. *Proc. Natl. Acad. Sci. USA* 85: 3213–3215, 1988.
10. DALSGAARD, C.-J. The sensory system. In: *Handbook of Chemical Neuroanatomy,* edited by A. Bjorklund, T. Hokfelt, and C. Owman. New York: Elsevier, 1988, vol. 6, p. 599–636.
11. DAVENPORT, A. P., R. G. HILL, and J. HUGHES. Quantitative analysis of autoradiograms. In: *Regulatory Peptides,* edited by J. M. Polak. Basel: Birkhauser Verlag, 1989, p. 137–153.
12. DENNY, P., Q. HAMID, J. E. KRAUSE, J. M. POLAK, and S. LEGON. Oligoriboprobes: tools for in situ hybridisation. *Histochemistry* 89: 481–483, 1988.
13. FUXE, K., E. ANGGARD, K. LUNDGREN, A. CINTRA, L. F. AGNATI, S. GALTON, and J. VANE. Localization of [^{125}I]endothelin-1 and [^{125}I]endothelin-3 binding sites in the rat brain. *Acta Physiol. Scand.* 137: 563–564, 1989.
14. FUXE, K., A. CINTRA, B. ANDBJER, E. ANGGARD, M. GOLDSTEIN, and L. F. AGNATI. 1989. Centrally administered endothelin-1 produces lesions in the brain of the male rat. *Acta Physiol. Scand.* 137: 155–156, 1989.
15. GALL, J., and M. PARDUE. Formation and detection of RNA–DNA hybrid molecules in cytological preparations. *Proc. Natl. Acad. Sci. USA* 63: 378–383, 1969.
16. GIAID, A., S. J. GIBSON, N. B. N. IBRAHIM, S. LEGON, S. R. BLOOM, M. YANAGISAWA, T. MASAKI, I. M. VARNDELL, and J. M. POLAK. Endothelin 1, an endothelium-derived peptide, is expressed in neurones of the human spinal cord and dorsal root ganglia. *Proc. Natl. Acad. Sci. USA* 86: 7634–7638, 1989.
17. GIAID, A., S. J. GIBSON, S. LEGON, T. HERRERO, D. UWANGHO, N. B. N. IBRAHIM, M. YANAGISAWA, T. MASAKI, S. R. BLOOM, and J. M. POLAK. Expression of endothelin immunoreactivity and mRNAs and its co-localisation with other peptides in the human brain. *J. Pathol.* 160: 177, 1990.
18. GIBSON, S. J., and J. M. POLAK. Neurochemistry of the spinal cord. In: *Immunocytochemistry: Modern Methods and Applications* (2nd ed.), edited by J. M. Polak and S. Van Noorden. Bristol: John Wright and Sons, 1986, p. 360–390.
19. GIBSON, S. J., and J. M. POLAK. Principles and applications of complementary RNA probes. In: *Modern Methods in Pathology: In Situ Hybridization,* edited by J. M. Polak, and J. McGee. Oxford: Oxford University Press.
20. GIBSON, S. J., J. M. POLAK, S. R. BLOOM, I. M. SABATE, P. M. MULDERRY, M. A. GHATEI, J. F. B. MORRISON, J. S. KELLY, R. EVANS, and M. G. ROSENFELD. Calcitonin gene–related peptide (CGRP) immunoreactivity in the spinal cord of man and of eight other species. *J. Neurosci.* 4: 3101–3111, 1984.
21. GIBSON, S. J., J. M. POLAK, S. R. BLOOM, and P. D. WALL. The distribution of nine peptides in rat spinal cord with special emphasis on the substantia gelatinosa and on the area around the central (lamina X). *J. Comp. Neurol.* 201: 65–79, 1981.
22. GIBSON, S. J., J. M. POLAK, A. GIAID, Q. A. HAMID, S. KAR, P. M. JONES, P. DENNY, S. LEGON, S. G. AMARA, R. K. CRAIG, S. R. BLOOM, R. J. A. PENKETH, C. RODEK, N. B. N. IBRAHIM, and A. DAWSON. Calcitonin gene–related peptide messenger RNA is expressed in sensory neurones of the dorsal root ganglia and also in spinal motoneurones in man and rat. *Neurosci. Lett.* 91: 282–288, 1988.
23. GIBSON, S. J., J. M. POLAK, T. KATAGIRI, H. SU, R. O. WELLER, D. B. BROWNELL, S. HOLLAND, J. T. HUGHES, S. KIKUYAMA, J. BALL, S. R. BLOOM, T. J. STEINER, J. DEBELLAROCHE, and F. CLIFFORD ROSE. A comparison of the distributions of eight peptides in spinal cord from normal controls and cases of motor neurone disease with special reference to Onuf's nucleus. *Brain Res.* 474: 255–278, 1988.
24. HAMILTON, M. G., R. FREW, and P. M. LUNDY. Effect of endothelin on Ca^{2+} influx, intracellular free Ca^{2+} levels and ligand binding to N and L type Ca^{2+} in rat brain. *Biochem. Biophys. Res. Commun.* 128: 1333–1338, 1989.
25. HENRY, J. L. Relation of substance P to pain transmission: neurophysiological evidence. In: *Substance P in the Nervous System* (CIBA Foundation Symposium 91), edited by R. Porter and M. O'Connor. London: Pitman, 1982, p. 206–224.
26. HIRANO, A., and M. IWATA. Pathology of motor neurons with special reference to amyotrophic lateral sclerosis and related diseases. In: *Amyotrophic Lateral Sclerosis* (Japan Medical Research Foundation Publication No 8), edited by T. Tsubaki and Y. Toykura. Baltimore: University Park Press, 1979, p. 107–135.
27. HOKFELT, T., C. POST, J. FREEDMAN, and J. M. LUNDBERG. Endothelin induces spinal lesions after intrathecal administration. *Acta Physiol. Scand.* 137: 555–556, 1989.
28. HOYER, D., C. WAEBER, and J. M. PALACIOS. [^{125}I]endothelin-1 binding sites: autoradio-

graphic studies in the brain and periphery of various species including humans. *J. Cardiovas. Pharmacol.* 13(Suppl. 5): S162–S165, 1989.
29. INOUE, A., M. YANAGISAWA, S. KIMURA, Y. KASUYA, T. MIYAUCHI, K. GOTO, and T. MASAKI. The human endothelin family: three structurally and pharmacologically distinct isopeptides predicted by three separate genes. *Proc. Natl. Acad. Sci. USA* 86: 2863–2867, 1989.
30. ITOH, Y., C. KIMURA, H. ONDA, and M. FUJINO. Canine endothelin-2: cDNA sequence for the mature peptide. *Nucleic Acids Res.* 17: 5389, 1989.
31. JONES, C. R., C. R. HILEY, J. T. PELTON and M. MOHR. Autoradiographic visualization of the binding sites for [^{125}I]endothelin in rat and human brain. *Neurosci. Lett.* 97: 276–279, 1989.
32. KITAMURA, K., T. TANAKA, J. KATO, T. ETO, K. TANAKA. Regional distribution of immunoreactive endothelin in porcine tissues: abundance in inner medulla of kidney. *Biochem. Biophys. Res. Commun.* 161: 348–352, 1989.
33. KITAMURA, K., T. TANAKA, J. KATO, T. OGAWA, T. ETO, and K. TANAKA. Immunoreactive endothelin in rat kidney inner medulla: marked decrease in spontaneously hypertensive rats. *Biochem. Biophys. Res. Commun.* 162: 38–44, 1989.
34. KLOOG, Y., I. AMBAR, E. KOCHVA, Z. WOLLBERG, A. BDOLAH, and M. SOKOLOVSKY. Sarafotoxin receptors mediate phosphoinositide hydrolysis in rat brain regions. *Fed. Eur. Biochem. Soc.* 242: 387–390, 1989.
35. KOSEKI, C., M. IMAI, Y. HIRATA, M. YANAGISAWA, and T. MASAKI. Autoradiographic distribution in rat tissues of binding sites for endothelin: a neuropeptide? *Am. J. Physiol.* 256 *(Regulatory Integrative Comp. Physiol.* 25): R858–R866, 1989.
36. KURASHI, Y., T. NANAYAMA, H. OHNO, M. MINAMI, and M. SATOL. Antinociception induced in rats by intrathecal administration of antiserum against calcitonin gene–related peptide. *Neurosci. Lett.* 92: 325–329, 1988.
37. LAUFER, R., and J. P. CHANGEUX. Calcitonin gene–related peptide elevated cyclic AMP levels in chick skeletal muscle: possible neurotrophic role for a coexisting messenger. *EMBO J.* 6: 901–906, 1987.
38. LAUWERYNS, J. M., L. V. VAN RANST, V. R. LLOYD, and D. T. O'CONNOR. Chromogranin in bronchopulmonary neuroendocrine cells: immunocytochemical detection in human, monkey and pig respiratory musoca. *J. Histochem. Cytochem.* 35: 113–118, 1987.
39. LEE, I., V. E. GOULD, R. MOLL, B. WIEDENMANN, and W. W. FRANKE. Synaptophysin expressed in the bronchopulmonary tract: neuroendocrine cells, neuroepithelial bodies, and neuroendocrine neoplasms. *Differentiation* 34: 115–125, 1987.
40. MACCUMBER, M. W., C. A. ROSS, and S. H. SNYDER. Endothelin-visualisation of messenger RNAs by in situ hybridisation provides evidence for local action. *Proc. Natl. Acad. Sci. USA* 86: 2785–2789, 1989.
41. MATSUMOTO, H., N. SUZUKI, H. ONDA, and M. FUJINO. Abundance of endothelin-3 in rat intestine, pituitary gland and brain. *Biochem. Biophys. Res. Commun.* 164: 74–80, 1989.
42. MATTEOLI, M., C. HAIMANN, F. TORRI-TARELLI, J. M. POLAK, B. CECCARELLI, and P. DECAMILII. Differential effect of α-latrotoxin on exocytosis from small synaptic vesicles and from large dense core vesicles containing calcitonin gene–related peptide at the frog neuromuscular junction. *Proc. Natl. Acad. Sci. USA* 85: 7366–7370, 1988.
43. MCBRIDE, J. T., D. R. SPRINGALL, R. J. D. WINTER, and J. M. POLAK. Quantitative immunocytochemistry shows calcitonin gene–related peptide-like immunoreactivity in lung neuroendocrine cells is increased by chronic hypoxia in the rat. *Am. J. Respir. Cell Mol. Biol.* 3: 587–593, 1990.
44. MIGNERY, G. A., T. C. SUDHOF, K. TAKEI, and P. DECAMILLI. Putative receptor for inositol 1,4,5-triphosphate similar to ryanodine receptor. *Nature* 342: 192–194, 1989.
45. MORA, M., M. MARCHI, S. J. GIBSON, and J. M. POLAK. Calcitonin gene–related peptide immunoreactivity at the human neuromuscular junction. *Brain Res.* 492: 404–407, 1989.
46. MOSER, P. C., and J. T. PELTON. Behavioural effects of centrally administered endothelin in the rat. *Br. J. Pharmacol.* 10: 124–126, 1990.
47. POLAK, J. M., and S. R. BLOOM. Distribution and tissue localization of VIP in the central nervous system and in seven peripheral organs. In: *Vasoactive Intestinal Polypeptide*, edited by S. I. Said (ed). New York: Raven, 1982, p. 107–120.
48. POLAK, J. M., and S. VAN NOORDEN (EDS.). *Immunocytochemistry: Modern Methods and Applications* (2nd ed.). Bristol: John Wright and Sons, 1986.
49. POLAK, J. M., and I. M. VARNDELL (EDS.). *Immunolabelling for Electron Microscopy*. Amsterdam: Elsevier Science Publishers, 1984.
50. POWER, R. F., J. WHARTON, Y. ZHAO, S. R. BLOOM, and J. M. POLAK. Autoradiographic localization of endothelin-1 binding sites in the cardiovascular and respiratory systems. *J. Cardiovasc. Pharmacol.* 13: 550–556, 1989.

51. ROSENFELD, M. G., J. J. MEROD, S. J. AMARA, L. W. SWANSON, J. RIVIER, W. W. VALE, and R. M. EVANS. Production of a novel neuropeptide encoded by the calcitonin gene via tissue specific RNA processing. *Nature* 304: 129–135, 1983.
52. ROZENGURT, N., D. R. SPRINGALL, and J. M. POLAK. Endothelin-like immunoreactivity is localised in airway epithelium of rat and mouse lung. *J. Pathol.* 160: 5–8, 1990.
53. SAID, S. I., and V. MUTT. Polypeptide with broad biological activity isolation from small intestine. *Science* 169: 1217–1218, 1970.
54. SALT, T. E., and R. G. HILL. Neurotransmitter candidates of somatosensory primary afferent fibres. *Neuroscience* 10: 1083–1103, 1983.
55. SCHICHIRI, M., Y. HIRATA, K. KANNO, K. OHTA, T. EMORI, and F. MARUMO. Effect of endothelin-1 on release of arginine-vasopressin from perfused rat hypothalamus. *Biochem. Biophys. Res. Commun.* 163: 1332–1337, 1989.
56. SHINMI, O., S. KIMURA, T. YOSHIZAWA, T. SAWAMURA, Y. UCHIYAMA, Y. SUGITA, I. KANAZAWA, M. YANAGISAWA, K. GOTO, and T. MASAKI. Presence of endothelin-1 in porcine spinal cord: isolation and sequence determination. *Biochem. Biophys. Res. Commun.* 162: 340–346, 1989.
57. STEEL, J. H., D. J. O'HALLORAN, P. M. JONES, S. VAN NOORDEN, W. W. CHIN, S. R. BLOOM, and J. M. POLAK. Combined use of immunocytochemistry and in situ hybridisation to study β-thyroid-stimulating hormone gene expression in pituitaries of hypothyroid rats. *Mol. Cell. Probes* 4: 385–398, 1990.
58. STEPHENSON, T. J., D. W. R. GRIFFITHS, and P. M. MILLS. Comparison of *Ulex europeaus* I lectin binding and Factor VIII–related antigen as markers of vascular endothelium in follicular carcinoma of the thyroid. *Histopathology* 10: 251–260, 1986.
59. SPINDEL, E. R., M. E. SUNDAY, H. HOEFLER, H. J. WOLFE, and J. F. HABENER. Transient elevation of messenger RNA encoding gastrin-releasing peptide, a putative pulmonary growth factor in human fetal lung. *J. Clin. Invest.* 80: 1172–1179, 1987.
60. TERENGHI, G., J. M. POLAK, Q. HAMID, E. O'BRIEN, P. DENNY, S. LEGON, J. E. DIXON, C. D. MINTH, S. L. PALAY, G. YASARGIL, and V. CHAN-PALAY. Localization of neuropeptide Y mRNA in neurons of human cerebral cortex by means of in situ hybridisation with a complementary RNA probe. *Proc. Natl. Acad. Sci. USA* 84: 7315–7317, 1987.
61. THOMPSON, R. J., J. F. DORAN, P. JACKSON, A. P. DHILLON, and J. RODE. PGP 9.5—a new marker for vertebrate neurones and neuroendocrine cells. *Brain Res.* 278: 224–228, 1983.
62. TSUJIMOTO, T., and M. KUNO. Calcitonin gene–related peptide prevents disuse-induced sprouting of rat motor nerve terminals. *J. Neurosci.* 8: 3951–3957, 1988.
63. VAN NOORDEN, S., and J. M. POLAK. Immunocytochemistry of regulatory peptides. In: *Techniques in Immunocytochemistry*, edited by G. R. Bullock and P. Petrusz. New York: Academic, 1985, vol. 3, p. 116–154.
64. WALL, P. D., and M. FITZGERALD. If substance P fails to fulfill the criteria as a sensory transmitter, what might its function be? In: *Substance P in the Nervous System* (CIBA Foundation Symposium 91), edited by R. Porter and M. O'Connor. London: Pitman, 1982, p. 249–266.
65. WATKINS, W. B. Immunohistochemical study of the hypothalamoneurohypophyseal system. I. Localisation of neurosecretory neurons containing neurophysin-I and neurophysin-II in domestic pig. *Cell Tissue Res.* 175: 165–181, 1976.
66. WIKLUND, N. P., A. OHLEN, C. U. WIKLUND, B. CEDERQVIST, P. HEDQVIST, and S. O. GUSTAFSSON. Neuromuscular actions of endothelin on smooth, cardiac and skeletal muscle from guinea pig, rat and rabbit. *Acta Physiol. Scand.* 137: 399–407, 1989.
67. WOOLF, C., and Z. WEISENFELD-HALLIN. Substance P and calcitonin gene–related peptide synergistically modulate the gain of the nociceptive flexor withdrawal reflex in the rat. *Neurosci. Lett.* 66: 226–230, 1986.
68. YANAGISAWA, M., H. KURIHARA, S. KIMURA, Y. TOMOBE, M. KOBAYASHI, Y. MITSUI, Y. YAZAKI, K. GOTO, and T. MASAKI. A novel potent vasoconstrictor peptide produced by vascular endothelial cells. *Nature* 332: 411–415, 1988.
69. YOSHIZAWA, I., S. KIMURA, I. KAMAZAWA, Y. UCHIYAMA, M. YANAGISAWA, and T. MASAKI. Endothelin localizes in the dorsal horn and acts on the spinal neurones: possible involvement of dihydropyridine-sensitive calcium channels and substance P release. *Neurosci. Lett.* 102: 179–184, 1989.
70. YOSHIZAWA, T., O. SHINMI, A. GIAID, M. YANAGISAWA, S. J. GIBSON, S. KIMURA, Y. UCHIYAMA, J. M. POLAK, T. MASAKI, and I. KANAZAWA. Endothelin: a novel peptide in the posterior pituitary system. *Science* 247: 462–463, 1990.
71. YOUNG, W. S. III, and M. J. KUHAR. A new method for receptor autoradiography: [^{3}H]opioid receptors in rat brain. *Brain Res.* 179: 255–270, 1979.
72. ZAMBONI, L., and C. DE MARTINO. Buffered picric acid formaldehyde: a new rapid fixative for electron microscopy. *J. Cell Biol.* 35: 148A, 1967.

13

Endothelin in Human Disease. I: Essential Hypertension, Vasospastic Angina, Acute Myocardial Infarction, and Chronic Renal Failure

YOSHIHIKO SAITO, KAZUWA NAKAO, AND HIROO IMURA

Following the discovery of endothelin-1 in culture medium of porcine aortic endothelial cells,[1] several hundred papers have been published concerning the biological properties of endothelin-1. Studies were performed in experimental animals and isolated vessels and show that endothelin-1 has both vasoconstricting and vasodilating activities.[1–5] Vasoconstricting activity results from the increase in the intracellular calcium concentration in vascular smooth muscle cells,[1–4] and vasodilation is related to the production of endothelium-derived relaxing factor and prostanoids by endothelin-1.[3,5] It was also reported that endothelin-1 causes proliferation of vascular smooth muscle cells[6] and mediates the release of hormones involved in the control of water, electrolytes, and blood pressure, such as atrial natriuretic peptide,[7,8] aldosterone,[9] and renin.[10] Experimental results indicate that endothelin-1 is involved in the control of cardiovascular homeostasis and in the pathophysiology of cardiovascular disorders (see Chapter 16).

To assess the pathophysiological significance of endothelin-1 in human disease, the evaluation of endothelin-1 synthesis in endothelin-1–producing vessels and organs is essential. To date, however, there is no information available on this matter. Recently, we and others developed specific radioimmunoassays (RIAs)[11–14] or sandwich-enzyme immunoassays (EIA)[15] for endothelin-1 and demonstrated its presence in human plasma. A few papers have been published about the plasma level of endothelin-1–like immunoreactivity in human cardiovascular diseases. In this chapter, we review the plasma endothelin-1–like immunoreactivity levels in patients with essential hypertension, vasospastic angina, acute myocardial infarction, and chronic renal failure based on recent literature and on our observations.

ASSAY SYSTEMS FOR MEASURING PLASMA ENDOTHELIN-1–LIKE IMMUNOREACTIVITY IN HUMANS

Before reviewing the plasma levels of endothelin-1–like immunoreactivity in patients with various cardiovascular disorders, we summarize assay systems for endothelin-1 developed in several laboratories because the assay methods

(RIA or EIA), extraction procedures, and specificities of antibodies used differ among laboratories (Table 13.1). Four laboratories established RIAs,[11-14] and Suzuki et al.[15] established sandwich EIAs. We and Suzuki et al.[15] developed mouse monoclonal antibodies for endothelin-1, and others used rabbit polyclonal antibodies. No assay systems are available for the direct measurement of the plasma endothelin-1–like immunoreactivity levels, and those that are available require plasma extraction. All laboratories except ours adopted a commercially available minicolumn such as Sep-Pak C_{18} and Spe C_8 for extraction of endothelin-1 from plasma. We developed a new extraction method for endothelin-1 using monoclonal antibody–coated beads.[12,16] The major advantage of this method is the specificity in comparison with other extraction methods. Endothelin-1 contains three acidic amino acids and one basic amino acid, and its carboxy-terminal portion is very hydrophobic. In addition, endothelin-1 is known to be very sticky. These chemical features should be considered in the assay procedure, especially in endothelin-1 extraction.

PLASMA ENDOTHELIN-1–LIKE IMMUNOREACTIVITY LEVELS IN
NORMAL SUBJECTS

The plasma endothelin-1–like immunoreactivity levels in normal subjects have been reported by several laboratories and range from 0.26 to 19.9 pg/ml (Table 13.1).[11-15] We developed six kinds of monoclonal antibodies (KY-endothelin-1 I to VI) and three kinds of polyclonal antibodies (endothelin-F5, endothelin-F9, endothelin-F10) with different specificities.[13,16] With these antibodies, the normal plasma endothelin-1–like immunoreactivity levels ranged from 9.9 ± 0.5 pg/ml (RIA with endothelin-F5) to 19.1 ± 1.1 pg/ml (RIA with KY-endothelin-1 I; (Table 13.1).[13] The difference mainly results from the presence of multiple forms of endothelin-1–like immunoreactivity with large and small molecular weights in human plasma and from different cross-reactivities with them in each assay system. We examined the molecular forms of endothelin-1–like immunoreactivity in human plasma using gel permeation chromatography (GPC) and reported that they consist of three components, endothelin-1, big endothelin-1, and another component with a molecular weight of 6 kd.[12] Taking the GPC profile and the plasma immunoreactivity levels into account, the plasma levels of endothelin-1, big endothelin-1, and the 6 kd endothelin are 3.9 ± 0.3, 5.9 ± 0.4, and 9.3 ± 0.6 pg/ml (mean ± SE), respectively, although the total endothelin-1–like immunoreactivity level in human plasma is 19.1 ± 1.1 pg/ml in our RIA with KY-endothelin-1 I. Suzuki et al.[15] reported that the plasma endothelin-1 level is 1.59 ± 0.32 pg/ml (mean ± SD) in sandwich EIA, which does not recognize human big endothelin-1. Miyauchi et al.[17] developed the specific sandwich EIA for human big endothelin-1 and reported that the plasma level of big endothelin-1 is ~3.5 pg/ml. Therefore, our results are essentially the same as those of Suzuki et al.[15] and Miyauchi et al.[17] The cross-reactivities with endothelin-2, endothelin-3, and big endothelin-1 were not described in the remaining RIAs from three other laboratories. Considering the relatively low level of plasma endothelin-1–like immunoreactivity measured by these RIAs (Table 13.1), these assay systems may not recognize

TABLE 13.1. Characteristics of assay systems for plasma endothelin (ET) levels reported in the literature

| References | Assay system | Plasma ET-1-like immunoreactivity levels in normal subjects (pg/ml, mean ± SD) | Sensitivity (pg/tube) | Cross-reactivity with | | | Extraction procedure | Recovery rate (%) |
				ET-2	ET-3	Big ET-1		
Ando et al.[11]	RIA	1.5 ± 0.5	0.5	N.D.[a]	27	N.D.	Spe C_8	61
Suzuki et al.[15]	Sandwich EIA	1.59 ± 0.32	0.2[b]	160	0.2	0.3	Sep-Pak C_{18}	90
Saito et al.[12]	RIA	19.1 ± 3.1[c] 9.9 ± 1.4[d]	0.5 0.5	100 100	60 2	100 100	Affinity beads	52
Cernacek et al.[13]	RIA	0.25 ± 0.24	0.12	N.D.	N.D.	N.D.	Sep-Pak C_{18}	51
Xuan et al.[15]	RIA	2.0 ± 1.4	0.62	N.D.	N.D.	N.D.	Sep-Pak C_{18}	78

[a]N.D., not described.
[b]pg/well.
[c]Value is determined by the RIA with KY-ET-1 I and is corrected by the recovery rate.
[d]Value is determined by the RIA with ET-F5 and is corrected the recovery rate.

big endothelin-1 or other endothelin-1–containing components. In the Second Meeting on Endothelin held in Tsukuba, Japan, in 1990, Japanese researchers have agreed that the plasma level of endothelin-1 is 1–4 pg/ml.

Regarding the primary structure of preproendothelin-1, Yanagisawa et al.[1] proposed that big endothelin-1 cleaved from proendothelin-1 at paired basic amino acids, typical processing signals, and then big endothelin-1 is processed to endothelin-1. However, a GPC profile in our study raises the possibility that an endothelin-1–containing peptide exists with a larger molecular weight than big endothelin-1. To determine the precise plasma level and exact molecular isoforms of endothelin-1–like immunoreactivity in human plasma, the precise biosynthetic pathway of endothelin-1 must be clarified. It also remains to be clarified whether endothelin-2 and endothelin-3 and their big forms are present in human plasma.

ESSENTIAL HYPERTENSION

The intravenous administration of endothelin-1 induces sustained elevation of arterial pressure with an initial hypotensive response in experimental animals.[1,2] It has also been reported that endothelin-1 produces vasoconstriction in various isolated arteries and veins in vitro. These observations prompted us to study the plasma endothelin-1–like immunoreactivity levels in essential hypertension.

We studied 20 patients with essential hypertension before drug treatment.[16,18] They had normal renal function and negative tests for urine protein. As shown in Figure 13.1, the plasma immunoreactivity level was significantly higher in hypertensives than in age-matched control subjects (30.1 ± 1.4 vs. 18.5 ± 0.9 pg/ml, determined by the RIA with KY-endothelin-1 I; mean ± SE), suggesting the possible involvement of endothelin-1 in the development and/or maintenance of essential hypertension. The patients in (WHO classification) stages II and III showed higher plasma endothelin-1–like immunoreactivity levels than did those in stage I (Fig. 13.2, left). In addition, when the patients were divided into two groups with plasma immunoreactivity levels below and above the average value (30 pg/ml), patients with the high level had organ complications more frequently (Fig. 13.2, right). There was no significant correlation between the plasma immunoreactivity level and systolic blood pressure ($r = 0.11$) or diastolic blood pressure ($r = -0.13$).[16] The plasma immunoreactivity levels in patients with essential hypertension were not correlated with plasma levels of atrial natriuretic peptide, aldosterone, norepinephrine, or epinephrine or with plasma renin activity.[16] These observations suggest that the elevated plasma level of the peptide is related to organ complications, including atherosclerosis, rather than with high blood pressure itself. Although Miyauchi et al.[19] reported no significant changes in the plasma immunoreactivity levels in patients with essential hypertension, they observed a positive correlation between the plasma endothelin-1–like immunoreactivity level and age. It is well known that systolic and diastolic blood pressures are correlated with age and that atherosclerotic changes in vessels are more frequently found in older than in younger people.

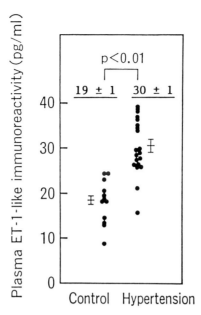

FIGURE 13.1. Plasma endothelin (ET)-1–like immunoreactivity level in patients with essential hypertension. Values are means ± SE.

The role of the increased plasma endothelin-1–like immunoreactivity level in essential hypertension is not clear at present. Komuro et al.[6] reported that endothelin-1 has a proliferative effect on vascular smooth muscle cells. It may be possible therefore that the increased secretion of endothelin-1 from endothelial cells plays an obligatory role in the development and aggravation of atherosclerotic changes in underlying intima and media and thereby is related to organ complications. A greater rise in the plasma endothelin-1–like immu-

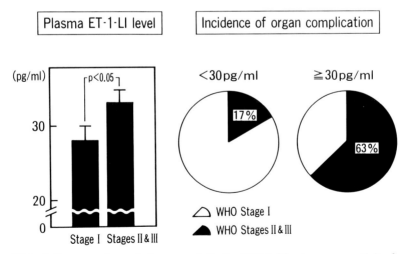

FIGURE 13.2. *Left:* comparison of plasma endothelin (ET)-1–like immunoreactivity levels of patients in stages II and III and those in stage I. *Right:* relation between plasma endothelin-1–like immunoreactivity level and incidence of organ complications.

noreactivity level was observed in hypertensives with renal hypofunction, which suggests the relation of endothelin-1 to renal complications in essential hypertension. Further studies are necessary to elucidate the mechanism whereby the plasma endothelin-1–like immunoreactivity level in essential hypertension is elevated.

VASOSPASTIC ANGINA

It is of great interest to know whether endothelin-1 participates in the pathogenesis of coronary artery spasm. Kurihara et al.[20] reported that intracoronary injection of endothelin-1 into dogs induces a significant reduction in coronary blood flow in association with ST-segment elevation and proposed the possible involvement of endothelin-1 in vasospasm.

We studied 14 patients with vasospastic angina. The diagnosis was made by the typical chest pain in association with ST-segment elevation. The basal plasma endothelin-1–like immunoreactivity levels in these patients were increased by 30% compared with the normal value ($P < 0.01$; Fig. 13.3). Toyooka et al.[21] also observed elevation of the basal plasma level of endothelin-1–like immunoreactivity in patients with vasospastic angina. These observations suggest that the elevation is related to endothelial disfunction or injury.

In spite of the elevation, no significant change was observed during spontaneous or hyperventilation-induced anginal attack. The average value of the plasma endothelin-1–like immunoreactivity levels did not significantly change during spontaneous attacks or 5 min after the disappearance of chest pain (before attack, 28.1 ± 2.6 pg/ml; during attack, 27.3 ± 1.7 pg/ml; after attack, 26.8 ± 2.4 pg/ml), but individual changes during the attack varied from case

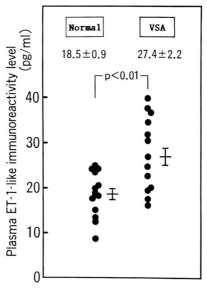

FIGURE 13.3. Plasma endothelin (ET)-1–like immunoreactivity levels in patients with vasospastic angina (VSA). Values are means ± SE.

to case (Fig. 13.4). The plasma immunoreactivity level was increased in about one-third of the patients and decreased in one third. In some patients, the level increased after the disappearance of chest pain. Although the change was not correlated with any clinical parameters in this small population of patients, measurements of the plasma endothelin-1–like immunoreactivity levels during attacks might be useful for the classification of the nature and severity of coronary vasospasm. Matsuyama et al.[22] observed that the plasma immunoreactivity level in the coronary sinus was elevated in patients with increased lactate production during acetylcholine-induced vasospasm, whereas it was not increased in patients without increased lactate production during the vasospasm.

ACUTE MYOCARDIAL INFARCTION

Thrombin and transforming growth factor β (TGF-β) are reported to increase the endothelin-1 mRNA level in cultured endothelial cells.[20,23] It is possible therefore that endothelin-1 has a pathophysiological significance in coronary artery thorombosis and acute myocardial infarction. Miyauchi et al.[17] reported that the endothelin-1–like immunoreactivity level in plasma collected within 1 day of onset of acute myocardial infarction is doubled in comparison with the normal level. The concentration then gradually declines over the next 4–6 days and returns to normal by day 13 (Fig. 13.5). We also observed a 1.6-fold increase in the plasma immunoreactivity level (32.0 ± 2.5 pg/ml in the RIA with KY-endothelin-1 I) in 10 patients with acute myocardial infarction within a few days after the onset. Using the specific sandwich EIAs for endothelin-1 and big endothelin-1 in acute myocardial infarction, Miyauchi et al.[17] discovered parallel increases in endothelin-1 and big endothelin-1.

Neither the mechanism responsible for the elevation or the major site of

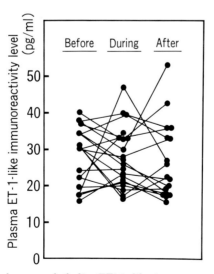

FIGURE 13.4. Changes in plasma endothelin (ET)-1–like immunoreactivity levels before, during, and after spontaneous anginal attacks.

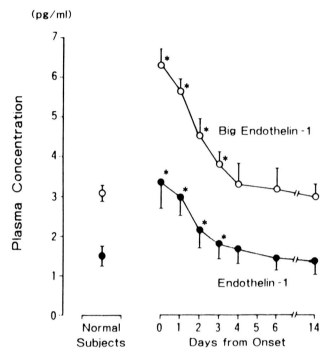

FIGURE 13.5. Changes in plasma endothelin-1–like immunoreactivity levels in patients with acute myocardial infarction. (From Miyauchi et al.[17])

endothelin-1 production are clear at present. As mentioned above, thrombin and TGF-β stimulate the expression of the endothelin-1 gene. It is possible therefore that endothelin-1 synthesis is increased in the coronary vessels in association with thrombus formation in acute myocardial infarction. However, in our preliminary study there was no significant difference in the plasma immunoreactivity level between the coronary sinus and the ascending aorta in two patients with acute myocardial infarction and one with impending infarction (Table 13.2). Cernacek and Stewart[13] and Hirata et al.[24] reported that cardiogenic shock and surgical stress respectively increase the plasma immunoreactivity level. It is possible therefore that the elevation of the im-

TABLE 13.2. Plasma endothelin (ET) levels in acute myocardial infarction

		Plasma ET-1–like immunoreactive levels (pg/ml)		
	Diagnosis	CS	Ao	FV
Patient 1	AMI	33.3	29.4	28.7
Patient 2	AMI	25.1	28.4	25.7
Patient 3	IMI	21.3	24.2	20.1

AMI, acute myocardial infarction; MI, myocardial infarction; IMI, impending myocardial infarction; CS, coronary sinus; Ao, ascending aorta; FV, femoral vein.

munoreactivity level in acute myocardial infarction is an acute-phase reaction to severe physical stress.

Watanabe et al.,[25] using monoclonal antibody for endothelin-1, made an interesting finding concerning the role of endothelin-1 in acute myocardial infarction. Acute myocardial infarction was induced in rat hearts by 1 h ligation of the coronary artery and followed by reperfusion. The administration of the antibody reduced the infarction size in a dose-dependent manner. An increase in the endothelin-1 level in the infarcted ventricle was also observed.

CHRONIC RENAL FAILURE

The renal artery is one of the vessels most sensitive to endothelin-1. Intrarenal infusion of endothelin-1 induces a significant decrease in renal blood flow and consequently causes acute renal failure in rats.[26] In this context, it was of interest to measure plasma endothelin-1–like immunoreactivity levels in renal failure.

We studied nine patients with chronic renal failure (CRF) caused by chronic glomerulonephritis, diabetic nephropathy, and chronic pyelonephritis. The plasma endothelin-1–like immunoreactivity levels in patients with CRF were about five times higher than those in normal subjects (Fig. 13.6) and is consistent with the fivefold increase recently reported.[27] The rise in CRF is much greater than that observed in essential hypertension, vasospastic angina, and acute myocardial infarction. In our study, the plasma immunoreactivity level was significantly correlated with the serum creatinine level, and there was no significant difference in the plasma concentration between CRF patients with and without hypertension. Shichiri et al.[28] have reported a small rise only in CRF patients with hypertension. Although Koyama et al.[27] reported that plasma endothelin-1–like immunoreactivity levels significantly

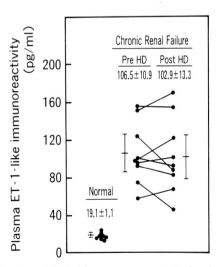

FIGURE 13.6. Plasma endothelin (ET)-1–like immunoreactivity levels in patients with chronic renal failure before and after hemodialysis (HD). Values are means ± SE.

decreased after hemodialysis, we did not observe a significant change between levels before and after hemodialysis (Fig. 13.6).

It is not clear at present whether the increased plasma level level in CRF is due to the augmentation of endothelin-1 synthesis or to the reduced clearance of the peptide in the kidney. Our preliminary observations indicate that large molecular forms of endothelin-1–like immunoreactivity (big endothelin-1 and the 6 kd endothelin) are predominantly increased in CRF. It is possible therefore that the elevation of the plasma immunoreactivity level can be explained by the reduced clearance of big endothelin-1 and the 6 kd endothelin in the kidney. Recent studies show the presence of endothelin-1–like immunoreactivity in rat kidney[29] and secretion of endothelin-1 or endothelin-2 from several renal epithelial cell lines.[30,31] It is not completely ruled out therefore that endothelin-1 systhesis is increased in the kidney in CRF.

Kon et al.[32,33] reported that anti-endothelin-1 serum ameliorates the vasoconstriction characteristic of postischemic nephrons and that antiendothelin-1 serum prevents the cyclosporin-induced deterioration of renal function. These observations suggest that endothelin-1 aggravates the renal function in CRF.

CONCLUSION

In this chapter, we reviewed the existing data on plasma endothelin-1–like immunoreactivity levels in patients with essential hypertension, vasospastic angina, acute myocardial infarction, and chronic renal failure. The measurement of plasma endothelin-1–like immunoreactivity is a clue with which to assess the pathophysiological significance of endothelin-1 in various human diseases, although its usefulness may be limited because endothelin-1 probably acts as a paracrine, or local, regulator rather than as a hormone, or systemic, regulator. The relatively smaller change in the immunoreactivity compared with that of the classic hormones is consistent with the idea that endothelin-1 is a local regulator. To determine the best sampling point for measurement of the plasma endothelin-1–like immunoreactivity level, the main vessels or organs of endothelin-1 production should be clarified. However, the physiological and pathophysiological significance of endothelin-1 as a circulating hormone cannot be ruled out at present, because the plasma level of endothelin-1 (10^{-12} M) is known to exert biological actions. The precise elucidation of the pathophysiological significance of endothelin-1 in human disease must await the development of specific antagonists for endothelin-1.

REFERENCES

1. YANAGISAWA, M., H. KURIHARA, S. KIMURA, Y. TOMOBE, M. KOBAYASHI, Y. MITSUI, Y. YAZAKI, K. GOTO, and T. MASAKI. A novel potent vasoconstrictor peptide produced by vascular endothelial cells. *Nature* 332: 411–415, 1988.
2. YANAGISAWA, M., and T. MASAKI. Endothelin, a novel endothelium-derived peptide. *Biochem. Pharmacol.* 38: 1877–1883, 1989.
3. SIMONSON, M. S., and M. J. DUNN. Endothelin—pathways of transmembrane signaling. *Hypertension* 15(Suppl. I): I-5–I-12, 1990.
4. HIRATA, Y., H. YOSHIMI, S. TAKATA, T. X. WATANABE, S. KUMAGAI, K. NAKAJIMA, and S. SAK-

AKIBARA. Cellular mechanism of action by a novel vasoconstrictor endothelin in cultured rat vascular smooth muscle cells. *Biochem. Biophys. Res. Commun.* 154: 868–875, 1988.
5. DE NUCCI, G., R. THOMAS, P. D'ORLEANS-JUSTE, E. ANTUNES, C. WALDER, T. D. WARNER, and J. R. VANE. Pressor effects of circulating endothelin are limited by its removal in the pulmonary circulation and by the release of prostacyclin and endothelium-derived relaxing factor. *Proc. Natl. Acad. Sci. USA* 85: 9797–9800, 1988.
6. KOMURO, I., H. KURIHARA, T. SUGIYAMA, F. TAKAKU, and Y. YAZAKI. Endothelin stimulates c-fos and c-myc expression and proliferation of vascular smooth muscle cells. *FEBS Lett.* 238: 249–252, 1988.
7. FUKUDA, Y., Y. HIRATA, H. YOSHIMI, T. KOJIMA, Y. KOBAYASHI, M. YANAGISAWA, and T. MASAKI. Endothelin is a potent secretagogue for atrial natriuretic peptide in cultured rat atrial myocytes. *Biochem. Biophys. Res. Commun.* 155: 167–172, 1988.
8. SHIRAKAMI, G., T. MAGARIBUCHI, T. YAMADA, H. ITOH, H. ARAI, K. HOSODA, S. SUGA, K. MORI, and H. IMURA. Inhibitory action of physiological level of endothelin-1 (ET-1) on ANP secretion, abstracted. *Circulation* 80(Suppl. II): II-585, 1989.
9. COZZA, E. N., C. E. GOMEZ-SANCHEZ, M. F. FOECKING, and S. CHIOU. Endothelin binding to cultured calf adrenal zona glomerulosa cells and stimulation of aldosterone secretion. *J. Clin. Invest.* 84: 1032–1035, 1989.
10. RAKUGI, H., M. NAKAMURA, H. SAITO, J. HIGAKI, and T. OGIHARA. Endothelin inhibits renin release from isolated rat glomeruli. *Biochem. Biophys. Res. Commun.* 155: 1244–1247, 1988.
11. ANDO, K., Y. HIRATA, M. SHICHIRI, T. EMORI, and F. MARUMO. Presence of immunoreactive endothelin in human plasma. *FEBS Lett.* 245: 164–166, 1989.
12. SAITO, Y., K. NAKAO, H. ITOH, T. YAMADA, M. MUKOYAMA, H. ARAI, K. HOSODA, G. SHIRAKAMI, S. SUGA, M. JOUGASAKI, S. MORICHIKA, and H. IMURA. Endothelin in human plasma and culture medium of aortic endothelial cells—detection and characterization with radioimmunoassay using monoclonal antibody. *Biochem. Biophys. Res. Commun.* 161: 320–326, 1989.
13. CERNACEK, P., and D. J. STEWART. Immunoreactive endothelin in human plasma: marked elevations in patients in cardiogenic shock. *Biochem. Biophys. Res. Commun.* 161: 562–567, 1989.
14. XUAN, Y. T., A. R. WHORTON, E. SHEARER-POOR, J. BOYD, and W. D. WATKINS. Determination of immunoreactive endothelin in medium from cultured endothelial cells and human plasma. *Biochem. Biophys. Res. Commun.* 164: 326–332, 1989.
15. SUZUKI, N., H. MATSUMOTO, C. KITADA, T. MASAKI, and M. FUJINO. A sensitive sandwich-enzyme immunoassay for human endothelin. *J. Immunol. Methods* 118: 245–250, 1989.
16. SAITO, Y., K. NAKAO, M. MUKOYAMA, G. SHIRAKAMI, H. ITOH, T. YAMADA, H. ARAI, K. HOSODA, S. SUGA, M. JOUGASAKI, Y. OGAWA, S. NAKAJIMA, M. UEDA, and H. IMURA. Application of monoclonal antibodies for endothelin to hypertensive research. *Hypertension* 15: 734–738, 1990.
17. MIYAUCHI, T., M. YANAGISAWA, T. TOMIZAWA, Y. SUGISHITA, N. SUZUKI, M. FIJINO, R. AJISAKA, K. GOTO, and T. MASAKI. Increased plasma concentrations of endothelin-1 and big endothelin-1 in acute myocardial infarction. *Lancet* 2: 53–54, 1989.
18. SAITO, Y., K. NAKAO, M. MUKOYAMA, and H. IMURA. Increased plasma endothelin level in patients with essential hypertension. *N. Engl. J. Med.* 322: 205, 1990.
19. MIYAUCHI, T., M. YANAGISAWA, N. SUZUKI, K. IIDA, Y. SUGISHITA, M. FUJINO, T. SAITO, K. GOTO, and T. MASAKI. Venous plasma concentrations of endothelin in normal and hypertensive subjects, abstracted. *Ciculation* 80(Suppl.): II-573, 1989.
20. KURIHARA, H., M. YOSHIZUMI, T. SUGIYAMA, K. YAMAOKI, R. NAGAI, F. TAKAKU, H. SATOH, J. INUI, M. YANAGISAWA, T. MASAKI, and Y. YAZAKI. The possible role of endothelin-1 in the pathogenesis of coronary vasospasm. *J. Cardiovasc. Pharmacol.* 13(Suppl. 5): S132–S137, 1989.
21. TOYO-OKA, T., T. AIZAWA, N. SUZUKI, Y. HIRATA, T. MIYAUCHI, M. YANAGISAWA, and T. MASAKI. Contribution of endothelin (ET) and atrial natriuretic factor (ANF) to coronary spasm in patients with vasospastic angina (VSA), abstracted. *Circulation* 80(Suppl.): II-126, 1989.
22. MATSUYAMA, K., H. YASUE, K. OKUMURA, Y. SAITO, K. NAKAO, G. SHIRAKAMI, and H. IMURA. Increased plasma level of endothelin-1-like immunoreactivity during coronary spasm in patients with coronary spastic angina. *Am. J. Cardiol.* 1991 (in press).
23. KURIHARA, H., M. YOSHIZUMI, T. SUGIYAMA, F. TAKAKU, M. YANAGISAWA, T. MASAKI, M. HAMAOKI, H. KATO, and Y. YAZAKI. Transforming growth factor-β stimulates the expression of endothelin mRNA by vascular endothelial cells. *Biochem. Biophys. Res. Commun.* 159: 1435–1440, 1989.

24. HIRATA, Y., K. ITOH, K. ANDO, M. ENDO, and F. MARUMO. Plasma endothelin levels during surgery. *N. Engl. J. Med.* 321: 1686, 1989.
25. WATANABE, T., N. SUZUKI, N. SHIMOMOTO, M. FUJINO, and A. IMADA. Endothelin in myocardial infarction. *Nature* 344: 114, 1989.
26. FIRTH, J. D., P. J. RATCLIFFE, A. E. G. RAINE, and J. G. G. LEDINGHAM. Endothelin: an important factor in acute renal failure? *Lancet* 2: 1179–1181, 1988.
27. KOYAMA H., T. TABATA, Y. NISHIZAWA, T. INOUE, H. MORI, and T. YAMAJI. Plasma endothelin levels in patients with uraemia. *Lancet* 1: 991–992, 1989.
28. SHICHIRI, M., Y. HIRATA, K. ANDO, T. EMORI, K. OHTA, S. KIMOTO, A. INOUE, and F. MARUMO. Plasma endothelin levels in patients with hypertension and end-stage renal failure, abstracted. *Circulation* 80(Suppl.): II-125, 1989.
29. KITAMURA, K., T. TANAKA, J. KATO, T. OGAWA, T. ETO, and K. TANAKA. Immunoreactive endothelin in rat kidney inner medulla: marked decrease in spontaneously hypertensive rats. *Biochem. Biophys. Res. Commun.* 162: 38–44, 1989.
30. KOSAKA, T., N. SUZUKI, H. MATSUMOTO, Y. ITOH, T. YASUHARA, H. ONDA, and M. FUJINO. Synthesis of the vasoconstrictor peptide endothelin in kidney cells. *FEBS Lett.* 249: 42–46, 1989.
31. SHICHIRI, M., Y. HIRATA, T. EMORI, K. OHTA, T. NAKAJIMA, K. SATO, A. SATO, and F. MARUMO. Secretion of endothelin and related peptides from renal epithelial cell lines. *FEBS Lett.* 253: 203–206, 1989.
32. KON, V., T. YOSHIOKA, A. FOGO, and I. ICHIKAWA. Glomerular actions of endothelin in vivo. *J. Clin. Invest.* 83: 1762–1767, 1989.
33. KON, V., M. SUGIURA, T. INAGAMI, and I. ICHIKAWA. Cyclosporine causes endothelin-dependent acute renal failure, abstracted. *Circulation* 80(Suppl.): II-483, 1989.

14

Endothelin in Human Disease. II: Shock, Pulmonary Hypertension, and Congestive Heart Failure

DUNCAN J. STEWART

Since the identification of endothelin by Yanagisawa and coworkers less than four years ago, there has been considerable interest in defining the molecular biology of endothelin production and its biological properties. It is perhaps indicative of the times that our understanding of the physiological role of endothelin has lagged considerably behind our knowledge of its molecular biology.[29] Although described in detail in accompanying chapters, some crucial aspects of endothelin's biological activity relating to its possible physiological and pathophysiological importance are briefly reviewed here. Unless otherwise specified, the endothelin referred to in this chapter is endothelin-1, the principal endothelin product of vascular endothelium. When appropriate, endothelin-2 and endothelin-3 will be specified.

Endothelin is a potent vasoconstrictor,[107] perhaps the most potent vasoconstrictor agent yet identified. Indeed, it was because of this property that a vasoconstrictor principle in endothelial cell–conditioned media first came to notice.[25,28] In vitro, endothelin has been shown to contract vessels from a wide variety of animal species,[22,97] including man.[9,12,13,26,108] In animal experiments, systemic administration of endothelin in vivo results in a complex hemodynamic response. Bolus administration of endothelin intravenously produces an initial decrease in systemic vascular resistance[32,45,107] possibly because of the release of prostacyclin[64,67,91] and endothelium-derived relaxing factor (EDRF)[82,101] by endothelin or via other mechanisms.[23] This is followed by a sustained, and often considerable, increase in vascular resistance and blood pressure as a result of the direct action of endothelin on vascular smooth muscle. In man, infusion of endothelin in the forearm bed[10] or systemically[98] also resulted in prolonged vasoconstriction.

The overall effects of circulating endothelin on the regional distribution of blood flow will depend on the differential responsiveness of the various vascular beds. The renal bed appears to be particularly sensitive to the constrictor actions of endothelin.[4,27,37,47,52,61] Endothelin produces marked reductions in renal plasma flow and glomerular filtration rate[7,39,48] and has a direct inhibitory action on tubular Na^+–K^+ exchange.[95,111] High concentrations of endothelin can compromise renal function,[39] raising the possibility that under certain conditions endothelin might contribute to (or mediate) renal failure.[18,92] Many other vascular beds also show an important vasoconstrictor responses to endothelin, e.g., mesenteric,[30,88] coronary,[11,16,20,35,42,82] cerebral,[2,3,17] and pulmonary[46,59] beds. However, in some preparations endothelin vasoconstriction appears to be less pronounced, and regional vasodilation may predominate.[19,46,82,100] To a large extent, the magnitude and direction of endothelin vasomotor response may be modified by the native endothelial layer.[36,82] In isolated, perfused vessel segments the intimal layer appears to present a near-complete barrier to endothelin,[63,76] preventing endothelin administered intraluminally from gaining access to the smooth muscle layer and producing contraction. In the coronary bed,[81,82] inhibition of EDRF or functional impairment of the endothelium[55,81,82] greatly potentiated the vasoconstrictor response to endothelin. Similar results were obtained in the isolated mesenteric bed[14,100]; however, here prostacyclin also appeared to play an important role. Therefore, endothelial damage, as occurs in a variety of pathophysiological states, may greatly enhance the constrictor action of circulating endothelin.

In addition to vasoconstriction, endothelin displays a plethora of other potent biological actions. Endothelin can contract a variety of different types of smooth muscle, including vesicular[49] and bronchial[50,70,94,96] It appears to be a growth modulating factor stimulating proliferation of smooth muscle cells,[15,38,58] fibroblasts,[6,90] and mesangial cells[78] in culture. It increases markedly the contractility of atrial and ventricular cardiac muscle[34,44,73,74,99] and has direct actions on the renal glomerulus[39] and tubules.[95,111] Endothelin also modulates the secretion of a variety of neurohumoral agents such as atrial natriuretic peptide,[21,33,105] renin,[65,89] and aldosterone.[57,60] In addition, it has been suggested that both endothelin-1 and endothelin-3 may act as neuropeptides[24,40,77,110] and modify neurotransmission[87,103,104,110] (for further details see Chapter 16).

Although far from complete, this brief summary illustrates the wide range of biological effects of the endothelin family of peptides. At present it is not known whether any of endothelin's diverse actions are physiologically relevant or, on the other hand, whether endothelin may contribute to pathophysiological states characterized by vasospasm and/or hypoperfusion. The clinical arena provides an ideal opportunity to determine the potential importance of endothelin in health and disease. With the recent advent of techniques to measure the plasma concentrations of endothelin, considerable evidence is now accumulating that will help provide a clearer picture of the biological relevance of endothelin in man.

EFFECT OF PHYSIOLOGICAL STIMULI ON PLASMA ENDOTHELIN LEVELS IN NORMAL SUBJECTS

Plasma endothelin concentration can be determined by radioimmunoassay using polyclonal or monoclonal antibodies to endothelin. Care must be taken to ascertain the specificity of the antibody, because in some assays there may be considerable cross-reactivity between the various endothelin peptides or between endothelin and big endothelin, its inactive precursor form. Because big endothelin appears in plasma in at least a twofold predominance over the active peptide,[54] this cross-reactivity can substantially contaminate the measurement and may account for some of the discrepancy between normal endothelin plasma levels reported by different groups. Most laboratories report normal concentrations of endothelin in the range of 1–3 pg/ml.[1,8,54,62,72,75,86,92,93,102,106] Figure 14.1 shows the normal range of immunoreactive endothelin measured in our laboratory. Miyauchi and coworkers[53] reported a slight but significant sex difference, with modestly higher levels in males, and a tendency for endothelin levels to increase with increasing age.

We have begun to examine the influence of physiological stimuli on plasma endothelin levels in normal subjects. Because it has been suggested that endothelin may contribute to the regulation of blood pressure and vascular resistance,[107] we have studied its response to a hypotensive stress induced by postural change.[82a] Figure 14.2 shows the effect of 60° head-up tilt on plasma endothelin concentration. In nearly all cases there is a rapid rise in endothelin levels, reaching a peak after 5 min. This increase, however, was not sustained throughout the 30 min period of tilting, but drifted back toward baseline at 10 and 15 min. A late increase in plasma endothelin occurred after 30 min of tilt, again reaching the level of statistical significance. This biphasic response of the immunoreactive endothelin level may reflect an initial hemodynamic stress at the onset of tilting and a second stress induced by prolonged tilt. The finding that endothelin plasma concentration is responsive to a physiological stimulus (i.e., posture) is in agreement with another report[75] and supports a role for endothelin in hemodynamic regulation in man. The circulating levels measured in normal subjects, even during postural stress, may be much too low to influence vascular tone directly. However, they may reflect a sub-

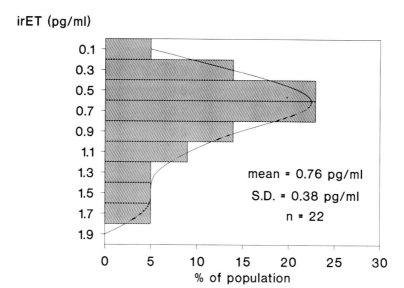

FIGURE 14.1. The distribution of immunoreactive endothelin-1 (irET-1) concentration in plasma from a group of normal subjects. Histograms indicate the percentage of the population within the ranges of endothelin concentration shown on the ordinate.

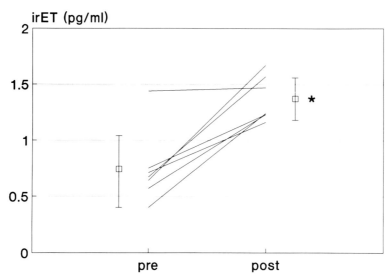

FIGURE 14.2. Response of plasma immunoreactive endothelin (irET) to postural stress induced by 60° upright tilt in seven healthy subjects. * = p < 0.05.

stantial increase in local endothelin production in the vasculature that could have a strong effect at the site of its release. It is unclear from these initial experiments what the stimulus inducing endothelin release into the bloodstream might be. Whether small changes in blood pressure can directly increase endothelin release by a baroreflex mechanism or whether the release is mediated secondarily by other neurohumoral factors (i.e., norepinephrine, angiotensin, and so forth) is not known. Further experiments are needed to establish the potential importance of other stimuli, such as exercise, increased (or decreased) blood flow, and hypoxia, among many others, in the control of endothelin release in vivo.

PLASMA ENDOTHELIN LEVELS IN PATHOPHYSIOLOGICAL STATES

The pathophysiological conditions that have been reported to be associated with elevated plasma endothelin levels are summarized in Table 14.1. We have examined plasma endothelin concentrations in a variety of clinical conditions characterized by profound abnormalities of vascular tone. The results, modified from our initial study[8] are shown in Figure 14.3. Several pathological states were found to be associated with an increase in plasma endothelin concentration well above the upper limit of normal. For the most part these were conditions characterized by a limitation in cardiac output (i.e., congestive heart failure or pulmonary hypertension) or by a tendency toward significant hypotension (i.e., cardiogenic shock). By far the highest levels of endothelin were observed in patients in cardiogenic shock, again consistent with a role for this vasoconstrictor peptide in the maintenance of blood pressure. However,

TABLE 14.1. Conditions with increased levels of endothelin in plasma

Shock (cardiogenic, septic, endotoxin)[8,56,62,85]
Acute and chronic renal failure[92,93,101a,102]
Systemic hypertension?[53,71]
Pulmonary hypertension[8,43,84]
Myocardial infarction[54,55,109]
Severe congestive heart failure (see text)[82a]
Subarachnoid hemorrhage[51,85a]
Major surgery[31]
Physiological stress: orthostatic (see text)[75,82a]

patients with other conditions such as chronic renal failure also exhibited modest but definite elevations in endothelin concentration. In the following sections, some specific diagnostic categories are examined in more detail in the search for clues concerning the stimuli controlling endothelin release in vivo and/or the importance of circulating endothelin as a neurohumoral factor.

Shock

Cardiogenic shock was chosen for study because it is a well-defined condition with a relatively straightforward pathophysiological mechanism, i.e., "pump failure." We reasoned that if circulating endothelin had any role whatsoever in the control of peripheral vascular resistance and perfusion pressure, then levels should be increased in this condition. Although a complete lack of response

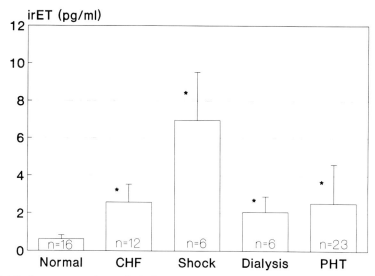

FIGURE 14.3. Immunoreactive endothelin (irET) in plasma of normal subjects and patients with congestive heart failure (CHF), cardiogenic shock, chronic renal failure on dialysis, and pulmonary hypertension (PHT). * = $p < 0.05$.

in plasma endothelin levels might exclude a role for circulating endothelin, the converse does not necessarily prove one. As is discussed below, it is by no means clear whether the circulating levels of endothelin exert a direct vasomotor (or other) effect or merely reflect an "overflow" into the bloodstream of endothelin produced and acting locally in the vasculature. It is also not known whether increased plasma levels of endothelin would have a beneficial effect in shock states, e.g., by increasing perfusion pressure, or be harmful, e.g., leading to renal failure.[18,92]

Are increases in circulating endothelin confined to shock occurring as a result of cardiac dysfunction, or is it a general feature of the shock state? As is discussed below, myocardial ischemia, a common component of cardiogenic shock, may in itself be a stimulus for endothelin release from the coronary circulation.[54,109] Therefore, we have begun to examine endothelin concentrations in patients with shock from diverse causes. Although as yet these studies are incomplete, it appears that plasma endothelin is consistently elevated in shock states regardless of cause. Indeed, patients with *septic shock* exhibit some of the highest venous endothelin levels that we have measured (16 ± 4 pg/ml, $N = 4$). This is in keeping with the limited data available from in vivo laboratory studies. Endotoxin, a mediator of many of the manifestations of septic shock, was found to be a potent stimulus for endothelin release in vivo and in vitro.[56,62,85]

Congestive Heart Failure

Heart failure, regardless of cause, results in activation of a variety of neurohumoral compensatory mechanisms. In its end stage, marked elevations in circulating catecholamines, renin, angiotensin, and atrial natriuretic peptide (among others) have been demonstrated. In our initial studies, we found no significant increase in endothelin concentration in the venous plasma of a small group of patients with congestive heart failure (CHF).[8] Other neurohumoral agents were not assessed in these patients. In collaboration with Dr. Jean L. Rouleau, we subsequently examined a larger group of patients with more severe CHF (NYHA classes III and IV) and have in addition measured corresponding levels of neurohumoral factors.[82a] The results, summarized in Figure 14.4, show that, although these patients were clinically normotensive, all demonstrated substantial increases in plasma endothelin that precisely paralleled the elevations in vasopressin, renin, aldosterone and norepinephrine. Therefore, a shock state is not a prerequisite for increased plasma endothelin.

It is difficult to determine in the clinical setting the exact mechanism underlying these increased levels or their importance with respect to hemodynamic compensation in heart failure. For this reason it is important to establish animal models in which controlled studies can be performed. The rapid pacing model of CHF in the dog provides many advantages over other animal models of heart failure and closely mimics the hemodynamic progression of abnormalities found in man. Figure 14.5 shows preliminary measurements[55] of plasma endothelin levels in dogs before and after induction of CHF by ventricular pacing at 250 beats per minute for approximately 6 weeks. In all ani-

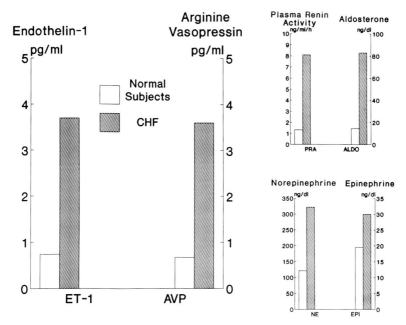

FIGURE 14.4. Plasma endothelin levels in normal subjects and patients with congestive heart failure (CHF) compared with other neurohumoral vasoactive factors: arginine vasopressin (AVP), plasma renin activity (PRA), aldosterone (ALDO), norepinephrine (NE), and epinephrine (EPI).

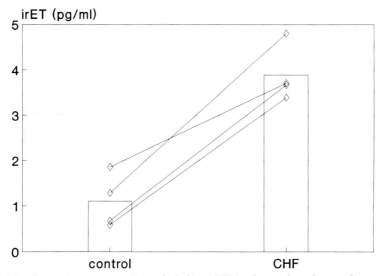

FIGURE 14.5. Plasma immunoreactive endothelin (irET) in plasma from dogs under control conditions and following development of congestive heart failure (CHF) induced by overdrive ventricular pacing.

mals there was a marked rise in plasma endothelin to levels similar to those observed in clinical heart failure in man. Therefore, the overdrive ventricular pacing dog model provides a convenient laboratory tool with which to address some crucial questions concerning mechanisms and significance of circulating endothelin in CHF.

Myocardial Infarction and Ischemia

It has been reported that endothelin levels are increased following myocardial infarction (MI).[54,72,109] There may be several possible mechanisms for the increased levels. Although it is tempting to conclude that endothelin is released from the coronary bed by ischemia, it is also possible that hemodynamic embarrassment following extensive infarction could result in reduced cardiac output and release from the peripheral circulation in a manner analogous to patients in shock.[8] Therefore, we have measured plasma endothelin levels in patients with well-tolerated infarction uncomplicated by heart failure, hypotension, rhythm disturbance, or recurrent ischemia.[41,82b] A schematic representation of our findings is given in Figure 14.6. An increase in plasma endothelin was a consistent feature of infarction, despite lack of any alteration in hemodynamic variables. This increase preceded (by about 10 h) the increase in creatinine phosphokinase, the standard biochemical marker of MI, raising the possibility that plasma endothelin might have diagnostic value. However, the peak level of endothelin concentration did not correlate with peak creatinine phosphokinase, suggesting that they may be markers of different events. Indeed, in several patients suffering spontaneous episodes of significant myocardial ischemia associated with reversible ST-segment depression but no rise in

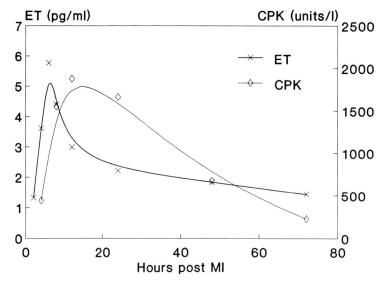

FIGURE 14.6. Time-course of plasma endothelin (ET) and creatinine phosphokinase (CPK) following onset of myocardial infarction (MI).

creatinine phosphokinase, plasma endothelin concentration was found to be increased to a similar extent as in MI. Therefore, plasma endothelin may reflect myocardial ischemia rather than infarction.

Pulmonary Hypertension

The lung has the greatest endothelial surface of any organ in the body.[69] Therefore, diseases of the lung more than of other organs may involve alterations in endothelin uptake or release that may subsequently alter pulmonary vascular resistance. Indeed, patients with severe pulmonary hypertension were found to have elevated plasma endothelin levels.[8] Although there was no significant difference in venous endothelin levels between groups of patients with various secondary causes of pulmonary hypertension and patients with primary pulmonary hypertension, there was a tendency for patients with pulmonary hypertension from lung disease or of "primary" etiology to have higher plasma concentrations.

From the limited laboratory data available to date, it appears that the normal lung may serve to clear endothelin from the circulation,[14,68,79] and single-pass clearances of ~50% have been reported.[68] In pulmonary hypertension there is a decrease in the number and calibre of the pulmonary vascular channels,[66,69] resulting in a decrease in the surface area of pulmonary endothelium. Therefore such a change may alter the ability of the lung to clear endothelin from the circulation. For this reason we have measured the endothelin gradient across the lung in normal subjects and in patients with pulmonary hypertension by measuring endothelin concentration in the systemic venous and arterial blood.[43,84] In normal subjects, arterial concentration was consistently lower than venous, giving an artery-to vein (A/V) ratio of considerably less than 1 (Fig. 14.7). In pulmonary hypertension of diverse secondary etiology, the A/V ratio was close to 1, suggesting that a decrease in endothelin clearance by the lung might contribute to the increased circulating levels seen in these patients. Finally, in patients with "primary" pulmonary hypertension, the arterial concentrations of endothelin were in nearly all cases greater than the venous levels, yielding an A/V ratio of significantly greater than 1. This is consistent with increased local production of endothelin in the pulmonary bed of these patients, which might contribute to, or even underlie, the abnormalities of pulmonary hemodynamics observed in this condition. This view is supported by observations demonstrating increased expression of preproendothelin-1 in an animal model of idiopathic pulmonary hypertension[80] preceding the development of increased pulmonary vascular resistance.

Other Conditions

In their initial report, Yanagisawa and coworkers[107] suggested that endothelin might contribute to increased vascular resistance in essential hypertension. However, there is as yet conflicting data concerning the role of circulating endothelin in these patients. Dr. Saito, who reported that plasma endothelin levels were elevated in patients with idiopathic systemic hypertension,[71] discusses this question in more detail in Chapter 13.

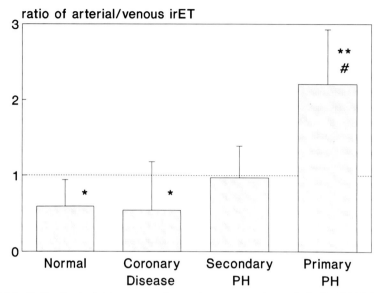

FIGURE 14.7. Ratio of arterial to venous plasma immunoreactive endothelin (irET) in normal subjects, patients with coronary disease but no pulmonary hypertension (PH), patients with secondary causes of PH, and patients with primary PH. * = $p < 0.05$ compared with unity; = $p < 0.05$ compared with secondary PH.

There is more agreement concerning the possible role of endothelin in renal failure. Numerous reports have appeared demonstrating increased plasma endothelin levels in patients with chronic renal failure (Table 14.1) and is reviewed in detail in Chapter 13.

SUMMARY AND CONCLUSIONS

Considerable evidence now exists that endothelin plasma levels fluctuate in a predictable manner in relation to a variety of pathophysiological conditions and also in response to physiological stress. However, what meaning or importance these changes in circulating endothelin concentration may have in terms of hemodynamic regulation and control of regional distribution of blood flow is not known. The highest endothelin levels measured in plasma in severe shock states only approach the threshold concentrations for vasoconstriction of isolated vessels in vitro,[107] including human artery and vein.[34,108] However, a recent report examining the effects of endothelin in man[98] demonstrated that infusions resulting in plasma concentrations <10 pmol/liter (<40 pg/ml), induced a significant pressor response. This is close to the range reported in human disease by ourselves and others (Table 14.1), especially taking into consideration differences in the other assay techniques that yielded a normal plasma range two- to threefold higher than ours (Fig. 14.1).

Thus the question of whether circulating endothelin may exert a "hormonal" action remains open. As well, more work is required to determine if some vascular beds or regions may be more sensitive to low concentrations of endothelin (i.e., veins vs. arteries) or if continuous exposure to endothelin, as occurs in vivo, might produce vasoconstriction at concentrations that do not necessarily produce short-term effects. Finally, the native endothelial lining of the vasculature has been reported to modify greatly the response to circulating endothelin.[36,82] Therefore, it might be anticipated that the vasoconstrictor action of circulating endothelin may be strongly potentiated in areas of endothelial damage (e.g., atherosclerosis), or as may occur in ischemia and reperfusion.[81]

Regardless of the functional importance of its circulating levels, endothelin is a product of vascular endothelium, and therefore its local concentration within the vessel wall at the site of its release may be orders of magnitude greater than that measured in plasma. Like other endothelial-derived vasoactive factors (i.e., prostacyclin and EDRF), endothelin might be best considered as a vascular autacoid, having its principal actions in tissues adjacent to its site of production. This would be particularly true if endothelin was released in a polarized manner toward the smooth muscle layers of the vessel wall, as is the case with other endothelial products.[4a] Thus endothelin levels found in plasma may result more from an "overflow" of local intravascular production into the bloodstream rather than from direct systemic secretion. In this case, circulating endothelin might be a marker of local vascular production, and therefore measurement of plasma levels provides important information about its regional release in various disease states.

However, the question then arises about where in the vasculature endothelin is produced and under what conditions. It is unlikely that the endothelium releases endothelin in a uniform manner in all beds and at all levels of the vascular tree. As well, if endothelin release has any relevance to the local control of vessel tone, one would expect that its production in any particular bed would be responsive to changes in circulatory conditions. Only limited information about the regional production of endothelin by various vascular beds in man is as yet available. However, the initial experience summarized here, and in accompanying chapters, would support heterogeneous release of endothelin by certain organ systems under specific physiological and pathophysiological circumstances. Thus the lung may act to both clear and release endothelin, and its net effect on circulating levels will depend on the balance of these two processes, which presumably may be altered in health and disease. As well, it appears that the heart reacts to an ischemic insult by releasing large amounts of endothelin into the circulation, which may contribute importantly to abnormalities in local control of vascular tone, leading to aggravation of ischemia and progression of infarction. However, the ability in man to determine precisely the mechanisms underlying abnormalities in endothelin production and clearance are limited by obvious technical and ethical considerations. More basic research is required using in vivo and in vitro models. Indicator-dilution techniques can better resolve the balance between production and clearance of endothelin across a vascular bed.[68] Studies using immunohistochemical techniques and in situ cDNA hybridization[24,40,107,110] (see Chapter 12) can define the macro- and microscopic distributions of endothelin message and mature product within organs and tissues. Experiments using these (and other) approaches in animal models of human disease may help to clarify some important questions being raised by the current observations in man.

The true test of the potential importance of endothelin in human disease awaits the development of specific antagonists of endothelin action and inhibitors of its production. At least two classes of agents are on the horizon that may greatly augment our ability to study the physiological and pathophysiological relevance of endothelin and even have therapeutic potential in the management of cardiovascular disorders. By analogy with other receptor-mediated processes, endothelin action might be interfered with by competitive inhibition at the level of its receptor. Although there has been considerable interest in developing and testing various analogs of endothelin, the search for a true endothelin blocker has as yet been unrewarding.

Alternatively, it might be possible to interfere with endothelin actions by blocking its synthesis or release. In their original report, Yanagisawa et al.[107] suggested that the conversion of the inactive precursor big endothelin to the mature peptide might be accomplished by the action of a specific "endothelin-converting enzyme" (ECE). Recently, promising candidates for the ECE have been identified, one of which is a neutral endopeptidase selectively inhibited by phosphoramidon.[51a] As well, two other mechanisms of regulation of endothelin production have been proposed. Boulanger and Lüscher[5] have suggested that endothelium-derived nitric oxide (i.e., EDRF) directly inhibits the release of endothelin by a cGMP-dependent mechanism, and work from our laboratory[83] suggests that vascular smooth muscle cells may release an inhibitor of endothelin production. The possible relevance of these and other inhibitory mechanisms for the future investigation of endothelin biology and perhaps for the treatment of disorders of regulation of vascular tone remains to be determined.

REFERENCES

1. ANDO, K., Y. HIRATA, M. SHICHIRI, T. EMORI, and F. MARUMO. Presence of immunoreactive endothelin in human plasma. *FEBS Lett.* 245: 164–166, 1989.
2. ARMSTEAD, W. M., R. MIRRO, C. W. LEFFLER, and D. W. BUSIJA. Influence of endothelin on piglet cerebral microcirculation. *Am. J. Physiol.* 257 (*Heart Circ. Physiol.* 26): H707–H710, 1989.
3. ASANO, T., I. IKEGAKI, Y. SUZUKI, S.-I. SATOH, and M. SHIBUYA. Endothelin and the production of cerebral vasospasm in dogs. *Biochem. Biophys. Res. Commun.* 159: 1345–1351, 1989.
4. BADR, K. F., J. J. MURRAY, M. D. BREYER, K. TAKAHASHI, T. INAGAMI, R. C. HARRIS, M. SCHWARTZBERG, and J. EBERT. Mesangial cell, glomerular and renal vascular responses to endothelin in the rat kidney: elucidation of signal transduction pathways. *J. Clin. Invest.* 83: 336–342, 1989.
4a. BASSENGE, E., R. BUSSE, and U. POHL. Abluminal release and assymetric response of rabbit arterial wall to endothelium-derived relaxing factor. *Circ. Res.* 61 (Suppl. 5): 68–73, 1987.
5. BOULANGER, C., and T. F. LÜSCHER. Release of endothelin from the porcine aorta. Inhibition by endothelium-derived nitric oxide. *J. Clin. Invest.* 85: 587–590, 1990.
6. BROWN, K. D., and C. J. LITTLEWOOD. Endothelin stimulates DNA synthesis in Swiss 3T3 cells: synergy with polypeptide growth factors. *Biochem. J.* 263: 977–980, 1989.
7. CAIRNS, H. S., M. E. ROGERSON, L. D. FAIRBANKS, G. H. NEILD, and J. WESTWICK. Endothelin induces an increase in renal vascular resistance and a fall in glomerular filtration rate in the rabbit isolated perfused kidney. *Br. J. Pharmacol.* 98: 155–160, 1989.

8. CERNACEK, P., and D. J. STEWART. Immunoreactive endothelin in human plasma: marked elevations in patients in cardiogenic shock. *Biochem. Biophys. Res. Commun.* 161: 562–567, 1989.
9. CHESTER, A. H., M. R. DASHWOOD, J. G. CLARKE, S. W. LARKIN, G. J. DAVIES, S. TADJKARIMI, A. MASERI, and M. H. YACOUB. Influence of endothelin on human coronary arteries and localization of its binding sites. *Am. J. Cardiol.* 63: 1395–1398, 1989.
10. CLARKE, J. G., N. BENJAMIN, S. W. LARKIN, D. J. WEBB, G. J. DAVIES, and A. MASERI. Endothelin is a potent long-lasting vasoconstrictor in men. *Am. J. Physiol.* 257 (*Heart Circ. Physiol.* 26): H2033–H2035, 1989.
11. CLOZEL, J.-P., and M. CLOZEL. Effects of endothelin on the coronary vascular bed in open-chest dogs. *Circ. Res.* 65: 1193–1200, 1989.
12. COCKS, T. M., N. L. FAULKNER, K. SUDHIR, and J. ANGUS. Reactivity of endothelin-1 on human and canine large veins compared with large arteries in vitro. *Eur. J. Pharmacol.* 171: 17–24, 1989.
13. COSTELLO, K.B., STEWART, D. J., and BAFFOUR, R. Endothelin is a potent constrictor of human vessels used in coronary revascularization surgery. *Eur. J. Pharmacol.* 186: 311–314, 1990.
14. DE NUCCI, G., R. THOMAS, P. D'ORLEANS-JUSTE, E. ANTUNES, C. WALDER, T. D. WARNER, and J. R. VANE. Pressor effects of circulating endothelin are limited by its removal in the pulmonary circulation and by the release of prostacyclin and endothelium-derived relaxing factor. *Proc. Natl. Acad. Sci. USA* 85: 9797–9800, 1988.
15. DUBIN, D., R. E. PRATT, J. P. COOKE, and V. J. DZAU. Endothelin, a potent vasoconstrictor, is a vascular smooth muscle mitogen. *Clin. Res.* 37: 517A, 1989.
16. EZRA, D., R. E. GOLDSTEIN, J. F. CZAJA, and G. Z. FEUERSTEIN. Lethal ischemia due to intracoronary endothelin in pigs. *Am. J. Physiol.* 257 (*Heart Circ. Physiol.* 26): H339–H343, 1989.
17. FARACI, F. M. Effects of endothelin and vasopressin on cerebral blood vessels. *Am. J. Physiol.* 257 (*Heart Circ. Physiol.* 26): H799–H803, 1989.
18. FIRTH, J. D., P. J. RATCLIFFE, A. E. G. RAINE, and J. G. G. LEDINGHAM. Endothelin: an important factor in acute renal failure. *Lancet* 2: 1179–1182, 1988.
19. FOLTA, A., I. G. JOSHUA, and R. C. WEBB. Dilator actions of endothelin in coronary resistance vessels and the abdominal aorta of the guinea pig. *Life Sci.* 45: 2627–2635, 1989.
20. FUKUDA, K., S. HORI, M. KUSUHARA, T. SATOH, S. KYOTANI, S. HANDA, Y. NAKAMURA, H. OONO, and K. YAMAGUCHI. Effect of endothelin as a coronary vasoconstrictor in the Langendorff-perfused rat heart. *Eur. J. Pharmacol.* 165: 301–304, 1989.
21. FUKUDA, Y., Y. HIRATA, H. YOSHIMI, Y. KOJIMA, Y. KOBAYASHI, M. YANAGISAWA, and T. MASAKI. Endothelin is a potent secretagogue for atrial natriuretic peptide in cultured rat atrial myocytes. *Biochem. Biophys. Res. Commun.* 155: 167–172, 1988.
22. FURCHGOTT, R. F., and P. M. VANHOUTTE. Endothelium-derived relaxing and contracting factors. *FASEB J.* 3: 2007–2018, 1989.
23. GARDINER, S. M., A. M. COMPTON, T. BENNETT, R. M. J. PALMER, and S. MONCADA. N^G-monomethyl-L-arginine does not inhibit the hindquarters vasodilator action of endothelin-1 in conscious rats. *Eur. J. Pharmacol.* 171: 237–240, 1989.
24. GIAID, A., S. J. GIBSON, N. B. N. IBRAHIM, S. LEGON, S. R. BLOOM, M. YANAGISAWA, T. MASAKI, I. M. VARNDELL, and J. M. POLAK. Endothelin-1, an endothelium-derived peptide, is expressed in neurons of the human spinal cord and dorsal root ganglia. *Proc. Natl. Acad. Sci. USA* 86: 7634–7638, 1989.
25. GILLESPIE, M. N., J. O. OWASOYO, I. F. MCMURTRY, and R. F. O'BRIEN. Sustained vasoconstriction by a peptidergic substance released from endothelial cells in culture. *J. Pharmacol. Exp. Ther.* 236: 339–343, 1986.
26. GODFRAIND, T., D. MENNIG, G. BRAVO, C. CHALANT, and P. JAUMIN. Inhibition by amlodipine of activity evoked in isolated human coronary arteries by endothelin, prostaglandin $F_{2\alpha}$, and depolarization. *Am. J. Cardiol.* 64: 58I–64I, 1989.
27. GOETZ, K. L., B. C. WANG, J. B. MADWED, J. L. ZHU, and R. J. LEADLEY, JR. Cardiovascular, renal, and endocrine responses to intravenous endothelin in conscious dogs. *Am. J. Physiol.* 255 (*Regulatory Integrative Comp. Physiol.* 24): R1064–R1068, 1988.
28. HICKEY, K. A., G. M. RUBANYI, R. J. PAUL, and R. F. HIGHSMITH. Characterization of a coronary vasoconstrictor produced by cultured endothelial cells. *Am. J. Physiol.* 248 (*Cell Physiol.* 17): C550–C556, 1985.
29. HILEY, C. R. Functional studies on endothelin catch up with molecular biology. *TIPS* 10: 47–49, 1989.
30. HILEY, C. R., S. A. DOUGLAS, and M. D. RANDALL. Pressor effects of endothelin-1 and some analogs in the perfused superior mesenteric arterial bed of the rat. *J. Cardiovasc. Pharmacol.* 13(Suppl. 5): S197–S199, 1989.

31. HIRATA, Y., K. ITOH, K. ANDO, M. ENDO, and F. MARUMO. Plasma endothelin levels during surgery. *N. Engl. J. Med.* 321: 1686–1686, 1989.
32. HOFFMAN, A., E. GROSSMAN, K. P. ÖSHMAN, E. MARKS, and H. R. KEISER. Endothelin induces an initial increase in cardiac output associated with selective vasodilation in rats. *Life Sci.* 45: 249–255, 1989.
33. HU, J. R., U. G. BERNINGER, and R. E. LANG. Endothelin stimulates atrial natriuretic peptide (ANP) release from rat atria. *Eur. J. Pharmacol.* 158: 177–178, 1988.
34. ISHIKAWA, T., M. YANAGISAWA, S. KIMURA, K. GOTO, and T. MASAKI. Positive inotropic action of novel vasoconstrictor peptide endothelin on guinea pig atria. *Am. J. Physiol.* 255 *(Heart Circ. Physiol.* 24): H970–H973, 1988.
35. KARWATOWSKA-PROKOPCZUK, E., and Å. WENNMALM. Effects of endothelin on coronary flow, mechanical performance, oxygen uptake, and formation of purines and on outflow of prostacyclin in the isolated rabbit heart. *Circ. Res.* 66: 46–54, 1990.
36. KAUSER, K., G. M. RUBANYI, and D. R. HARDER. Endothelium-dependent modulation of endothelin-induced vasoconstriction and membrane depolarization in cat cerebral arteries. *J. Pharmacol. Exp. Ther.* 252: 93–97, 1990.
37. KING, A. J., B. M. BRENNER, and S. ANDERSON. Endothelin: a potent renal and systemic vasoconstrictor peptide. *Am. J. Physiol.* 256 *(Renal Fluid Electrolyte Physiol.* 25): F1051–F1058, 1989.
38. KOMURO, I., H. KURIHARA, T. SUGIYAMA, F. TAKAKU, and Y. YAZAKI. Endothelin stimulates c*fos* and c-*myc* expression and proliferation of vascular smooth muscle cells. *FEBS Lett.* 238: 249–252, 1988.
39. KON, V., T. YOSHIOKA, A. FOGO, and I. ICHIKAWA. Glomerular actions of endothelin in vivo. *J. Clin. Invest.* 83: 1762–1767, 1989.
40. KOSEKI, C., M. IMAI, Y. HIRATA, M. YANAGISAWA, and T. MASAKI. Autoradiographic distribution in rat tissues of binding sites for endothelin: a neuropeptide. *Am. J. Physiol.* 256 *(Regulatory Integrative Comp. Physiol.* 25): R858–R866, 1989.
41. KUBAC, G., P. CERNACEK, F. MOHAMED, and D. J. STEWART. Plasma endothelin (ET) is elevated in the early hours following myocardial infarction (MI): possible diagnostic and prognostic implications, abstracted. *J. Hypertens.* 8(Suppl. 3): S4, 1990.
42. LARKIN, S. W., J. G. CLARKE, B. E. KEOGH, L. ARAUJO, C. RHODES, G. J. DAVIES, K. M. TAYLOR, and A. MASERI. Intracoronary endothelin induces myocardial ischemia by small vessel constriction in the dog. *Am. J. Cardiol.* 64: 956–958, 1989.
43. LEVY, R. D., D. LANGLEBEN, P. CERNACEK, and D. J. STEWART. Increased plasma levels of the endothelium-derived vasoconstrictor endothelin in pulmonary hypertension: marker or mediator of disease?, abstracted. *Am. Rev. Respir. Dis.* 141: A889, 1990.
44. LI, K., STEWART, D. J., and ROULEAU, J.-L. Myocardial contractile actions of endothelin-1 in rat and rabbit papillary muscles. *Circ. Res.* 69 (in press), 1991.
45. LIPPTON, H., J. GOFF, and A. HYMAN. Effects of endothelin in the systemic and renal vascular beds in vivo. *Eur. J. Pharmacol.* 155: 197–199, 1988.
46. LIPPTON, H. L., T. A. HAUTH, W. R. SUMMER, and A. L. HYMAN. Endothelin produces pulmonary vasoconstriction and systemic vasodilation. *J. Appl. Physiol.* 66: 1008–1012, 1989.
47. LOUTZENHISER, R., M. EPSTEIN, K. HAYASHI, and C. HORTON. Direct visualization of effects of endothelin on the renal microvasculature. *Am. J. Physiol.* 258 *(Renal Fluid Electrolyte Physiol.* 27): F61–F68, 1990.
48. LÓPEZ-FARR, A., I. MONTAN S., I. MILLÁS, AND J. M. LÓPEZ-NOVOA. Effect of endothelin on renal function in rats. *Eur. J. Pharmacol.* 163: 187–189, 1989.
49. MAGGI, C. A., S. GIULIANI, R. PATACCHINI, P. SANTICIOLI, D. TURINI, G. BARBANTI, and A. MELI. Potent contractile activity of endothelin on the human isolated urinary bladder. *Br. J. Pharmacol.* 96: 755–757, 1989.
50. MAGGI, C. A., R. PATACCHINI, S. GIULIANI, and A. MELI. Potent contractile effect of endothelin in isolated guinea-pig airways. *Eur. J. Pharmacol.* 160: 179–182, 1989.
51. MASAOKA, H., R. SUZUKI, Y. HIRATA, T. EMORI, F. MARUMO, and K. HIRAKAWA. Raised plasma endothelin in aneurysmal subarachnoid haemorrhage. *Lancet* 2: 1402–1402, 1989.
51a. MCMAHON, E. G., M. A. PALOMO, W. M. MOORE, J. F. MCDONALD, and M. K. STERN. Phosphoramidon blocks the pressor activity of porcine big endothelin-1-(1–39) in vivo and conversion of big endothelin-1-(1–39) to endothelin-1-(1–21) in vitro. *Proc. Natl. Acad. Sci. USA* 88: 703–707, 1991.
52. MILLER, W. L., M. M. REDFIELD, and J. C. BURNETT, JR. Integrated cardiac, renal, and endocrine actions of endothelin. *J. Clin. Invest.* 83: 317–320, 1989.
53. MIYAUCHI, T., M. YANAGISAWA, N. SUZUKI, K. IIDA, Y. SUGISHITA, M. FUJINO, T. SAITO, K. GOTO, and T. MASAKI. Venous plasma concentrations of endothelin in normal and hypertensive subjects, abstracted. *Circulation* 80: II-573–II-573, 1989.
54. MIYAUCHI, T., M. YANAGISAWA, T. TOMIZAWA, Y. SUGISHITA, N. SUZUKI, M. FUJINO, R. AJI-

SAKA, K. GOTO, and T. MASAKI. Increased plasma concentrations of endothelin-1 and big endothelin-1 in acute myocardial infarction. *Lancet* 2: 53–54, 1989.
55. MOGHTADER, S., P. BELICHARD, A. CALDERONE, J. L. ROULEAU, and D. J. STEWART. Endothelin-induced coronary vasoconstriction is enhanced by experimental heart failure in dogs, abstracted. *FASEB J.* 4: A960, 1990.
56. MOREL, D. R., J. S. LACROIX, A. HEMSEN, D. A. STEINIG, J.-F. PITTET, and J. M. LUNDBERG. Increased plasma and pulmonary lymph levels of endothelin during endotoxin shock. *Eur. J. Pharmacol.* 167: 427–428, 1989.
57. MORISHITA, R., J. HIGAKI, and T. OGIHARA. Endothelin stimulates aldosterone biosynthesis by dispersed rabbit adreno-capsular cells. *Biochem. Biophys. Res. Commun.* 160: 628–632, 1989.
58. NAKAKI, T., M. NAKAYANA, S. YAMAMOTO, and R. KATO. Endothelin-mediated stimulation of DNA synthesis in vascular smooth muscle. *Biochem. Biophys. Res. Commun.* 158: 880–883, 1989.
59. OHTSUKA, M., Y. UCHIDA, M. SAOTOMA, N. NINOMIYA, Y. ISHII, and S. MASUGAWA. Endothelin exerts to human pulmonary artery as a vasoconstrictor in vitro, abstracted. *Am. Rev. Respir. Dis.* 139A: 51–51, 1989.
60. OTSUKA, A., H. MIKAMI, K. KATAHIRA, T. TSUNETOSHI, K. MINAMITANI, and T. OGIHARA. Changes in plasma renin activity and aldosterone concentration in response to endothelin injection in dogs. *Acta Endocrinol. (Copenh)* 121: 361–364, 1989.
61. PERNOW, J., J.-F. BOUTIER, A. FRANCO-CERECEDA, J. S. LACROIX, R. MATRAN, and J. M. LUNDBERG. Potent selective vasoconstrictor effects of endothelin in the pig kidney in vivo. *Acta Physiol. Scand.* 134: 573–574, 1988.
62. PERNOW, J., A. HEMSÉN, and J. M. LUNDBERG. Increased plasma levels of endothelin-like immunoreactivity during endotoxin administration in the pig. *Acta Physiol. Scand.* 137: 317–318, 1989.
63. POHL, U., and R. BUSSE. Differential vascular sensitivity to luminally and adventitially applied endothelin-1. *J. Cardiovasc. Pharmacol.* 13(Suppl. 5): S188–S190, 1989.
64. RAE, G. A., M. TRYBULEC, G. DE NUCCI, and J. R. VANE. Endothelin-1 releases eicosanoids from rabbit isolated perfused kidney and spleen. *J. Cardiovasc. Pharmacol.* 13(Suppl. 5): S89–S92, 1989.
65. RAKUGI, H., M. NAKAMARU, H. SAITO, J. HIGAKI, and T. OGIHARA. Endothelin inhibits renin release from isolated rat glomeruli. *Biochem. Biophys. Res. Commun.* 155: 1244–1247, 1988.
66. REEVES, J. T., B. M. GROVE, and D. TURIKEVICH. The case for treatment of selected patients with primary pulmonary hypertension. *Am. Rev. Respir. Dis.* 134: 342–346, 1986.
67. REYNOLDS, E. E., L. L. S. MOK, and S. KUROKAWA. Phorbol ester dissociates endothelin-stimulated phosphoinositide hydroylysis and arachidonic acid release in vascular smooth muscle cells. *Biochem. Biophys. Res. Commun.* 160: 868–873, 1989.
68. RIMAR, S., and C. N. GILLIS. Differential uptake of endothelin by rabbit coronary and pulmonary circulations, abstracted. *Circulation* 80-II: 213–213, 1989.
69. RYAN, U. S., and J. FROKJAGER-JENSEN. Pulmonary endothelium and processing of pulmonary solutes: structure and function. In: *The Pulmonary Circulation and Acute Lung Injury*, edited by S. I. Said. Mount Kisco, NY: Futura, p. 37–60, 1985.
70. RYAN, U. S., M. K. GLASSBERG, and K. B. NOLOP. Endothelin-1 from pulmonary artery and microvessels acts on vascular and airway smooth muscle. *J. Cardiovasc. Pharmacol.* 13(Suppl. 5): S57–S62, 1989.
71. SAITO, Y., K. NAKAO, M. MUKOYAMA, and H. IMURA. Increased plasma endothelin level in patients with essential hypertension. *N. Engl. J. Med.* 322: 205–205, 1990.
72. SALMINEN, K., I. TIKKANEN, O. SAIJONMAA, M. NIEMINEN, F. FYHRQUIST, and M. H. FRICK. Modulation of coronary tone in acute myocardial infarction by endothelin. *Lancet* 2: 747–747, 1989.
73. SCHOMISCH-MORAVEC, C., E. E. REYNOLDS, R. W. STEWART, and M. BOND. Endothelin is a positive inotropic agent in human and rat heart in vitro. *Biochem. Biophys. Res. Commun.* 159: 14–18, 1989.
74. SHAH, A. M., M. J. LEWIS, and A. H. HENDERSON. Inotropic effects of endothelin in ferret ventricular myocardium. *Eur. J. Pharmacol.* 163: 365–367, 1989.
75. SHICHIRI, M., Y. HIRATA, K. ANDO, K. KANNO, T. EMORI, K. OHTA, and F. MARUMO. Postural change and volume expansion affect plasma endothelin levels. *JAMA* 263: 661–661, 1990.
76. SHIGENO, T., T. MIMA, K. TAKAKURA, M. YANAGISAWA, A. SAITO, K. GOTO, and T. MASAKI. Endothelin-1 acts in cerebral arteries from the adventitial but not from the luminal side. *J. Cardiovasc. Pharmacol.* 13(Suppl. 5): S174–S176, 1989.
77. SHINMI, O., S. KIMURA, T. SAWAMURA, Y. SUGITA, T. YOSHIZAWA, Y. UCHIYAMA, M. YANAGI-

SAWA, K. GOTO, T. MASAKI, and I. KANAZAWA. Endothelin-3 is a novel neuropeptide: isolation and sequence determination of endothelin-1 and endothelin-3 in porcine brain. *Biochem. Biophys. Res. Commun.* 164: 587–593, 1989.
78. SIMONSON, M. S., S. WANN, P. MEN, G. R. DUBYAK, M. KESTER, Y. NAKAZATO, J. R. SEDOR, and M. J. DUNN. Endothelin stimulates phospholipase C, Na^+/H^+ exchange, c-*fos* expression, and mitogenesis in rat mesangial cells. *J. Clin. Invest.* 83: 708–712, 1989.
79. SIRVIO, M.-L., K. METSARINNE, O. SAIJONMAA, and F. FYHRQUIST. Tissue distribution and half-life of ^{125}I-endothelin in the rat: importance of pulmonary clearance. *Biochem. Biophys. Res. Commun.* 167: 1191–1195, 1990.
80. STELZNER, T. J., M. YANAGISAWA, T. SAKURAI, K. SATO, S. WEBB, M. ZAMORA, I. F. MCMURTRY, J. H. FISHER, and R. F. O'BRIEN. Increased lung preproendothelin (preET-1) mRNA expression in rats with pulmonary hypertension, abstracted. *FASEB J.* 4: A1147–A1147, 1990.
81. STEWART, D. J., and R. BAFFOUR. Ischemia-reperfusion potentiates endothelin-induced constriction in the coronary resistance bed. In: *Endothelium-Derived Contracting Factors,* edited by G. M. Rubanyi, and P. M. Vanhoutte. Basel: Karger, 1990, p. 212–220.
82. STEWART, D. J., and R. BAFFOUR. Functional state of the endothelium determines the response to endothelin in the coronary circulation. *Cardiovasc. Res.* 24: 7–12, 1990.
82a. STEWART, D. J., P. CERNACEK, and J.-L. ROULEAU. High endothelin levels in heart failure and loss of normal response to postural change, abstracted. *Circulation* 82: III 725, 1990.
82b. STEWART, D. J., G. KUBAC, K. B. COSTELLO, AND P. CERNACEK. Increased plasma endothelin-1 in the early hours of acute myocardial infarction. *J. Am. Coll. Cardiol.* 18: (in press), 1991.
83. STEWART, D. J., D. LANGLEBEN, P. CERNACEK, and K. CIANFLONE. Endothelin release is inhibited by coculture of endothelial cells with cells of vascular media. *Am. J. Physiol.* 259 (*Heart Circ. Physiol.* 28): H1928–1932, 1990.
84. STEWART, D. J., R. D. LEVY, P. CERNACEK, and D. LANGLEBEN. Increased plasma endothelin-1 in pulmonary hypertension: marker or mediator of disease. *Ann. Intern. Med.* 114: 464–469, 1991.
85. SUGIURA, M., T. INAGAMI, and V. KON. Endotoxin stimulates endothelin-release in vivo and in vitro as determined by radioimmunoassay. *Biochem. Biophys. Res. Commun.* 161: 1220–1227, 1989.
85a. SUZUKI H., S. SATO, Y. SUZUKI, M. OKA, T. TSUCHIYA, I. LINO, T. YAMANAKA, N. ISHIHARA, and S. SHIMODA. Endothelin immunoreactivity in cerebrospinal fluid of patients with subarachnoid haemorrhage. *Ann. Med.* 22: 233–236, 1990.
86. SUZUKI, N., H. MATSUMOTO, C. KITADA, T. MASAKI, and M. FUJINO. A sensitive sandwich-enzyme immunoassay for human endothelin. *J. Immunol. Methods* 118: 245–250, 1989.
87. TABUCHI, Y., M. NAKAMARU, H. RAKUGI, M. NAGANO, H. MIKAMI, and T. OGIHARA. Endothelin inhibits presynaptic adrenergic neurotransmission in rat mesenteric artery. *Biochem. Biophys. Res. Commun.* 161: 803–808, 1989.
88. TABUCHI, Y., M. NAKAMARU, H. RAKUGI, M. NAGANO, and T. OGIHARA. Endothelin enhances adrenergic vasoconstriction in perfused rat mesenteric arteries. *Biochem. Biophys. Res. Commun.* 159: 1304–1308, 1989.
89. TAKAGI, M., H. TSUKADA, H. MATSUOKA, and S. YAGI. Inhibitory effect of endothelin on renin release in vitro. *Am. J. Physiol.* 257 (*Endocrinol. Metab.* 20): E833–E838, 1989.
90. TAKUWA, N., Y. TAKUWA, M. YANAGISAWA, K. YAMASHITA, and T. MASAKI. A novel vasoactive peptide endothelin stimulates mitogenesis through inositol lipid turnover in Swiss 3T3 fibroblasts. *J. Biol. Chem.* 264: 7856–7861, 1989.
91. THIEMERMANN, C., P. S. LIDBURY, G. R. THOMAS, and J. R. VANE. Endothelin-1 releases prostacyclin and inhibits ex vivo platelet aggregation in the anesthetized rabbit. *J. Cardiovasc. Pharmacol.* 13(Suppl. 5): S138–S142, 1989.
92. TOMITA, K., K. UJIIE, T. NAKANISHI, S. TOMURA, O. MATSUDA, K. ANDO, M. SHICHIRI, Y. HIRATA, and F. MARUMO. Plasma endothelin levels in patients with acute renal failure. *N. Engl. J. Med.* 321: 1127–1127, 1989.
93. TOTSUNE, K., T. MOURI, K. TAKAHASHI, M. OHNEDA, M. SONE, T. SAITO, and K. YOSHINAGA. Detection of immunoreactive endothelin in plasma of hemodialysis patients. *FEBS Lett.* 249: 239–242, 1989.
94. TOUVAY, C., B. VILAIN, F. PONS, P.-E. CHABRIER, J. M. MENCIA-HUERTA, and P. BRAQUET. Bronchopulmonary and vascular effect of endothelin in the guinea-pig. *Eur. J. Pharmacol.* 176: 23–33, 1990.
95. TSUCHIYA, K., M. NARUSE, T. SANAKA, K. NARUSE, Y. KATO, Z. P. ZENG, K. NITTA, K. SHIZUME,

H. DEMURA, and N. SUGINO. Effects of endothelin on renal hemodynamics and excretory functions in anesthetized dogs. *Life Sci.* 46: 59–65, 1990.
96. UCHIDA, Y., H. NINOMIYA, M. SAOTOME, A. NOMURA, M. OHTSUKA, M. YANAGISAWA, K. GOTO, T. MASAKI, and S. HASEGAWA. Endothelin, a novel vasoconstrictor peptide, as potent bronchoconstrictor. *Eur. J. Pharmacol.* 154: 227–228, 1988.
97. VANHOUTTE, P. M. Endothelium dependent contractions. *Blood Vessels* 28: 74–83, 1991.
98. VIERHAPPER, H., O. WAGNER, P. NOWOTNY, and W. WALDHAEUSL. Effect of endothelin-1 in man. *Circulation* 81: 1415–1418, 1990.
99. VIGNE, P., M. LAZDUNSKI, and C. FRELIN. The inotropic effect of endothelin-1 on rat atria involves hydrolysis of phosphatidylinositol. *FEBS Lett.* 249: 143–146, 1989.
100. WARNER, T. D., G. DE NUCCI, and J. R. VANE. Rat endothelin is a vasodilator in the isolated perfused mesentery of the rat. *Eur. J. Pharmacol.* 159: 325–326, 1989.
101. WARNER, T. D., J. A. MITCHELL, G. DE NUCCI, and J. R. VANE. Endothelin-1 and endothelin-3 release EDRF from isolated perfused arterial vessels of the rat and rabbit. *J. Cardiovasc. Pharmacol.* 13(Suppl. 5): S85–S88, 1989.
101a. WARRENS, A. N., M. J. D. CASSIDY, K. TAKAHASHI, M. A. GHATEI, and S. R. BLOOM. Endothelium in renal failure. *Nephrol. Dial. Transplant.* 5: 418–422, 1990.
102. WEBB, D. J., and J. R. COCKCROFT. Plasma immunoreactive endothelin in uraemia. *Lancet* 1: 1211–1211, 1989.
103. WIKLUND, N. P., A. ÖHL N, and B. CEDERQVIST. Inhibition of adrenergic neuroeffector transmission by endothelin in the guinea-pig femoral artery. *Acta Physiol. Scand.* 134: 311–312, 1988.
104. WIKLUND, N. P., A. ÖHL N, C. U. WIKLUND, B. CEDERQVIST, P. HEDQVIST, and L. E. GUSTAFSSON. Neuromuscular actions of endothelin on smooth, cardiac and skeletal muscle from guinea-pig, rat and rabbit. *Acta Physiol. Scand.* 137: 399–407, 1989.
105. WINQUIST, R. J., A. L. SCOTT, and G. P. VLASUK. Enhanced release of atrial natriuretic factor by endothelin in atria from hypertensive rats. *Hypertension* 14: 111–114, 1989.
106. XUAN, Y.-T., A. R. WHORTON, E. SHEARER-POOR, J. BOYD, and W. D. WATKINS. Determination of immunoreactive endothelin in medium from cultured endothelial cells and human plasma. *Biochem. Biophys. Res. Commun.* 164: 326–332, 1989.
107. YANAGISAWA, M., H. KURIHARA, S. KIMURA, Y. TOMOBE, M. KOBAYASHI, Y. MITSUI, Y. YAZAKI, K. GOTO, and T. MASAKI. A novel potent vasoconstrictor peptide produced by vascular endothelial cells. *Nature* 332: 411–415, 1988.
108. YANG, Z., F. R. BÜHLER, D. DIEDERICH, and T. F. LUSCHER. Different effects of endothelin-1 on cAMP-and cGMP-mediated vascular relaxation in human arteries and veins: comparison with norepinephrine. *J. Cardiovasc. Pharmacol.* 13(Suppl. 5): S129–S131, 1989.
109. YASUDA, M., M. KOHNO, A. TAHARA, H. ITAGANE, I. TODA, K. AKIOKA, M. TERAGAKI, H. OKU, K. TAKEUCHI, and T. TAKEDA. Circulating immunoreactive endothelin in ischemic heart disease. *Am. Heart J.* 119: 801–806, 1990.
110. YOSHIZAWA, T., S. KIMURA, I. KANAZAWA, Y. UCHIYAMA, M. YANAGISAWA, and T. MASAKI. Endothelin localizes in the dorsal horn and acts on the spinal neurones: possible involvement of dihydropyridine-sensitive calcium channels and substance P release. *Neurosci. Lett.* 10: 179–184, 1989.
111. ZEIDEL, M. L., H. R. BRADY, B. C. KONE, S. R. GULLANS, and B. M. BRENNER. Endothelin, a peptide inhibitor of Na^+-K^+-ATPase in intact renal tubular epithelial cells. *Am. J. Physiol.* 257 (*Cell Physiol.* 26): C1101–C1107, 1989.

15

Endothelin and the Homeostatic Function of the Endothelial Cell

REGINA M. BOTTING AND JOHN R. VANE

Cultured porcine endothelial cells release a peptide into the culture medium that causes slow, long-lasting contractions of isolated vascular strips.[17] This peptide was isolated, characterized, and synthesized in 1988 by Yanagisawa et al.,[66] who named it "endothelin." The endothelin from porcine endothelial cells consists of a chain of 21 amino acids held together by two disulfide bridges and is generated from a precursor molecule by a previously unknown protease. It was subsequently discovered that three isomers of endothelin, expressed by three different genes, are probably present in all mammalian species.[22] Endothelin-1 is the only one made by endothelial cells and was first called "porcine" or "human" endothelin. Endothelin-2 is more potent than endothelin-1 and has a longer duration of action on blood pressure. It is not known where it is made except perhaps in kidney cells.[27] Endothelin-3, first called "rat" endothelin, may be associated with nervous tissue.[67] The structure of endothelin-3 differs from that of endothelin-1 by 6 of the 21 amino acid residues and consists of (Thr^2, Phe^4, Thr^5, Tyr^6, Lys^7, Tyr^{14})-substituted endothelin-1. Endothelin-2 differs from endothelin-1 by two amino acids and corresponds to (Trp^6, Leu^7)-substituted endothelin-1.

CARDIOVASCULAR EFFECTS OF ENDOTHELINS

Yanagisawa et al.[66] initially demonstrated the striking vasopressor properties of endothelin-1 on the blood pressure of a chemically denervated rat. Endothelin-1 was ten times more potent as a vasoconstrictor than angiotensin II, and the hypertensive effect of a single dose lasted for over 1 h. The potent pressor action of endothelin-1, endothelin-2, and endothelin-3 has been confirmed in most species[4,7,14,19,21,25,29,30,35,37,40,44,47,48,52] using pithed, anesthetized, or conscious animal models and also in humans. In the majority of experiments a transient, dose-related depressor response occurred immediately after a bolus injection and preceded the prolonged pressor effect. This vasodilator effect was not seen, however, when a slow infusion as opposed to a bolus injection of endothelin-1 was given[19] and could not be elicited in anesthetized guinea pigs.[30]

The short lasting fall in blood pressure was particularly prominent in conscious cats,[33] in anesthetized rats with a high resting blood pressure,[8,20,28,29] and in spontaneously hypertensive rats.[62,65] The depressor responses were smaller when endothelin-1 was injected intravenously than when it was administered into the left ventricle. This difference in effect indicated partial removal of endothelin-1 by the lung circulation, a finding that was confirmed in perfused lung preparations.[8]

Although endothelin-3 had only one-tenth to one-third of the activity of endothelin-1 in raising the blood pressure of anesthetized cat or rat, the peptides were equipotent in their ability to lower blood pressure.[53] The receptors for endothelin that mediate smooth muscle contraction must therefore be different from those that trigger relaxation. The relative proportions of these receptors probably differ in different vascular beds, since endothelin-1 dilates most prominently the blood vessels of skeletal muscle and skin and has only a weak dilator action in the mesenteric vascular bed.[7,20,25,41,42,48,52]

Interestingly, the depressor effects of endothelin-1 in anesthetized rats or cats were not attenuated by cyclooxygenase inhibition with indomethacin[61] or sodium meclofenamate,[40] showing that these were not caused by release of vasodilator prostanoids. Another labile product of the endothelium, endothelium-derived relaxing factor (EDRF; now known to be nitric oxide [NO]) mediated these depressor responses as shown in perfused mesenteric preparations of the rat.[8] The abolition of the depressor effect of endothelin-1 by the specific inhibitor of EDRF/NO synthesis N^G-monomethyl-L-arginine (L-NMMA) and the reversal of the block by L-arginine but not by D-arginine confirmed EDRF/NO release.[60] However, in less well-controlled experiments, L-NMMA did not inhibit the depressor action of endothelin-1 in conscious rats.[13]

Although the depressor action of endothelins was not mediated by release of eicosanoids, some evidence has accrued that the long-lasting pressor effect may be modified by stimulation of the release of cyclooxygenase products. In a pithed rat preparation[57] or in rabbits anesthetized with sodium pentobarbitone[32] administration of indomethacin caused significant augmentation of the pressor responses elicited by endothelin-1 or endothelin-3 (Fig. 15.1). In support of this, it was shown that indomethacin augmented the pressor activity of endothelin-1 in the blood-perfused rat mesentery[18,51] and potentiated the contractile responses to endothelin-1 in cat isolated cerebral arteries.[24] However, the potentiation of the pressor action of endothelin-1 with indomethacin could not be confirmed in other experiments using pithed rats[28,29] or in chemically denervated rats.[10] These differences may be due to the different protocols used. Thus, when indomethacin was injected into rats between two doses of endothelin-1, a substantial increase in the second pressor effect was observed (Fig. 15.2). However, when indomethacin was given before the first dose, the pressor effect of endothelin-1 was similar to that in control rats.[8]

The pressor effect of the endothelin-1 precursor big endothelin-1 (human) in anesthetized rabbits was also potentiated by indomethacin, thus suggesting that cyclooxygenase products modulate the effects of big endothelin-1 (Fig. 15.3). It is not clear whether big endothelin-1; itself released eicosanoids or whether its conversion to endothelin-1 was responsible for the prostanoid release.[9] Thus, vasoconstrictor doses of endothelins release dilator prostanoids that attenuate the pressor action.

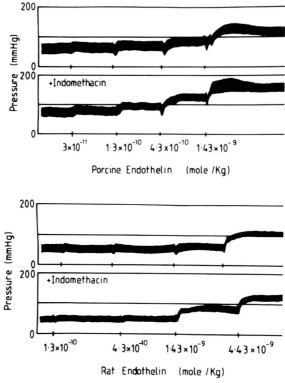

FIGURE 15.1. Cumulative dose–response curves for the effects of endothelin-1 (porcine endothelin) or endothelin-3 (rat endothelin) in a pithed rat prior to and following indomethacin (15 μmol/kg i.v.). The responses to both endothelin-1 and endothelin-3 were enhanced by indomethacin, and endothelin-3 was less potent than endothelin-1 as a pressor agent in the pithed rat. (From Walder et al.[57])

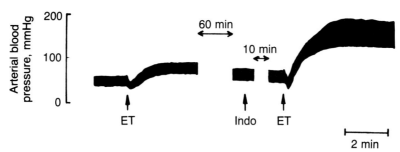

FIGURE 15.2. Indomethacin potentiates the pressor response to endothelin-1 (ET) in the pithed rat. Overall, in six experiments, endothelin-1 (1 nmol/kg) injected i.v. induced an increase in blood pressure of 43.3 ± 6.1 mm Hg (N = 6). Indomethacin (Indo, 14 μmol/kg) was administered after the blood pressure returned to control values. A second dose of endothelin-1 (1 nmol/kg) induced an increase in blood pressure of 78.3 ± 9.2 mm Hg, approximately twice that induced by the first challenge (N = 6). Similar results were obtained with piroxicam (15 μmol/kg; N = 4). (From de Nucci et al.[8])

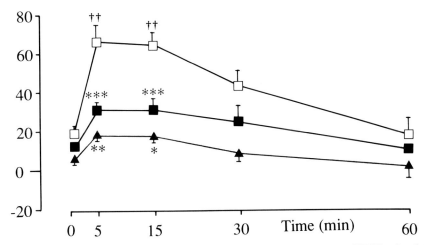

FIGURE 15.3. Time-course for changes in left ventricular systolic pressure (LVSP) after injection of big endothelin (h) at either 1 nmol/kg (▲) or 3 nmol/kg (■). Note that with big endothelin (h, 3 nmol/kg) after indomethacin treatment (□) there was a substantial potentiation of the pressor effect. ($*P < 0.05$; $** P < 0.01$; $*** P < 0.001$) (From D'Orleans-Juste et al.[9])

RELEASE OF EICOSANOIDS FROM PERFUSED ISOLATED ORGANS

Generation of vasodilator eicosanoids by the lungs,[8] spleen, kidney,[49] and heart[23] may contribute importantly to the overall hemodynamic effects of endothelin-1. The spleen and kidney are important determinants of cardiovascular homeostasis, the former as a blood reservoir mobilized during exercise, shock, or hemorrhage and the second in regulation of blood pressure and hydroelectrolytic balance.

Isolated Lungs

Guinea Pig
Endothelin-1 (1 or 10 nM) infused for 3 min induced a sustained release of prostacyclin and thromboxane A_2 (TXA; Fig. 15.4) from guinea pig lungs perfused through the pulmonary artery with Krebs-Ringer solution. The effluent was analyzed by radioimmunoassay for 6-oxo-prostaglandin $F_{1\alpha}$ (6-oxo-PGF$_{1\alpha}$) as a measure of prostacyclin release and TXB_2 as a measure of TXA_2 release.[8]

Rat
Both endothelin-1 (10 nM) and endothelin-3 (10 nM) released prostacyclin from isolated rat lung preparations, with little or no release of TXA_2.[8]

Rabbit
A significant increase above basal levels in the release of prostacyclin but not TXA_2 was elicited from perfused rabbit lungs by endothelin-1 (10 nM; Fig. 15.5).

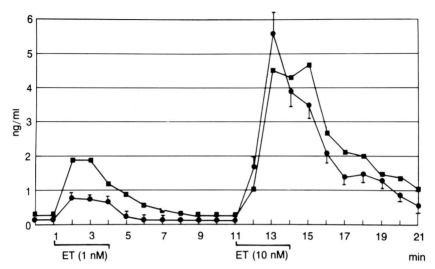

FIGURE 15.4. Endothelin-1 (ET) releases prostacyclin and thromboxane A_2 (TXA_2) from isolated lungs of guinea pigs. Endothelin-1 (1 or 10 nM) infused for 3 min induced dose-dependent release of both prostacyclin and TXA_2 (measured by radioimmunoassay as 6-oxo-$PGF_{1\alpha}$ [■] and TXB_2 [●]. The release of both eicosanoids continued after the infusion was terminated (N = 3). The SEM values (±5%–15%) for the 6-oxo-$PGF_{1\alpha}$ concentrations were omitted for the sake of clarity. The release of eicosanoids was confirmed by bioassay. (From de Nucci et al.[8])

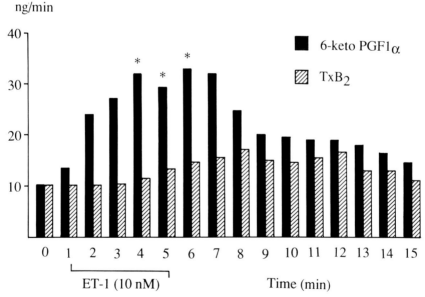

FIGURE 15.5. Endothelin-1 (ET-1 10 nM) increases release of prostacyclin but not thromboxane A_2 (TXA_2) from isolated perfused lungs of rabbits. Prostacyclin was measured by radioimmunoassay as 6-oxo-$PGF_{1\alpha}$ and TXA_2 by radioimmunoassay as TXB_2. The release of prostacyclin continued after the infusion of endothelin-1 was terminated. The SEM values were omitted for the sake of clarity.

Rabbit Isolated Spleen

Endothelin-1 given as a bolus injection released prostaglandin E_2 (PGE_2), prostacyclin, and TXA_2 from the rabbit isolated perfused spleen identified by differential bioassay using the Vane cascade.[49] An infusion of endothelin-1 for 5 min caused a sustained release, over minutes, of PGE_2, prostacyclin, and TXA_2 identified by radioimmunoassay (Fig. 15.6). Treatment of the spleen with indomethacin for 30 min prevented eicosanoid release.

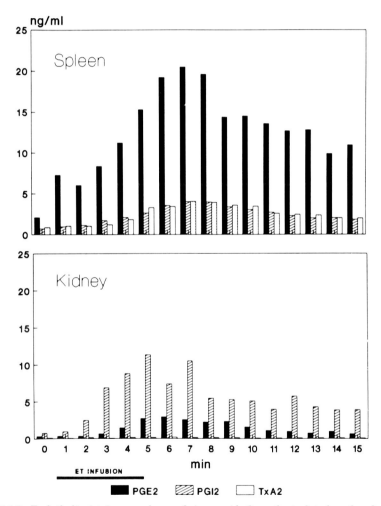

FIGURE 15.6. Endothelin-1 triggers release of eicosanoids from the isolated perfused spleen (N = 4) and kidney (N = 3) of the rabbit. Concentrations of PGE_2, 6-oxo-$PGF_{1\alpha}$ (prostacyclin), and TXB_2 (TXA_2) were assessed by radioimmunoassay of 1 min aliquots of the organ effluent before (0 min), during a 5 min infusion of endothelin-1 at 10 nM (1–5 min), or over the 10 min following the end of the infusion (6–15 min). The SEM ($\pm 5\%$–25%) were omitted for the sake of clarity. (From Rae et al.[49])

Release From Kidney

Kidneys treated in vitro or in vivo with endothelin-1 released PGE_2 and prostacyclin. Bolus injections of endothelin-1 (2.5–50 pmol) caused a release of prostacyclin and PGE_2 into the effluent from rabbit isolated kidneys perfused with Krebs-Ringer solution.[48] The eicosanoids were identified by means of a cascade of bioassay tissues. Infusions of 10 nM endothelin-1 for 5 min increased the concentrations of prostacyclin and PGE_2 identified by radioimmunoassay (Fig. 15.6) in the effluent. Release of the eicosanoids was abolished by treatment of the kidneys with indomethacin for 30 min.

Infusion of endothelin-1 (5 ng/kg/min) into the renal artery of anesthetized dogs induced renal vasoconstriction that was potentiated by inhibition of cyclooxygenase with aspirin. Plasma concentrations of PGE_2 and prostacyclin measured by radioimmunoassay increased during the infusion, and this increase was also abolished by pretreatment with aspirin. Thus prostaglandins modulate the renal actions of endothelin-1.[43]

Isolated Perfused Heart

Rabbit isolated hearts were perfused with Tyrodes solution by the method of Langendorff, and the effects of endothelin-1 (0.1–10 nM) on various parameters, including coronary flow and outflow of 6-oxo-$PGF_{1\alpha}$, were studied. Coronary flow was dose dependently decreased while the concentration of 6-oxo-$PGF_{1\alpha}$ in the perfusing fluid was elevated to 14 times the control levels by 10 nM endothelin-1. Indomethacin (50 μM) almost completely abolished this outflow of the prostacyclin metabolite.[23]

However, neither 6-oxo-$PGF_{1\alpha}$ nor TXB_2 could be detected in the effluent from perfused isolated hearts of the rat after injection of low doses (1–100 pmol) of endothelin-1, which caused a transient decrease followed by an increase in coronary tone. The cyclooxygenase inhibitor flurbiprofen did not alter either the vasodilator or the vasoconstrictor responses to endothelin-1, indicating that in the rat heart neither were modulated by release of prostanoids.[3]

Rat Perfused Mesentery

Prostacyclin (measured as 6-oxo-$PGF_{1\alpha}$) was released from the perfused mesenteric bed of the rat by endothelin-1 (10 or 40 pmol) and endothelin-3. This release was abolished with indomethacin. Endothelin-3 stimulated the release of prostacyclin less potently than did endothelin-1.[50]

EDRF/NO RELEASE FROM ISOLATED PERFUSED ORGANS

Endothelin-1 and endothelin-3 released the potent vasodilator EDRF/NO[46] from endothelial cells of the rabbit luminally perfused isolated aorta and rat mesentery preparations.

Rabbit Isolated Aorta

The complete length of a rabbit aorta from the aortic arch to the left renal artery was removed and perfused through the lumen with Krebs solution containing indomethacin (5 µM) contracted with U46619 (11-dideoxy-9-methanoepoxy-PGF$_{2\alpha}$, 30 nM). EDRF/NO was characterized both by the typically rapid disappearance of its activity down the bioassay cascade and by potentiation of its survival by superoxide dismutase.[15] Endothelin-1 released EDRF/NO in 8 of 12 experiments (Fig. 15.7). The release of EDRF/NO induced by endothelin-1 was substantially potentiated by infusion of potassium chloride through the aorta, whereas that to other EDRF/NO releasers was not. The reason why potassium chloride[59] increased EDRF/NO release induced by endothelin-1 from the aorta is not clear. One interpretation is that potassium chloride partially depolarized the membranes of the endothelial cells and facilitated release by endothelin-1.

Rat Perfused Mesentery

The isolated arterial mesenteric vascular bed of male Wistar rats[8] was perfused with oxygenated Krebs solution containing indomethacin (5 µM), according to the method of McGregor.[38] Vasoconstrictions induced by intraarterial injections of endothelin-1 or endothelin-3 were recorded in preparations with a low resting perfusion pressure of 15–22 mm Hg. Endothelin-1 was more than ten times more active than endothelin-3 as a vasoconstrictor.[59]

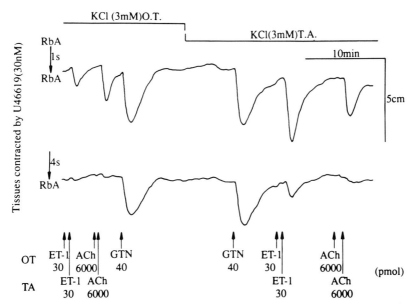

FIGURE 15.7. Potassium chloride (KCl) potentiates the release of EDRF from the luminally perfused aorta of the rabbit in response to endothelin-1 (ET-1, 30 pmol) but not to acetylcholine (ACh, 6000 pmol). The effluent from the aorta superfused two deendothelialized rabbit aortic strips. With KCl infused over the tissues (OT), ACh released more EDRF than endothelin-1; when KCl was infused through the aorta (TA), endothelin-1 released more EDRF than did ACh. GTN, glyceryl trinitrate. (From Warner et al.[59])

In mesenteric vascular beds preconstricted with methoxamine (perfusion pressure 35–120 mm Hg) both endothelin-1 (28 of 46 experiments) and endothelin-3 (14 of 19 experiments) caused dose-dependent vasodilations.[58] These vasodilations were due to the release of EDRF/NO, because they were abolished by removal of the endothelium with sodium deoxycholate and inhibited by methylene blue or oxyhemoglobin (Figs. 15.8, 15.9). Vasodilations induced by the direct-acting nitrovasodilator sodium nitroprusside were not inhibited either by removal of the endothelium or by oxyhemoglobin.

Low doses of endothelin-1 or endothelin-3 (0.1–3 pmol) were vasodilator, with endothelin-3 being slightly less potent than endothelin-1. However, high doses of endothelin-3 (10–300 pmol) were more effective vasodilators than were high doses of endothelin-1, because the initial vasodilation produced by endothelin-1 was largely overpowered by the subsequent vasoconstriction (Fig. 15.10).

The finding that the vasoconstrictor responses to endothelin-1 and endothelin-3 in the rat isolated perfused mesentery were potentiated by either removal of the endothelium or by the presence of hemoglobin or methylene blue[59] clearly demonstrates that the release of EDRF/NO can substantially limit the vasopressor action of endothelin. One particularly interesting finding is that endothelin-3 is equiactive with endothelin-1 in causing EDRF/NO release. This is intriguing because in the same system we have also confirmed

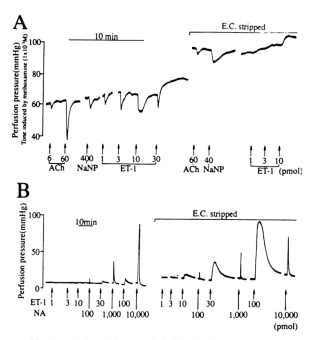

FIGURE 15.8. A: vasodilations induced by acetylcholine (Ach, 6–60 pmol), sodium nitroprusside (NaNP, 40–400 pmol), or endothelin-1 (ET-1, 30 pmol) in rat isolated perfused mesentery preconstricted with methoxamine (10 μM). Removal of the endothelial cells (E.C.) by perfusion with sodium deoxycholate (2.4 mM, 30 s) abolished all vasodilations except those to sodium nitroprusside. B: vasoconstrictions to endothelin-1 (1–100 pmol) but not norepinephrine (NA, 100–10,000 pmol) were potentiated by removal of the endothelial cells. (From Warner et al.[59])

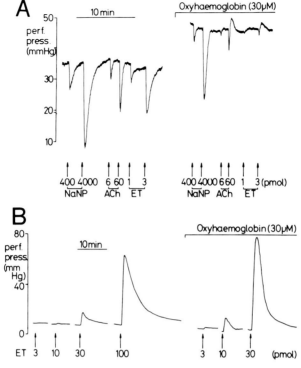

FIGURE 15.9. Endothelin-1 (ET) releases EDRF in the rat isolated perfused mesentery. *A:* to assess the contribution of EDRF to the vascular responses, oxyhemoglobin (30 μ*M*) was infused through the mesentery. Acetylcholine (ACh)-induced dose-dependent vasodilations were mediated via release of EDRF, for they were abolished by oxyhemoglobin. Endothelin-1 at lower doses also induced vasodilations, which were abolished by oxyhemoglobin. Responses to sodium nitroprusside (NaNP) were only slightly attenuated. *B:* in preparations in which no tonus was induced, endothelin-1 induced vasoconstrictions that were strongly potentiated by infusion of oxyhemoglobin. (From de Nucci et al.[8])

the observation by Inoue et al.[22] that as a vasoconstrictor endothelin-3 has less than one-tenth the activity of endothelin-1. Thus endothelin-3 is more selective than endothelin-1 as a vasodilator in the mesentery because at the higher doses the vasodilator effect is still evident, whereas for endothelin-1 it is overwhelmed by the stronger vasoconstrictor activity (Fig. 15.10). Clearly, then, in the rat isolated perfused mesentery endothelin-3 is principally vasodilator and endothelin-1 vasodilator or vasoconstrictor depending on the dose used.[58] Thus the putative endothelin receptors on the endothelial cells, stimulation of which leads to the release of EDRF/NO, may differ from those on the smooth muscle, which mediate vasoconstrictor responses.

Evidence for the presence of multiple receptors for endothelin has also been presented by Spokes et al.[53] and Maggi et al.,[36] based on different profiles of activity of endothelin analogs on isolated smooth muscle preparations. Fukuda et al.[12] showed in rat perfused mesenteric arteries that vasodilating effects of low doses of endothelin-3 were abolished with an infusion of 50 μ*M* L-NMMA, thus confirming the release of EDRF/NO.

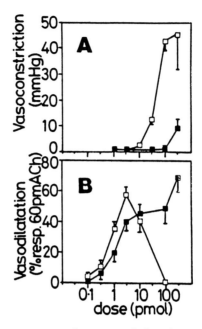

FIGURE 15.10. *A:* Dose–response curve of vasoconstrictions in response to endothelin-1 (N = 25, □) and endothelin-3 (N = 20, ■) in rat isolated perfused mesentery. Vasoconstrictor responses were recorded as rises in perfusion pressure (mm Hg). *B:* dose–response curve of vasodilations in response to endothelin-1 (N = 15, □) and endothelin-3 (N = 10, ■) Data are expressed as percentage of the vasodilator response to a control dose of acetylcholine (ACh, 60 pmol) in each preparation. Basal perfusion pressure was increased by infusion of methoxamine (30–100 μM). All data are expressed as means ± SEM. (From Warner et al.[58])

Release from Other Tissues

There is some evidence that endothelins are able to release EDRF/NO from the perfused kidney and liver but not from cultures of porcine endothelial cells or from canine blood vessels. Endothelin-1 (10^{-11}–10^{-9} M) weakly induced release of EDRF/NO in the rabbit isolated perfused kidney. The EDRF/NO inhibitors methylene blue or hemoglobin slightly augmented the increase in renal vascular resistance produced by endothelin-1.[6] The vasodilation observed after injection of endothelin-1 into the portal vein of the dog liver was not attenuated with indomethacin. This suggested that endothelin-1 was not releasing cyclooxygenase products, and the observed vasodilation was probably due to release of EDRF/NO from vascular endothelial cells in the liver.[63,64]

Endothelin-1 did not stimulate or modify the production of EDRF/NO from porcine aortic endothelial cells in culture. The content of cGMP in cultures of porcine endothelial cells was not altered by endothelin-1, whereas bradykinin induced a rapid and transient production of the nucleotide, indicating EDRF/NO release. A similar lack of release of EDRF/NO by endothelin-1 was seen in isolated canine femoral arteries and veins contracted with endothelin-1 in which the tissue content of cGMP was not altered.[56]

PLATELET AGGREGATION EX VIVO AND IN VIVO
IN THE ANESTHETIZED RABBIT

Ex Vivo Aggregation

Injections of endothelin-1, endothelin-3, or big endothelin-1 all inhibited ex vivo platelet aggregation in anesthetized rabbits. Measurements of cyclic nucleotides established that platelet levels of cAMP were also raised after administration of endothelin-1.

Rabbits anesthetized with pentobarbitone and artificially ventilated[54] had polyethylene catheters placed in the left ventricle, the left femoral vein, and the femoral artery for intraarterial injections, intravenous injections, and withdrawal of blood samples. Aggregation of platelets to adenosine diphosphate (ADP) was measured in a Payton aggregometer. At time 0, endothelin-1 at 1 nmol/kg was injected into the left ventricle of the rabbit, and ex vivo platelet aggregation was measured at 1, 5, 15, 30, and 60 min afterwards.

Three minutes after the induction of platelet aggregation with ADP, platelet-rich plasma was collected into trichloracetic acid for measurement of nucleotide levels. cAMP was measured by radioimmunoassay with an [^{125}I]cAMP RIA system and cGMP by radioimmunoassay with an acetylated [^{125}I]cGMP RIA system.

Intraarterially injected endothelin-1 caused an 83% ± 9% inhibition of ex vivo ADP-induced platelet aggregation within 5 min. This inhibition still amounted to 49% ± 10% of the control value at 30 min, but was no longer evident 60 min after injection of the peptide (Fig. 15.11). Pretreatment of the rabbits with indomethacin 20 min before injection of endothelin-1 abolished the antiaggregatory effects of the peptide at 5 min. However, even after pretreatment with indomethacin, endothelin-1 still reduced the ADP-induced platelet aggregation by about 40% 15–60 min after injection of the peptide. Injection of endothelin-1 was associated with a significant increase in platelet cAMP after 5 min. This increase in cAMP was no longer evident 30 and 60 min after injection of the peptide. Pretreatment with indomethacin abolished the endothelin-1 induced increase in cAMP (Fig. 15.12). No significant changes were observed in platelet cGMP levels.

Intraarterial endothelin-3 caused a 77% ± 10% inhibition of ADP-induced platelet aggregation 1 min after injection. This inhibition was no longer evident 30 min after injection of the peptide and was abolished by pretreatment with the cyclooxygenase inhibitor piroxicam.[31] The transient depressor response caused by endothelin-3 was not followed by an increase in blood pressure even after piroxicam. Neither endothelin-1 nor endothelin-3 had any effect on in vitro platelet aggregation.

Big endothelin-1 at 3 nmol/kg inhibited ex vivo platelet aggregation to a less significant degree than a smaller dose of endothelin-1 (Fig. 15.13). It is not certain whether this effect was due to conversion of big endothelin-1 to endothelin-1. It is also puzzling why big endothelin-1 was equipotent with endothelin-1 in raising blood pressure but much less potent at inhibiting the aggregation of platelets.[9]

Thus endothelin-1 injected either into the left ventricle or intravenously

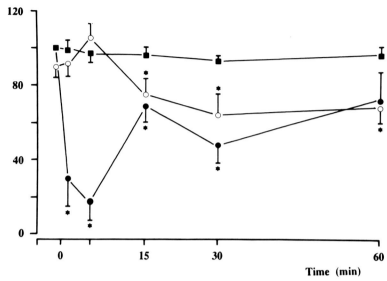

FIGURE 15.11. Endothelin-1 inhibits adenosine diphosphate (1.6 μg/ml)–induced platelet aggregation in the rabbit ex vivo. Different groups of animals received either vehicle only (■), endothelin-1 only (●, 1 nmol/kg i.a.), or endothelin-1 after indomethacin pretreatment (○, 0.5 mg/kg i.v.). Data are expressed as mean ± SEM of six observations. *$P < 0.05$ when compared with control. (From Thiemermann et al.[54])

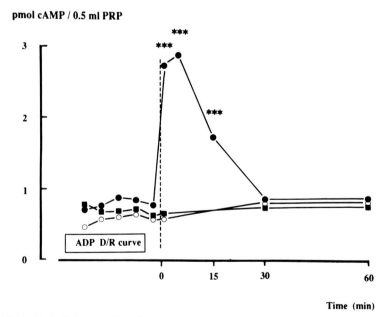

FIGURE 15.12. Endothelin-1–induced increase in ex vivo platelet cAMP levels in the rabbit is abolished by pretreatment with indomethacin. Different groups of animals received either vehicle only (■), endothelin-1 only (●, 1 nmol/kg i.a.), or endothelin-1 after pretreatment with indomethacin (○, 5 mg/kg i.v.). Data are expressed as mean values of six determinations ***$P < 0.001$ when compared with control. PRP, platelet-rich plasma; ADP, adenosine diphosphate; D/R, dose–response. (From Thiemermann et al.[54])

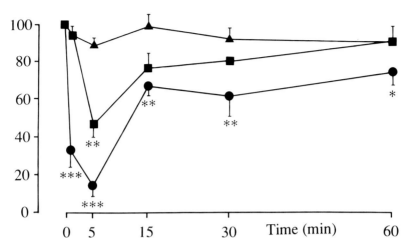

FIGURE 15.13. Inhibition of platelet aggregation ex vivo after injection of big endothelin (h) at either 1 nmol/kg (▲) or 3 nmol/kg (■) compared with inhibition after the injection of endothelin-1 (1 nmol/kg, ●; *$P < 0.05$; **$P < 0.01$; *** $P < 0.001$). (From D'Orléans-Juste et al.[9])

inhibited platelet function ex vivo.[54] This inhibition of platelet aggregation was associated with enhanced platelet cAMP levels, but not those of cGMP. Inhibition of platelet aggregation by prostacyclin also involves activation of adenylate cyclase, but not guanylate cyclase. The view that endothelin-1 inhibits platelet aggregation via release into the circulation of endogenous prostacyclin is strengthened by the results with indomethacin. Pretreatment of the rabbit with this cyclooxygenase inhibitor abolished both the endothelin-1–induced inhibition of platelet aggregation and the corresponding rise in platelet cAMP.[54] This rise in platelet cAMP was therefore most likely due to the release of prostacyclin as confirmed by measurements of plasma 6-oxo-PGF$_{1\alpha}$.[16] It is certainly unlikely to be due to release of EDRF, which also inhibits platelet function,[1,2] because EDRF does not survive in the bloodstream.

The selective inhibition of platelet aggregation by endothelin-3 without increasing systemic blood pressure[31] reinforces the hypothesis that the putative endothelin receptors on the endothelial cells differ from those on the smooth muscle. Thus synthetic endothelin analogs that lack pressor activity, but act selectively on endothelial cells to release prostacyclin and EDRF, may well be of therapeutic value in certain cardiovascular disorders.

In Vivo Aggregation

The in vivo platelet-inhibitory effects of endothelin-1 were investigated in anesthetized rabbits by scintigraphic determination of [^{111}indium] labeled platelet aggregation.[55] Platelets were aggregated with submaximal intravenous doses of ADP, collagen, platelet-activating factor, or thrombin.

Endothelin-1 inhibited ADP-stimulated platelet aggregation in vivo; a maximum inhibition of 78% of the control value was reached at 3 min, with 45% inhibition at 15 min, and a return to control values at 30 min after injection of the peptide. Endothelin-1 also inhibited in vivo platelet aggregation induced by collagen or platelet-activating factor by 86% and 52% respectively, but had no effect on thrombin-induced platelet aggregation. Indomethacin abolished the endothelin-1–induced inhibition of ADP-stimulated platelet aggregation and significantly potentiated and prolonged the pressor response brought about by endothelin-1. Thus endothelin-1 potently inhibited platelet aggregation in the anesthetized rabbit in vivo by releasing a hypotensive and antiaggregatory cyclooxygenase product, presumably prostacyclin, into the circulation.

RELEASE OF TISSUE PLASMINOGEN ACTIVATOR

The fibrinolytic system in rabbits is activated by intraarterial injections of endothelin-1, endothelin-2, and endothelin-3. When applied directly to euglobulin clots in vitro the peptides did not induce fibrinolysis, but blood samples taken from rabbits injected with endothelin-1, endothelin-2, or endothelin-3 showed a significant, biphasic reduction of the euglobulin clot lysis time. The first phase of this endothelin-induced enhancement of plasma fibrinolytic activity was associated with release of tissue plasminogen activator (tPA) into the circulation (Fig. 15.14). One minute after administration of endothelin-1, endothelin-2, or endothelin-3 there was a transient but significant increase in blood tPA antigen levels that returned to basal values during the course of the experiment. Endothelin-3 was weaker than either of the other two peptides both in reducing ex vivo clot lysis time and in releasing tPA.[32]

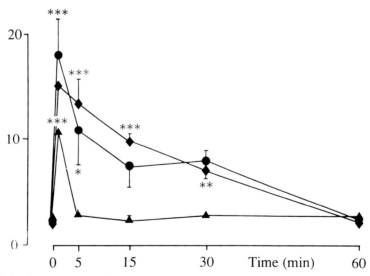

FIGURE 15.14. Stimulation of fibrinolysis by 1 nmol/kg endothelin-1 (●; N = 3), endothelin-2 (◆; N = 3), and endothelin-3 (▲; N = 3) as determined by ex vivo measurement of tissue plasminogen activator (t-PA) antigen levels. Results are expressed as mean ± SEM of N experiments. *$P < 0.05$; **$P < 0.01$; *** $P < 0.001$. (From Lidbury et al.[32])

A late-onset fibrinolytic effect that occurred between 30 and 60 min after injection of endothelin-3 matched the fibrinolysis observed after an injection of prostacyclin and was reversed by pretreatment of the rabbits with indomethacin. Thus this second phase of fibrinolysis was probably associated with release of prostacyclin.[26]

INTERACTIONS BETWEEN ENDOTHELIN-1 AND NO

Contractions evoked by endothelin-1 in rings (with and without endothelium) of canine femoral and coronary arteries and of canine femoral veins were modulated by release of endogenous EDRF/NO or by addition of exogenous NO. Removal of the endothelium and hence of EDRF/NO enhanced the sensitivity to endothelin-1 of veins, but not of arteries. In rings of arteries maximally contracted with endothelin-1, stimulation of the release of EDRF/NO with either acetylcholine or Ca^{2+} ionophore A23187 caused rapid relaxation. NO elicited concentration-dependent inhibition and eventually abolition of the contraction of the arteries evoked by endothelin-1. These observations implied that the release of EDRF/NO can overcome vasoconstrictor responses to endothelin-1.[56]

Curtailment of the constrictor effects of endothelin-1 by NO donors has been demonstrated in vivo on the microcirculation of the hamster[45] and the rat[11] and on the blood pressure of the anesthetized dog.[39] Arterioles of the hamster cheek pouch or exteriorized rat mesentery responded to topically applied endothelin-1 with dose-dependent constrictions. These could be overcome by applications of sodium nitroprusside, which decomposes to nitric oxide, or of acetylcholine, which releases nitric oxide from the vascular endothelium.[11] Constriction of the perfused coronary blood vessels in anesthetized dogs with endothelin-1 was reversed by an infusion of nitroglycerin. Thus, where effective concentrations of endothelin-1 and NO are present together, the dilator action of NO can override the constrictor effects of endothelin-1.[39]

A potentially important interaction between the two systems is shown by the inhibition by EDRF/NO of the production of endothelin-1 by endothelial cells. Preincubation of intact pig aortae with the inhibitor of EDRF/NO synthesis L-NMMA potentiated the thrombin-stimulated but not the basal release of endothelin-1.[5] Thus under normal physiological conditions EDRF/NO may suppress the elaboration of endothelin-1. However, impaired production of EDRF/NO as, for example, in hyperlipidemia, hypertension, or atherosclerosis could result in the unopposed release of endothelin-1 and pathological constriction of diseased blood vessels.[34]

CONCLUSIONS

Endothelin-1 has the characteristics of a local hormone that, when released by the endothelium, will constrict the underlying vascular smooth muscle. Endothelin-1 can also release vasodilator substances such as prostacyclin and EDRF/NO. One interpretation of these results is that endothelin-1 released abluminally by endothelial cells acts locally on the underlying smooth muscle, causing vasoconstriction, whereas the pressor activity of any endothelin-1 reaching the circulating blood is limited by the release of prostacyclin and EDRF/NO and by inactivation in the lungs.

ACKNOWLEDGMENTS

The William Harvey Research Institute is supported by Glaxo, Parke Davis, and the ONO Pharmaceutical Companies.

REFERENCES

1. ALHEID, U., J. C. FRÖLICH, and U. FÖRSTERMANN. Endothelium-derived relaxing factor from cultured human endothelial cells inhibits aggregation of human platelets. *Thromb. Res.* 47: 561–571, 1987.
2. AXUMA, H., M. ISHIKAWA, and S. SEKIZAKI. Endothelium-dependent inhibition of platelet aggregation. *Br. J. Pharmacol.* 88: 411–415, 1986.
3. BAYDOUN, A. R., S. H. PEERS, G. CIRINO, and B. WOODWARD. Vasodilator action of endothelin-1 in the perfused rat heart. *J. Cardiovasc. Pharmacol.* 15: 759–763, 1990.
4. BENNETT, T., A. M. COMPTON, and S. M. GARDINER. Regional haemodynamic effects of endothelin-2 and sarafotoxin-S6b in conscious rats. *J. Physiol.* 423 22P (abstract) 1990.
5. BOULANGER, C., and T. F. LÜSCHER. Release of endothelin from the porcine aorta: inhibition by endothelium-derived nitric oxide. *J. Clin. Invest.* 85: 587–590, 1990.
6. CAIRNS, H. S., M. E. ROGERSON, L. D. FAIRBANKS, G. H. NEILD, and J. WESTWICK. Endothelin induces an increase in renal vascular resistance and a fall in glomerular filtration rate in the rabbit isolated perfused kidney. *Br. J. Pharmacol.* 98: 155–160, 1989.
7. CLOZEL, M., and J.-P. CLOZEL. Effects of endothelin on regional blood flows in squirrel monkeys. *J. Pharmacol. Exp. Ther.* 250: 1125–1131, 1989.
8. DE NUCCI, G., R. THOMAS, P. D'ORLÉANS-JUSTE, E. ANTUNES, C. WALDER, T. D. WARNER, and J. R. VANE. Pressor effects of circulating endothelin are limited by its removal in the pulmonary circulation and by the release of prostacyclin and endothelium-derived relaxing factor. *Proc. Natl. Acad. Sci. USA* 85: 9797–9800, 1988.
9. D'ORLÉANS-JUSTE, P., P. S. LIDBURY, T. D. WARNER, and J. R. VANE. Intravascular big endothelin increases circulating levels of endothelin-1 and prostanoids in the rabbit. *Biochem. Pharmacol.* 39: R21–R22, 1990.
10. EGLEN, R. M., A. D. MICHEL, N. A. SHARIF, S. R. SWANK, and R. L. WHITING. The pharmacological properties of the peptide endothelin. *Br. J. Pharmacol.* 97: 1297–1307, 1989.
11. FORTES, Z. B., G. DE NUCCI, and J. GARCIA-LEME. Effect of endothelin-1 on arterioles and venules in vivo. *J. Cardiovasc. Pharmacol.* 13(Suppl. 5): S200–S201, 1989.
12. FUKUDA, N., Y. IZUMI, M. SOMA, Y. WATANABE, M. WATANABE, M. HATANO, I. SAKUMA, and H. YASUDA. L-N^G-momomethyl arginine inhibits the vasodilating effects of low dose of endothelin-3 on rat mesenteric arteries. *Biochem. Biophys. Res. Commun.* 167: 739–745, 1990.
13. GARDINER, S. M., A. M. COMPTON, T. BENNETT, R. M. J. PALMER, S. MONCADA. The effect of N^G-monomethyl-L-arginine (L-NMMA) on the haemodynamic actions of endothelin-1 in conscious Long-Evans rats. *Br. J. Pharmacol.* 98 626P (abstract), 1989.
14. GIVEN, M. B., R. F. LOWE, H. LIPPTON, A. L. HYMAN, G. E. SANDER, and T. D. GILES. Hemodynamic actions of endothelin in conscious and anesthetized dogs. *Peptides* 10: 41–44, 1989.
15. GRYGLEWSKI, R. J., R. M. J. PALMER, and S. MONCADA. Superoxide anion is involved in the breakdown of endothelium-derived vascular relaxing factor. *Nature* 320: 454–456, 1986.
16. HERMAN, F., K. MAGYAR, P.-E. CHABRIER, P. BRAQUET, and J. FILEP. Prostacyclin mediates antiaggregatory and hypotensive actions of endothelin in anaesthetized beagle dogs. *Br. J. Pharmacol.* 98: 38–40, 1989.
17. HICKEY, K. A., G. RUBANYI, R. J. PAUL, and R. F. HIGHSMITH. Characterization of a coronary vasoconstrictor produced by cultured endothelial cells. *Am. J. Physiol.* 248 (*Cell Physiol.* 17): C550–C556, 1985.
18. HILEY, C. R., S. A. DOUGLAS, and M. D. RANDALL. Pressor effects of endothelin-1 and some analogs in the perfused superior mesenteric arterial bed of the rat. *J. Cardiovasc. Pharmacol.* 13(Suppl. 5): S197–S199, 1989.

19. HINOJOSA-LABORDE, C., J. W. OSBORN, JR., and A. W. COWLEY, JR. Hemodynamic effects of endothelin in conscious rats. *Am. J. Physiol.* 256(*Heart Circ. Physiol.* 25): H1742–H1746, 1989.
20. HOFFMAN, A., E. GROSSMAN, K. P. ÖHMAN, E. MARKS, and H. R. KEISER. Endothelin induces an initial increase in cardiac output associated with selective vasodilation in rats. *Life Sci.* 45: 249–255, 1989.
21. HUGHES, A. D., S. A. M. THOM, N. WOODALL, M. SCHACHTER, W. M. HAIR, G. N. MARTIN, and P. S. SEVER. Human vascular responses to endothelin-1: observations in vivo and in vitro. *J. Cardiovasc. Pharmacol.* 13(Suppl. 5): S225–S228, 1989.
22. INOUE, A., M. YANAGISAWA, S. KIMURA, Y. KASUYA, T. MIYAUCHI, K. GOTO, and T. MASAKI. The human endothelin family: three structurally and pharmacologically distinct isopeptides predicted by three separate genes. *Proc. Natl. Acad. Sci. USA* 86: 2863–2867, 1989.
23. KARWATOWSKA-PROKOPCZUK, E., and Å WENNMALM. Effects of endothelin on coronary flow, mechanical performance, oxygen uptake, and formation of purines and on outflow of prostacyclin in the isolated rabbit heart. *Circ. Res.* 66: 46–54, 1990.
24. KAUSER, K., G. M. RUBANYI, D. R. HARDER. Endothelium-dependent modulation of endothelin-induced vasoconstriction and membrane depolarization in cat cerebral arteries. *J. Pharmacol. Exp. Ther.* 252: 93–97, 1989.
25. KNUEPFER, M. M., S. PING HAN, A. J. TRAPANI, K. F. FOK, and T. C. WESTFALL. Regional hemodynamic and baroreflex effects of endothelin in rats. *Am. J. Physiol.* 257 (*Heart Circ. Physiol.* 26): H918–H926, 1989.
26. KORBUT, R., P. LIDBURY, G. R. THOMAS, and J. R. VANE. Fibrinolytic activity of endothelin-3. *Thromb. Res.* 55: 797–799, 1989.
27. KOSAKA, T., N. SUZUKI, H. MATSUMOTO, Y. ITOH, T. YASUHARA, H. ONDA, and M. FUJINO. Synthesis of the vasoconstrictor peptide in kidney cells. *FEBS Lett.* 249: 42–46, 1989.
28. LE MONNIER DE GOUVILLE, A. C., and I. CAVERO. Haemodynamic profile of endothelin in pithed rats. *Br. J. Pharmacol.* 96: 95P, 1989.
29. LE MONNIER DE GOUVILLE, A. C., S. MONDOT, H. LIPPTON, A. HYMAN, and I. CAVERO. Hemodynamic and pharmacological evaluation of the vasodilator and vasoconstrictor effects of endothelin-1 in rats. *J. Pharmacol. Exp. Ther.* 252: 300–311, 1990.
30. LI, L., T. ISHIKAWA, T. MIYAUCHI, M. YANAGISAWA, S. KIMURA, K. GOTO, and T. MASAKI. Pressor response to endothelin in guinea pigs. *Jpn. J. Pharmacol.* 49: 549–550, 1989.
31. LIDBURY, P. S., C. THIEMERMANN, G. R. THOMAS, and J. R. VANE. Endothelin-3: selectivity as an antiaggregatory peptide in vivo. *Eur. J. Pharmacol.* 166: 335–338, 1989.
32. LIDBURY, P. S., C. THIEMERMANN, R. KORBUT, and J. R. VANE. Endothelins release tissue plasminogen activator and prostanoids. *Eur. J. Pharmacol.* 186: 205–212, 1990.
33. LIPPTON, H., J. GOFF, and A. HYMAN. Effect of endothelin in the systemic and renal vascular beds in vivo. *Eur. J. Pharmacol.* 155: 197–199, 1988.
34. LÜSCHER, T. F., Z. YANG, D. DIEDERICH, and F. R. BÜHLER. Endothelium-derived vasoactive substances: potential role in hypertension, atherosclerosis, and vascular occlusion. *J. Cardiovasc. Pharmacol.* 14(Suppl. 6): S63–S69, 1989.
35. MACLEAN, M. R., M. D. RANDALL, and C. R. HILEY. Effects of moderate hypoxia, hypercapnia and acidosis on haemodynamic changes induced by endothelin-1 in the pithed rat. *Br. J. Pharmacol.* 98: 1055–1065, 1989.
36. MAGGI, C. A., S. GIULIANI, R. PATACCHINI, P. SANTICIOLI, P. ROVERO, A. GIACHETTI, and A. MELI. The C-terminal hexapeptide, endothelin-(16–21), discriminates between different endothelin receptors. *Eur. J. Pharmacol.* 166: 121–122, 1989.
37. MCAULEY, M. A., I. M. MACRAE, and J. L. REID. Haemodynamic responses to endothelin after peripheral and central administration in the conscious rat. *Br. J. Pharmacol.* 98, 709P (abstract) 1989.
38. MCGREGOR, D. D. The effect of sympathetic nerve stimulation on vasoconstrictor responses in perfused mesenteric blood vessels of the rat. *J. Physiol. (Lond).* 177: 21–30, 1965.
39. MILLER, W. L., P. G. CAVERO, L. L. AARHUS, D. M. HEUBLEIN, and J. C. BURNETT, JR. Endothelin-mediated arterial vasoconstriction is heterogenous and attenuated by nitroglycerin. *Clin. Res.* 37: 930A, 1989.
40. MINKES, R. K., D. H. COY, W. A. MURPHY, D. B. MCNAMARA, and P. J. KADOWITZ. Effects of porcine and rat endothelin and an analog on blood pressure in the anaesthetized cat. *Eur. J. Pharmacol.* 164: 571–575, 1989.
41. MINKES, R. K., and P. J. KADOWITZ. Influence of endothelin on systemic arterial pressure and regional blood flow in the cat. *Eur. J. Pharmacol.* 163: 163–166, 1989.
42. MINKES, R. K., and P. J. KADOWITZ. Differential effects of rat endothelin on regional blood flow in the cat. *Eur. J. Pharmacol.* 165: 161–164, 1989.
43. MIURA, K., T. YUKIMARA, Y. YAMASHITA, T. SHIMMEN, M. OKUMURA, M. IMANISHI, and K. YA-

MAMOTO. Endothelin stimulates the renal production of prostaglandin E_2 and I_2 in anaesthetized dogs. *Eur. J. Pharmacol.* 170: 91–93, 1989.

44. MORTENSEN, L. H., and G. D. FINK. Hemodynamic effect of human and rat endothelin administration into conscious rats. *Am. J. Physiol.* 258 (*Heart Circ. Physiol.* 27): H362–H368, 1990.

45. ÖHLÉN, A., J. RAUD, P. HEDQVIST, and N. P. WIKLUND. Microvascular effects of endothelin in the rabbit tenuissimus muscle and hamster cheek pouch. *Microvasc. Res.* 37: 115–118, 1989.

46. PALMER, R. M. J., A. G. FERRIGE, and S. MONCADA. Nitric oxide release accounts for the biological activity of endothelium-derived relaxing factor. *Nature* 327: 524–526, 1987.

47. PERNOW, J., A. HEMSÉN, and J. M. LUNDBERG. Tissue specific distribution, clearance and vascular effects of endothelin in the pig. *Biochem. Biophys. Res. Commun.* 161: 647–653, 1989.

48. PING HAN, S., A. J. TRAPANI, K. F. FOK, T. C. WESTFALL, and M. M. KNUEPFER. Effects of endothelin on regional hemodynamics in conscious rats. *Eur. J. Pharmacol.* 159: 303–305, 1989.

49. RAE, G. A., M. TRYBULEC, G. DE NUCCI, and J. R. VANE. Endothelin-1 releases eicosanoids from rabbit isolated perfused kidney and spleen. *J. Cardiovasc. Pharmacol.* 13(Suppl. 5): S89–S92, 1989.

50. RAKUGI, H., M. NAKAMARU, Y. TABUCHI, M. NAGANO, H. MIKAMI, and T. OGIHARA. Endothelin stimulates the release of prostacyclin from rat mesenteric arteries. *Biochem. Biophys. Res. Commun.* 160: 924–928, 1989.

51. RANDALL, M. D., S. A. DOUGLAS, and C. R. HILEY. Vascular activities of endothelin-1 and some alanyl substituted analogues in resistance beds of the rat. *Br. J. Pharmacol.* 98: 685–699, 1989.

52. ROHMEISS, P., J. PHOTIADIS, S. ROHMEISS, and T. UNGER. Hemodynamic actions of intravenous endothelin in rats: comparison with sodium nitroprusside and methoxamine. *Am. J. Physiol.* 258(*Heart Circ. Physiol.* 27): H337–H346, 1990.

53. SPOKES, R. A., M. A. GHATEI, and S. R. BLOOM. Studies with endothelin-3 and endothelin-1 on rat blood pressure and isolated tissues: evidence for multiple endothelin receptor subtypes. *J. Cardiovasc. Pharmacol.* 13(Suppl. 5): S191–S192, 1989.

54. THIEMERMANN, C., P. S. LIDBURY, G. R. THOMAS, and J. R. VANE. Endothelin-1 releases prostacyclin and inhibits ex vivo platelet aggregation in the anesthetized rabbit. *J. Cardiovasc. Pharmacol.* 13(Suppl. 5): S138–S141, 1989.

55. THIEMERMANN, C., G. R. MAY, C. P. PAGE, and J. R. VANE. Endothelin-1 inhibits platelet aggregation in vivo: a study with [111]indium-labelled platelets. *Br. J. Pharmacol.* 99: 303–308, 1990.

56. VANHOUTE, P. M., W. AUCH-SCHWELK, C. BOULANGER, P. A. JANSSEN, Z. S. KATUSIC, K. KOMORI, V. M. MILLER, V. B. SCHINI, and M. VIDAL. Does endothelin-1 mediate endothelium-dependent contractions during anoxia? *J. Cardiovasc. Pharmacol.* 13(Suppl. 5): S124–S128, 1989.

57. WALDER, C. E., G. R. THOMAS, C. THIEMERMANN, and J. R. VANE. The hemodynamic effects of endothelin-1 in the pithed rat. *J. Cardiovasc. Pharmacol.* 13(Suppl. 5): S93–S97, 1989.

58. WARNER, T. D., G. DE NUCCI, and J. R. VANE. Rat endothelin is a vasodilator in the isolated perfused mesentery of the rat. *Eur. J. Pharmacol.* 159: 325–326, 1989.

59. WARNER, T. D., J. A. MITCHELL, G. DE NUCCI, and J. R. VANE. Endothelin-1 and endothelin-3 release EDRF from isolated perfused arterial vessels of the rat and rabbit. *J. Cardiovasc. Pharmacol.* 13(Suppl. 5): S85–S88, 1989.

60. WHITTLE, B. J. R., J. LOPEZ-BELMONTE, and D. D. REES. Modulation of the vasodepressor actions of acetylcholine, bradykinin, substance P and endothelin in the rat by a specific inhibitor of nitric oxide formation. *Br. J. Pharmacol.* 98: 646–652, 1989.

61. WHITTLE, B. J. R., A. N. PAYNE, and J. V. ESPLUGUES. Cardiopulmonary and gastric ulcerogenic actions of endothelin-1 in the guinea pig and rat. *J. Cardiovasc. Pharmacol.* 13(Suppl. 5): S103–S107, 1989.

62. WINQUIST, R. J., P. B. BUNTING, V. M. GARSKY, P. K. LUMMA, and T. L. SCHOFIELD. Prominent depressor response to endothelin in spontaneously hypertensive rats. *Eur. J. Pharmacol.* 163: 199–203, 1989.

63. WITHRINGTON, P. G., G. DE NUCCI, and J. R. VANE. The actions of endothelin on the hepatic arterial and portal vascular beds of the anaesthetised dog. *Br. J. Pharmacol.* 96, 165P (abstract) 1989.

64. WITHRINGTON, P. G., G. DE NUCCI, and J. R. VANE. Endothelin-1 causes vasoconstriction and vasodilation in the blood perfused liver of the dog. *J. Cardiovasc. Pharmacol.* 13(Suppl. 5): S209–210, 1989.

65. WRIGHT, C. E., and J. R. FOZARD. Regional vasodilation is a prominent feature of the haemodynamic response to endothelin in anaesthetized, spontaneously hypertensive rats. *Eur. J. Pharmacol.* 155: 201–203, 1988.
66. YANAGISAWA, M., H. KURIHARA, S. KIMURA, Y. TOMOBE, M. KOBAYASHI, Y. MITSUI, Y. YAZAKI, K. GOTO, and T. MASAKI. A novel potent vasoconstrictor peptide produced by vascular endothelial cells. *Nature* 332: 411–415, 1988.
67. YANAGISAWA, M., and T. MASAKI. Endothelin, a novel endothelium-derived peptide. *Biochem. Pharmacol.* 38: 1877–1883, 1989.

16

Hypothetical Role of Endothelin in the Control of the Cardiovascular System

GABOR M. RUBANYI AND JOHN T. SHEPHERD

The discovery that vascular endothelial cells can synthesize autocrine and paracrine substances that can alter the tone of the underlying smooth muscle and inhibit the aggregation of platelets revealed novel mechanisms that may contribute to cardiovascular homeostasis. This began with a series of studies that led to the discovery of prostacyclin[41] and was followed by the demonstration that endothelial cells also form a nonprostanoid substance that was termed "endothelium-derived relaxing factor" (EDRF).[21] Subsequent studies have shown that the major EDRF is nitric oxide or a labile nitroso compound[20,27] formed from L-arginine in the endothelial cell.[42]

DeMey and Vanhoutte[13] showed that arachidonic acid and thrombin evoked endothelium-dependent relaxations in segments of canine arteries, but endothelium-dependent contractions in vein segments. These contractions could be prevented by inhibitors of cyclooxygenase. Anoxia also can evoke endothelium-dependent contractions, an effect that is mediated by a diffusible substance(s) that is not a metabolite of arachidonic acid.[49] These studies indicated that, similar to EDRFs, the endothelium can produce vasoconstrictor autacoids (endothelium-derived contracting factors [EDCFs]).[45] The demonstration that endothelial cells in culture produce a potent peptidergic vasoconstrictor[25] and its isolation, sequencing, and identification as a 21-amino acid peptide termed "endothelin"[59] further emphasized the potential role of the endothelium in regulating vascular tone.

The discovery of the variety of endothelium-derived relaxing and contracting autacoids and paracoids that may be released with appropriate stimuli opens the question of their role in normal circulatory regulation. In this chapter, we speculate on the potential role of endothelin in circulatory control under physiological and pathological conditions. Most of the proposed functions of endothelin are merely hypothetical. They are based on existing data obtained with exogenously administered synthetic peptides and not by detection of local changes in endogenous endothelin production or by the use of selective inhibitors of endothelin biosynthesis or action. However, the collected thoughts in this chapter (some of which have already been addressed in more detail elsewhere in this book), may focus attention on certain areas of cardiovascular physiology and pathophysiology where this unique peptide may play a primary role.

LOCAL CONTROL OF VASCULAR TONE

Maintenance of Basal Tone

In some vascular beds, after all known neurohumoral control mechanisms have been inhibited, the vascular smooth muscle is not completely relaxed, indicating the presence of a basal vascular tone. The mechanism of maintenance of basal vascular tone is still unknown. One postulate is an intrinsic mechanism in the smooth muscle cells sensitive to changes in transmural pressure. Another possibility is a still unknown circulating substance. A third is that the endothelial cells respond to changes in transmural pressure by altering the ratio of EDRFs and EDCFs in favor of the latter. As examples, isolated canine basilar,[28] carotid,[45] and cat cerebral arteries,[23] when subjected to a sudden stretch or increase in transmural pressure, respond with an endothelium-mediated increase in tone. Because the response to both the onset and release of a quick stretch occurs rapidly, it could serve to effect moment-to-moment changes in vascular tone. Thus it is unlikely to involve endothelin, because this peptide is not released abruptly from endothelial cells[8,17] and when released causes a sustained contraction of the smooth muscle both in vitro and in vivo.[25,58] However, these very properties make endothelin a unique candidate for being the mediator of long-term modulation of vascular tone (Fig. 16.1).[47]

Endothelin is produced by cultured endothelial cells at a slow basal rate,[8,17] and, although its expression can be stimulated by various agonists,[59]

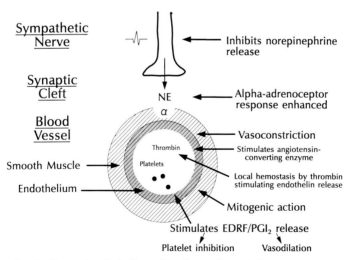

FIGURE 16.1. Local effects of endothelin on blood vessels, including actions on smooth muscle (contraction and mitogenic stimulation) adrenergic neurotransmission, angiotensin-converting enzyme (in endothelial cells), and release of prostanoids (e.g., prostacyclin; PGI_2) and EDRF from endothelium. It is postulated that these local effects may play a role in (*1*) maintenance of *basal* vascular tone, (*2*) inhibition of platelet aggregation (via prostacyclin and EDRF), and (*3*) hemostasis. For further details, see text.

facilitation of endothelin release is detectable only several hours after stimulation.[8] There is no evidence that endothelin is stored in endothelial cells, so its release may directly be connected to its de novo synthesis. Because of its high vasoconstrictor potency and long-lasting action, the continuous release of small amounts of endothelin from endothelial cells toward the underlying smooth muscle cells could contribute to the maintenance of vascular tone. An important feature of endothelin-induced vascular contraction is that it can be inhibited by most known vasodilator agents and by the potent endogenous vasodilator EDRF.[57] If basal tone is maintained by endothelin, it may be balanced by the tonic release of EDRF under physiological conditions, and it should be effectively antagonized by local vasodilator mechanisms in case of increased metabolic requirements. An imbalance between the production of endothelin and EDRF could lead to pathologically elevated vascular tone (vasospasm, hypertension). Thus the known properties of endothelin make it a potential candidate for long-term maintenance of basal vascular tone.

Interaction with Other Vasoactive Substances—Vasodilation and Antiplatelet Activities

In addition to a direct effect on vascular smooth muscle, endothelin can modulate blood vessel tone indirectly by modulating the release and/or action of other vasoactive substances (Fig. 16.1). Endothelin suppresses the liberation of norepinephrine (NE) from adrenergic nerve terminals in the blood vessel wall but potentiates the effect of NE on postjunctional α-adrenoceptors.[52] In some (but not all) vascular beds, endothelin apparently stimulates the release of prostacyclin and EDRF from endothelial cells,[14,56] which in turn will decrease blood vessel tone and inhibit platelet aggregation and adhesion (Fig. 16.2). This (indirect) mechanism may be responsible for *(1)* the vasodilator action of low doses of endothelin when injected into anesthetized animals[14] and *(2)* suppression of platelet aggregation.[56] It has been shown recently that EDRF (nitrous oxide [NO]) may suppress biosynthesis or release of endothelin (Fig. 16.2).[3,9] This raises the intriguing possibility of negative feedback regulation of not only the biological action, but also the biosynthesis or release of endothelin by other endothelium-derived substances.

Hemostasis and Vascular Growth

Thrombin is a potent stimulator of endothelin biosynthesis and release.[50,59] One may speculate that, in the event of blood coagulation, thrombin-induced release of endothelin and the consequent long-lasting vasoconstriction triggered by the peptide contributes to *hemostasis,* a role thus far attributed primarily to serotonin released from activated platelets. This mechanism may be especially effective in the case of endothelial dysfunction (i.e., loss of EDRF production), since thrombin is also a potent stimulant of EDRF release (see also Chapter 15).[13]

In addition to regulation of local vascular tone, endothelin may also modulate *vascular smooth muscle growth/proliferation.* Endothelin is a smooth muscle mitogen[15] and may therefore mediate endothelium-dependent growth

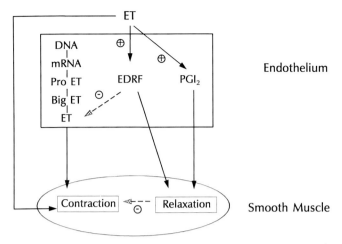

FIGURE 16.2. Interaction between endothelin (ET) and endothelium-derived relaxing substances (EDRF and PGI$_2$). Endothelin, a potent vascular smooth muscle constrictor, can stimulate the release of PGI$_2$, and EDRF from endothelial cells (see also Fig. 16.1). This autocrine mechanism can "balance" the vasoconstrictor effect of endothelin by (*1*) functional antagonism at the level of smooth muscle (PGI$_2$ and EDRF) and (*2*) inhibition of endothelin biosynthesis (EDRF).

and remodeling of the blood vessel wall.[16,36] Endothelin may stimulate smooth muscle proliferation indirectly by facilitating the (local) production of another potent mitogen, angiotensin II (see later under Renin–Angiotensin System). It is possible that, in addition to the pathological consequences of its excessive production (i.e., vasospasm, hypertension), the maintenance of basal vascular tone and structure, as well as contribution to hemostasis, represent important physiological roles of the peptide. Thus endothelin joins a host of other biologically active substances that modulate vascular tone or structure (Fig. 16.3). Under physiological conditions a delicate balance must exist between relaxing/antiproliferative (e.g., prostacyclin, EDRF [NO]; endothelium-derived hyperpolarizing factor [EDHF]) (Fig. 16.3, right) and contracting/proliferative factors (e.g., arachidonic acid metabolites or free radicals [EDCF1], hypoxia [EDCF2], and endothelin; Fig. 16.3, left). This *harmony* will serve the purpose of controlling local vascular homeostasis. An imbalance (endothelial dysfunction) can lead to increased tone (vasospasm) or vessel wall proliferation and remodeling (see also Chapter 9).

REFLEX CONTROL OF THE CIRCULATION

Central Nervous System

Endothelin-binding sites have been identified in the basal ganglia and brain stem.[11] Specific binding sites and immunoreactive endothelin are also present in the supraoptic and paraventricular nuclei of the hypothalamus. With chronic infusion of endothelin into the cerebral ventricles of conscious rats there is a progressive increase in arterial blood pressure accompanied by in-

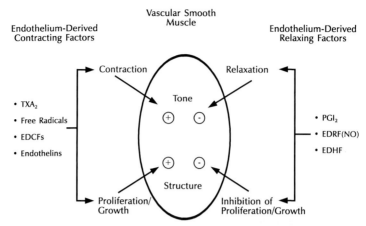

FIGURE 16.3. Endothelium-derived vasoactive factors and the local control of vascular tone/structure. Under physiological conditions a fine balance exists between endothelium-derived relaxing (PGI_2, EDRF, EDHF) and contracting (TXA_2, free radicals, EDCFs, endothelin) factors. The same substances may also modulate smooth muscle growth and proliferation in an opposite manner. Imbalance between these factors in favor of the contracting/proliferative factors (endothelial dysfunction) can lead to pathophysiological consequences such as vasospasm, hypertension, and vessel wall thickening.

creased urinary excretion of NE, epinephrine, and vasopressin,[54] probably reflecting activation of autonomic vasomotor centers. These observations suggest that endothelin produced in the central nervous system may modulate the central control of circulation (Fig. 16.4) via an as yet unidentified mechanism.

Resetting of Arterial Baroreceptors

The arterial mechanoreceptors in the carotid sinus and aortic arch respond to static and pulsatile changes in transmural pressure, with resultant changes in autonomic outflow. Resetting occurs with acute changes in pressure, as during the rise in arterial blood pressure during dynamic or static exercise, as well as in established hypertension.[56]

Resetting may be due to central modulation of the reflex arc or to changes at the site of the mechanoreceptors. Resetting at the latter site means that the afferent nerve activity, for example, in the glossopharyngeal nerve from the carotid sinus, is reduced at an equivalent arterial pressure and vascular strain.[8] Peripheral resetting can occur in a variety of ways:

1. It can occur as a consequence of structural changes in the vessel wall, for example, in chronic hypertension where the increase in collagen causes a decrease in compliance with decreased strain and hence decreased mechanoreceptor activity.[5] This would cause chronic resetting.
2. Activation of Na^+,K^+-ATPase may contribute to acute resetting during an arterial pressure increase as a consequence of the movement of Na^+ into the neurons activating the Na^+K^+-pump with a resultant hyperpolarization.[24]

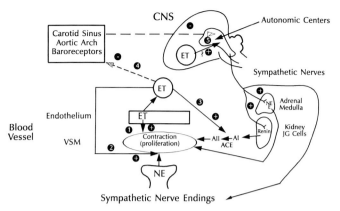

FIGURE 16.4. Potential sites where endothelin (ET) can contribute to the control of blood pressure (vascular tone). *Local actions:* (1) contraction of vascular smooth muscle; (2) potentiation of α-adrenergic vasoconstriction; (3) activation of angiotensin converting enzyme (ACE). *Systemic and central actions:* (4) resetting (suppression) of baroreceptors with consequent increase in sympathetic tone, the result of which is increased synthesis/release and circulating levels of norepinephrine (NE), epinephrine (E), vasopressin (not shown), and renin (see Fig. 16.7); and (5) direct activation (?) of vasomotor centers by locally produced endothelin in the central nervous system. AI, AII, angiotensin I and II, respectively. For further details, see text.

3. The discovery of EDRFs and EDCFs has indicated that these substances may also be involved in acute resetting of the arterial baroreflexes. For example, in rabbits treated with indomethacin the activity of the carotid baroreceptors is suppressed. This suggests that a prostanoid released during vascular stretch contributes in a paracrine manner to the activation of baroreceptors in normotensive animals.[6,39] Microcarrier beads coated with bovine aortic endothelium, when introduced into a carotid sinus denuded of endothelium, suppresses baroreceptor activity when the endothelial culture is activated with calcium ionophore.[8] Also an increase in flow in the absence of a change in transmural pressure causes endothelium-dependent relaxation of conduit arteries[48] as a consequence of the release of EDRF in response to the increase in wall shear stress.[30,55] An increase in flow, perhaps in the same way, could decrease the activity of the carotid sinus mechanoreflex. Chapleau et al.[6] have demonstrated the release of an "inhibitory factor" from cultured endothelial cells that suppresses baroreceptor activity. They state that the potential role of this "inhibitory factor" in chronic baroreceptor resetting remains to be determined.

These studies raise the possibility that the local release of endothelium-derived vasoactive factors at the site of the mechanoreceptors in the circulation, in both the carotid sinus and the aortic arch, and also those in the heart and lungs, could contribute to acute (PGI_2, EDRF) and chronic (endothelin) resetting. That endothelin may be involved in this mechanism is supported by the demonstration that the peptide suppresses the activity of the mechanoreceptors in the isolated canine carotid sinus during step increases in carotid sinus pressure (Fig. 16.4).[7] However, baroreflex modulation of heart rate remained intact after injection of endothelin to rats.[32]

Sympathetic Nervous System

Studies in the isolated, perfused mesenteric vascular bed of spontaneously hypertensive and Wistar-Kyoto rats suggest that endothelin inhibits presynaptic adrenergic neurotransmission (Fig. 16.1).[52] In contrast, endothelin stimulates the release of NE and epinephrine from bovine adrenal chromaffin cells in culture.[2] It has also been observed that subpressor doses of endothelin enhance the pressor response evoked by exogenous norepinephrine in perfused rat mesenteric arteries.[52] Ganglionic blockade attenuates the endothelin infusion-induced bradycardia and decrease in cardiac output, suggesting that reflex autonomic pathways are activated by the exogenously administered peptide. Thus endothelin may play a role in the control of vascular tone by a variety of direct and indirect mechanisms (Figs. 16.1–16.4): *(1)* there is modulation of central control mechanisms (probably via enhancing sympathetic nerve activity) by endothelin produced in the central nervous system; *(2)* endothelin released in the vicinity of baroreceptors (carotid sinus, aortic arch, and so forth) may suppress their activity; *(3)* modulation of adrenergic neurotransmission in chromaffin cells, in the adrenal medulla, and in the blood vessel wall can lead to an increased/decreased liberation of NE and augmentation of postsynaptic effect of the neurotransmitter at α-adrenoceptors; *(4)* there is local modulation of the synthesis/release of other endothelium-derived vasoactive substances (EDRF, PGI_2) and of angiotensin II; and *(5)* there is direct action on vascular smooth muscle.

HUMORAL CONTROL OF THE CIRCULATION

Aldosterone and Sodium Balance

Aldosterone, a steroid hormone produced by the zona glomerulosa of the adrenal cortex, stimulates Na^+ reabsorption by stimulating the unidirectional Na^+ transport mechanism in the collecting tubule/duct of the kidney (Fig. 16.5). Using the technique of tissue autoradiography, Koseki et al.[35] and Davenport et al.[11] demonstrated binding sites for endothelin in the adrenal gland. A high density of binding was found in the zona glomerulosa, suggesting that endothelin may modulate the biosynthesis of aldosterone.[10,34] Indeed, endothelin stimulates aldosterone biosynthesis directly in dispensed zona glomerulosa cells of rabbit and calf.[10,41] Pharmacological concentrations of endothelin stimulate the release of aldosterone in vivo.[22,40]

In addition to the direct stimulation of its biosynthesis, endothelin can facilitate the production of aldosterone indirectly via angiotensin II, a strong stimulant of aldosterone biosynthesis (Fig. 16.5). Endothelin may elevate angiotensin II levels by stimulating endothelial angiotensin-converting enzyme (ACE).[29]

The known effects of endothelin on renal function suggest a synergistic action with aldosterone (i.e., antinatriuresis) but via a different mechanism

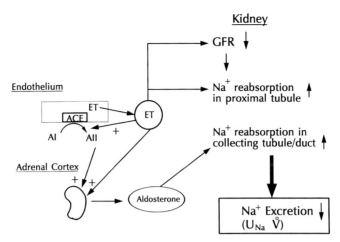

FIGURE 16.5. Potential role of endothelin (ET) in the control of sodium balance. The peptide can suppress renal sodium (Na$^+$) excretion by decreasing GFR and by direct action on Na$^+$ reabsorption in the proximal tubule. In addition, it stimulates the release of the antinatriuretic hormone aldosterone from the adrenal cortex either directly by action on zona glomerulosa cells or indirectly by elevation of sympathetic tone (not shown) and angiotensin II levels (by stimulating ACE). CNS, central nervous system; AI, AII, angiotensin I and II, respectively; GFR, glomerular filtration rate; JG, juxtaglomerular; VSM, vascular smooth muscle; U$_{Na}$, urinary sodium content; V, diuresis. For further details, see text.

(Fig. 16.5). Exogenous administration of endothelin causes a dose-dependent decrease in glomerular filtration rate (GFR).[22,37,40] This may be achieved through contraction of mesangial cells, leading to glomerular capillary vasoconstriction and a lower ultrafiltration coefficient (K_f)[1,31] and also by a decrease in glomerular filtration pressure via potent afferent arterial vasoconstriction. Corresponding reductions in urine flow and sodium excretion are observed, as well as a decrease in the fractional excretion of both sodium and lithium, the latter of which is a marker for the reabsorption of sodium by the proximal tubule. These observations are consistent with an endothelin-induced increase in the reabsorption of sodium and fluid from the proximal tubule. Endothelin can also directly inhibit tubular Na$^+$ reabsorption (Fig. 16–5).[18] However, natriuresis was observed after injection of subpressor doses of endothelin to rats, which do not compromise GFR.[31]

Antidiuretic Hormone and Water Balance

It was demonstrated that endothelin is present in paraventricular and supraoptic nuclear neurons and their terminals in the neurohypophysis of the rat.[60] Water deprivation depleted endothelin from these neurons. Because the same neurons are the site of biosynthesis of antidiuretic hormone (ADH; or vasopressin, a nine amino acid peptide that facilitates water absorption in the collecting tubule/duct by increasing water permeability) and water deprivation is the stimulus for its secretion, one can speculate that there are some interactions between endothelin and ADH in the hypothalamus—neurohypophysis system (Fig. 16.6). Although the direct effect of endothelin in ADH (vasopressin) secretion has not been reported, endothelin infusion significantly elevates circulating and urine levels of vasopressin.[40]

Pressor doses of endothelin increase hematocrit and plasma protein concentration, suggesting loss of water into the interstitium.[31,32] It is still unknown whether endothelin-induced hemoconcentration is due to increased vascular permeability and/or elevated capillary hydrostatic pressure. These findings suggest that endothelin may influence water balance (in opposite ways) at the level of kidney (water conservation) and also in systemic capillaries (water loss from plasma).

The exact role of endothelin in modulating these humoral systems needs to be determined. However, the results thus far suggest that endothelin may play some role in the central and peripheral regulation of water and electrolyte balance. This (still hypothetical) role of endothelin can contribute to the control of blood pressure by mechanisms (i.e., regulating plasma volume and/or osmolality) other than its direct actions on vascular smooth muscle or its modulation of the central and systemic control of the circulation.

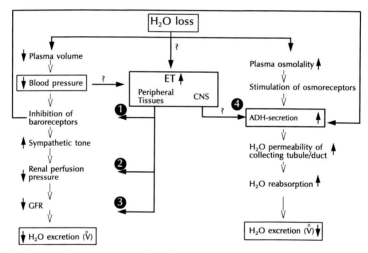

FIGURE 16.6. Potential role of endothelin (ET) in the regulation of water balance. Water loss stimulates compensatory mechanisms, which include increased sympathetic activity (via baroreceptor inhibition; left) and secretion of antidiuretic hormone (ADH) via stimulation of osmoreceptors in the central nervous system (right). Inhibition of baroreceptors can also stimulate the release of ADH. Both mechanisms will reduce water excretion (diuresis; V̊) in the kidney with resultant normalization of plasma volume. Local production of endothelin (either in peripheral tissues or the central nervous system) may contribute to these regulatory (homeostatic) pathways in several ways: 1, inhibition of baroreceptor function; 2, constriction of afferent arterioles; 3, decreasing GFR in the kidney; and 4, effects on autonomic centers (not shown) and hypothetically on ADH secretion (via a central mechanism or through baroreceptor resetting). Although an increase in circulating endothelin was reported in severe hypotension (e.g., cardiogenic, endotoxin, and hemorrhagic shock) and a release of endothelin from the neurohypophysis during water deprivation, a causal relationship between these homeostatic mechanisms and endothelin remains to be determined.

Renin–Angiotensin System: Synergism with Endothelin in the Control of Vascular Tone and Proliferation

Several studies report the inhibitory effect of endothelin on renin release in vitro.[38,44,53] Endothelin inhibited renin release in a dose-dependent fashion in a superfusion system of dispersed rat juxtaglomerular cells.[53] Similar findings have been reported on renin release from isolated rat glomeruli[44] and in rat renal cortical slices.[38] However, in vivo studies have demonstrated a significant rise in plasma renin activity following intravenous infusion of endothelin.[22,40]

This apparent discrepancy between the in vitro and in vivo observations can be explained if endothelin activated a mechanism in vivo that stimulates renin release and overrides the possible direct inhibitory action of the peptide. Based on the available information, the following mechanisms can lead to the observed increase in renin release in vivo (Fig. 16.7): *(1)* elevation of sympathetic activity (a known stimulus for renin release) by endothelin either directly at the neurotransmission level (postsynaptic facilitation of NE action on target tissues) or indirectly by suppression of baroreceptor reflexes; *(2)* constriction of afferent arterioles in the kidney, leading to decreased renal perfusion pressure, which stimulates renin secretion via the baroreceptor mechanism in the juxtaglomerular apparatus; and *(3)* decrease in sodium delivery to the macula densa, activating renin release via the osmoreceptor mechanism.

In cultured bovine pulmonary artery endothelial cells, endothelin stimu-

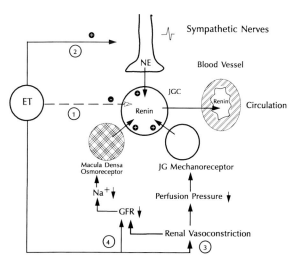

FIGURE 16.7. Potential mechanisms by which endothelin (ET) can modulate renin release from the juxtaglomerular cells (JGC) of the renal cortex: Direct action on JGC supresses renin release in vitro (1). However, infusion of endothelin raises circulating renin levels in vivo by the following putative mechanisms: increased sympathetic activity (2); renal (afferent–glomerular) vasoconstriction that decreases perfusion pressure, which stimulates renal mechanoreceptors in the JG apparatus (3); reduced Na^+ delivery, which activates osmoreceptors in the macula densa (4).

lates ACE activity (Fig. 16.4).[29] The increased renin release and activation of ACE may result in elevated angiotensin II levels. Angiotensin II then may act synergistically with endothelin in *(1)* increasing vascular tone and promoting cell proliferation and growth (Fig. 16.1) and *(2)* facilitating aldosterone biosynthesis (Fig. 16.5).

Atrial Natriuretic Peptide (ANP): Functional Antagonism with Endothelin

Endothelin can release ANP, a potent natriuretic, diuretic, and vasodilator peptide, from isolated rat cardiac myocytes[19] and from isolated rat atria.[26,51] Endothelin is a potent secretagogue for ANP release from spontaneously contracting atria obtained from hypertensive and normal rats.[58] A nearly identical pattern of receptor distribution has been observed in peripheral tissues for endothelin and ANP in vivo.[51] This observation is interesting, because some of the known physiological actions of ANP oppose those of endothelin (Fig. 16.8).

In the kidney, endothelin has a potent antinatriuretic effect (Fig. 16.5), while ANP evokes natriuresis. Similarly, endothelin and ANP have opposing effects on renal blood flow and on peripheral vascular resistance and blood pressure. ANP effectively antagonizes endothelin-induced vasoconstriction.[4] Based on these findings, one may speculate that ANP represents a negative feedback mechanism against endothelin-induced renal and vascular effects.

SUMMARY AND CONCLUSION

The various studies demonstrate that endothelin(s) can be formed in many cells of the body, as well as in the endothelium, and thus have the potential for

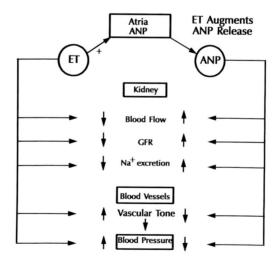

FIGURE 16.8. Functional antagonism between endothelin (ET) and atrial natriuretic peptide (ANP). For details, see text.

many of the effects suggested by in vitro and in vivo studies. Endothelin appears to function as a local rather than a circulating hormone.[14] These autocrine and paracrine actions demonstrate that endothelin has the potential to regulate vascular tone; to modulate cardiovascular reflexes and renal function; to affect renin, aldosterone, and ANP release; and to stimulate ACE activity. Although we speculated on its potential role in cardiovascular homeostasis, its role in health and disease remains to be established. Thus it joins the ranks of other potent vasoactive substances that are present in many of the body tissues, such as bradykinin, histamine, and serotonin, all of which have the capability of modulating cardiovascular behavior by central, systemic, and/or local actions, but whose specific role in the intact circulatory system remains to be determined.

We should repeatedly emphasize that this chapter is mostly hypothetical, and most (if not all) of the hypotheses put forward serve the purpose of stimulating thoughts and further studies in this exciting new field of cardiovascular physiology and pathophysiology. The test of the validity of these hypotheses can be accomplished only after the still-lacking necessary experimental tools (i.e., sensitive methods to detect local release of endothelin in vivo or selective inhibitors of its biosynthesis or action) become available.

REFERENCES

1. BADR, K., J. MURRAY, M. BREYER, K. TAKAHASHI, T. INAGAMI, and R. HARRIS. Mesangial cell, glomerular, and renal vascular responses to endothelin in the rat kidney. *J. Clin. Invest.* 83: 336–342, 1989.
2. BOARDER, M. R., and D. B. MARRIOTT. Characterization of endothelin-1 stimulation of catecholamine release from adrenal chromaffin cells. *J. Cardiovasc. Pharmacol.* 13(Suppl. 5): S223–S224, 1989.
3. BOULANGER, C., and T. F. LUSCHER. Release of endothelin from the porcine aorta—inhibition by endothelium-derived nitric oxide. *J. Clin. Invest.* 85: 587–590, 1990.
4. BONHOMME, M. C., M. CANTIN, and R. GARCIA. Relaxing effect of atrial natriuretic factor on endothelin-precontracted strips. *Proc. Soc. Exp. Biol. Med.* 191: 309–315, 1989.
5. BROWN, A. M. Receptors under pressure: an update on baroreceptors. *Circ. Res.* 46: 1–10, 1980.
6. CHAPLEAU, M. W., G. HAJDUCZOK and F. M. ABBOUD. Mechanisms of resetting of arterial baroreceptors: an overview. *Am. J. Med. Sci.* 295(4): 327–334, 1988.
7. CHAPLEAU, M. W., G. HAJDUCZOK, and F. M. ABBOUD. Endothelin suppresses baroreceptor activity: a new mechanism contributing to hypertension?, abstracted. *Hypertension* 14(3): 336, 1989.
8. CLOZEL, M., and W. FISCHLI. Human cultured endothelial cells do secrete endothelin-1 *J. Cardiovasc. Pharmacol.* 13(Suppl. 5): S229–S231, 1989.
9. COCKS, T. M., S. J. KING, R. A. WOODS, and J. A. ANGUS. Hemoglobin increases the amount of endothelin present in conditioned media of cultured endothelial cells. *Proc. Aust. Physiol. Pharmacol. Soc.* 20: 115, 1989.
10. COZZA, E. N., C. E. GOMEZ-SANCHEZ, M. F. FOECKING, and S. CHIOU. Endothelin binding to cultured calf adrenal zona glomerulosa cells and stimulation of aldosterone secretion. *J. Clin. Invest.* 84: 1032–1035, 1989.
11. DAVENPORT, A. P., D. J. NUNEZ, J. A. HALL, A. J. KAUMANN, and M. J. BROWN. Autoradiographical localization of binding sites for porcine [^{125}I]endothelin-1 in humans, pigs, and rats: functional relevance in humans. *J. Cardiovasc. Pharmacol.* 13(Suppl. 5): S166–S170, 1989.
12. DEMEY, J. G., M. CLAEYS, and P. M. VANHOUTTE. Endothelium-dependent inhibitory effects of acetylcholine, adenosine triphosphate, thrombin and arachidonic acid in the canine femoral artery. *J. Pharmacol. Exp. Ther.* 222: 166–173, 1983.
13. DEMEY, J. G., and P. M. VANHOUTTE. Heterogenous behavior of the canine arterial and venous wall. *Circ. Res.* 51: 439–447, 1982.

14. DE NUCCI, G. R., R. THOMAS, P. D'ORLÉANS-JUSTE, E. ANTUNES, T. D. WALDER, and J. R. VANE. Pressor effects of circulatory endothelin are limited by its removal in the pulmonary circulation and by the release of prostacyclin and endothelium-derived relaxing factor. *Proc. Natl. Acad. Sci. USA* 85: 9797–9800, 1988.
15. DUBIN, D., R. E. PRATT, J. P. COOKE, and V. J. DZAU. Endothelin, a potent vasoconstrictor, is a vascular smooth-muscle mitogen, abstracted. *Clin. Res.* 37: 517, 1989.
16. DZAU, V. J., and G. H. GIBBONS. Cell biology of vascular hypertension in systemic hypertension. *Am. J. Cardiol.* 62: 30G–35G, 1988.
17. EMORI, T., Y. HIRATA, K. OHTA, M. SHICHIRA, and F. MARUMO. Secretory mechanism of immunoreactive endothelin in cultured bovine endothelial cells. *Biochem. Biophys. Res. Commun.* 160: 93–100, 1989.
18. FERRARIO, R. G., R. FOULKES, P. SALVATI, C. PATRONS. Hemodynamic and tubular effects of endothelin and thromboxane in the isolated perfused rat kidney. *Eur. J. Pharmacol.* 171: 127–134, 1989.
19. FUKUDA, Y., Y. HIRATA, H. YSHIMI, T. KOJIMA, Y. KOBAYASHI, M. YANAGISAWA, and T. MASAKI. Endothelin is a potent secretagogue for atrial natriuretic peptide cultured in rat atrial myocytes. *Biochem. Biophys. Res. Commun.* 155(1): 167–172, 1988.
20. FURCHGOTT, R. F., M. T. KHAN, and D. JOTHIANANDAN. Comparison of endothelium dependent relaxation and nitric oxide induced relaxation in rabbit aorta, abstracted. *Federation Proc.* 46: 385, 1987.
21. FURCHGOTT, R. F., and J. V. ZAWADZKI. The obligatory role of endothelial cells in the relaxation of arterial smooth muscle by acetylcholine. *Nature* 288: 373–376, 1980.
22. GOETZ, K. L., B. C. WANG, J. B. MADWED, J. L. ZHU, R. J. LEADLEY, JR. Cardiovascular, renal and endocrine responses to intravenous endothelin in conscious dogs. *Am. J. Physiol.* 255 (*Regulatory Integrative Comp. Physiol.* 24): R1064–R1068, 1988.
23. HARDER, D. R. Pressure-induced myogenic activation of cat cerebral arteries is dependent on intact endothelim. *Circ. Res.* 60: 102–107, 1987.
24. HEESCH, C. M., F. M. ABBOUD, and M. D. THAMES. Acute resetting of carotid sinus baroreceptors. II. Possible involvement of electrogenic Na^+ pump. *Am. J. Physiol.* 247: (*Heart Circ. Physiol.* 16): H833–H834, 1984.
25. HICKEY, K. A., G. RUBANYI, R. J. PAUL, and R. F. HIGHSMITH. Characterization of a coronary vasoconstrictor produced by cultured endothelial cells. *Am. J. Physiol.* 248 (*Cell Physiol.* 17): C550–C556, 1985.
26. HU, R. J., U. G. BERNINGER, and R. E. LANG. Endothelin stimulates atrial natriuretic peptide (ANP) release from rat atria. *Eur. J. Pharmacol.* 158: 177–178, 1988.
27. IGNARRO, L. J., R. BYRNS, and K. S. WOOD. Pharmacological and biochemical properties of EDRF: evidence that EDRF is closely related to nitric oxide radical, abstracted. *Circ. Res.* 74(Suppl. II): II–287, 1986.
28. KATUSIC, Z. S., J. T. SHEPHERD, and P. M. VANHOUTTE. Endothelium-dependent contractions to stretch in canine basilar arteries, abstracted. *Federation Proc.* 45: 289, 1986.
29. KAWAGUCHI, H., K. ITO, and H. YASUDA. Endothelin stimulates angiotensin converting enzyme activity. Presented at *The International Society of Hypertension, 13th Scientific Meeting, Montreal, December 1990*.
30. KHAYUTIN, V. M., A. M. MELKUMYANTS, A. N. ROGOZA, E. S. VESELOVA, S. A. BALASHOV, and V. P. NIKOLSKY. Flow-induced control of arterial lumen, abstracted. *Acta Physiol. Hung.* 68: 334, 1989.
31. KING, A. J., B. M. BRENNER, and S. ANDERSON Endothelin: a potent renal and systemic vasoconstrictor peptide. *Am. J. Physiol.* 256 (*Renal Fluid Electrolyte Physiol.* 25): F1051–F1058, 1989.
32. KING, A. J., J. M. PFEFFER, M. A. PFEFFER, and B. M. BRENNER. Systemic hemodynamic effects of endothelin in the rat. *Am. J. Physiol.* 258: (*Heart Circ. Physiol.* 27): H787–H792, 1990.
33. KNUEPFER, M.M., S. P. HAN, A. J. TRAPANI, K. F. FOK, and T. C. WESTFALL. Regional hemodynamic and baroreflex effects of endothelin in rats. *Am. J. Physiol.* 257 (*Heart Circ. Physiol.* 26): H918–H926, 1989.
34. KOHZUKI, M., C. I. JOHNSTON, S. Y. CHAI, D. J. CASLEY, F. ROGERSON, and F. A. O. MENDELSOHN. Endothelin receptors in rat adrenal gland visualized by quantitative autoradiography. *Clin. Exp. Pharmacol. Physiol.* 14(4): 239–242, 1989.
35. KOSEKI, C., M. IMAI, Y. HIRATA, M. YANAGISAWA, and T. MASAKI. Autoradiographic distribution in rat tissues of binding sites for endothelin: a neuropeptide? *Am. J. Physiol.* (*Regulatory Integrative Comp. Physiol.* 25): R858–R866, 1989.
36. LANGILLE, B. L., and F. O'DONNELL. Reductions in arterial diameter produced by chronic decreases in blood flow are endothelium-dependent. *Science* 231: 405–407, 1986.
37. LOPEZ-FARRE, A., I. MONTANES, I. MILLAS, and J. LOPEZ-NOVOA. Effect of endothelin on renal function in rats. *Eur. J. Pharmacol.* 163: 187–189, 1989.

38. MATSUMURA, Y., K. NAKASE, R. IKEGAWA, K. HAYASHI, T. OHYAMA, and S. MORIMOTO. The endothelin-derived vasoconstrictor peptide endothelin inhibits renin release in vitro. *Life Sci.* 44(2): 149–157, 1989.
39. MCDOWELL, T. S., T. S. AXTELLE, M. W. CHAPLEAU, and F. M. ABBOUD. Prostaglandins in carotid sinus enhance baroreflex in rabbits. *Am. J. Physiol.* 257 (*Regulatory Integrative Comp. Physiol.* 26): R445–R450, 1989.
40. MILLER, W. L., M. M. REDFIELD, and J. C. BURNETT, JR. Integrated cardiac, renal, and endocrine actions of endothelin. *J. Clin. Invest.* 83: 317–320, 1989.
41. MONCADA, S., R. GRYGLEWSKI, S. BUNTING, and J. R. VANE. An enzyme isolated from arteries transforms prostaglandin endoperoxidase to an unstable substance that inhibits platelet aggregation. *Nature* 263: 663–665, 1976.
42. MORISHITA, R., J. HIGAKI, and T. OGIHARA. Endothelin stimulates aldosterone biosynthesis by dispersed rabbit adreno-capsular cells. *Biochem. Biophys. Res. Commun.* 160(2): 628–632, 1989.
43. PALMER, R. M. J., A. G. FERRIGE, and S. MONCADA. Nitric oxide release accounts for the biological activity of endothelium-derived relaxing factor. *Nature* 327: 524–526, 1987.
44. RAKUGI, H., M. NAKAMARU, H. SAITO, J. HIGAKI, and T. OGIGARA. Endothelin inhibits renin release from isolated rat glomeruli. *Biochem. Biophys. Res. Commun.* 155(3): 1244–1247, 1988.
45. RUBANYI, G. M. Endothelium-dependent pressure-induced contraction of isolated canine carotid arteries. *Am. J. Physiol.* 255 (*Heart Circ. Physiol.* 24): H783–H788, 1988.
46. RUBANYI, G. M. Endothelium-derived vasoconstrictor factors. In: *Endothelial Cell,* edited by U. S. Ryan. Cleveland, OH: CRC, 1985, vol. III, p. 61–74.
47. RUBANYI, G. M. Maintenance of "basal" vascular tone may represent a physiological role for endothelin. *J. Vasc. Med. Biol.* 6: 315–316, 1989.
48. RUBANYI, G. M., C. J. ROMERO and P. M. VANHOUTTE. Flow-induced release of endothelium-derived relaxing factor. *Am. J. Physiol.* 250 (*Heart Circ. Physiol.* 19): H1145–H1149, 1986.
49. RUBANYI, G. M., and P. M. VANHOUTTE. Hypoxia releases a vasoconstrictor substance from canine vascular endothelium. *J. Physiol.* 364: 45–56, 1985.
50. SCHINI, V. B., H. HENDRICKSON, D. M. HEUBLEIN, J. C. BURNETT, JR., and P. M. VANHOUTTE. Thrombin enhances the release of endothelin from cultured porcine aortic endothelial cells. *Eur. J. Pharmacol.* 165: 333–334, 1989.
51. STASCH, J. P., C. HIRTH-DIETRICH, S. KAZDA, D. NEUSER. Endothelin stimulates release of natriuretic peptides in vitro and in vivo. *Life Sci.* 45(10): 869–875, 1989.
52. TABUCHI, Y., M. NAKAMURU, H. RAKUGI, M. NAGANO, H. MIKAMI, and T. OGIHARA. Endothelin interacts with the neuroeffector junction in rat mesenteric artery, abstracted. *Hypertension* 14(3): 336, 1989.
53. TAKAGI, M., H. MATSUOKA, K. ATARASHI, and S. YAGI. Endothelin: a new inhibitor of renin release. *Biochem. Biophys. Res. Commun.* 157(3): 1164–1168, 1988.
54. TAKASHI, H., M. NISHIMURA, M. MATSUSAWA, I. IKEGAKI, M. SAKAMOTO, and M. YOSHIMURA. Centrally-induced cardiovascular effects of endothelin in conscious rats. Presented at *The International Society of Hypertension, 13th Scientific Meeting, Montreal, December 1990.*
55. TESFAMARIAM, B., and R. A. COHEN. Inhibition of adrenergic vasoconstriction by endothelial cell shear stress. *Circ. Res.* 63: 720–725, 1988.
56. THIEMERMANN, C., P. S. LIDBURY, G. R. THOMAS, and J. R. VANE. Endothelin-1 releases prostacyclin and inhibits ex vivo platelet aggregation in the anesthetized rabbit. *J. Cardiovasc. Pharmacol.* 13(Suppl. 5): S138–S141, 1989.
57. VANHOUTTE, P. M., W. AUCH-SCHWELK, and C. BOULANGER. Does endothelin-1 mediate endothelium-dependent contractions during anoxia? *J. Cardiovasc. Pharmacol.* 13 (Suppl. 5): S124–S128, 1989.
58. WINQUIST, R. J., A. L. SCOTT, and G. P. VLASUK. Enhanced release of atrial natriuretic factor by endothelin in atria from hypertensive rats. *Hypertension* 14(1): 111–114, 1989.
59. YANAGISAWA, M., H. KURIHARA, S. KIMURA, Y. TOMOBE, M. KABAYASHI, Y. MITSUI, Y. YAZAKI, K. GOTO, and T. MASAKI. A novel potent vasoconstrictor peptide produced by vascular endothelial cells. *Nature* 332: 411–415, 1988.
60. YASHIZAWA T, O. SHINMI, A. GIAID, M. YANAGISAWA, J. GIBSON, S. KIMURA, Y. OCHIYAMA, J. M. POLAK, T. MASAKI, and I. KANAZAWA. Endothelin is a novel peptide in the posterior pituitary system. *Science* 247: 462–464, 1990.

Index

A 23187. *See* Calcium ionophore
Abluminal release, 25
Acetylcholine, 5, 8, 13, 129, 130
Acetylsalicylic acid, 4
Action potential, 27
Affinity-labeling technique, 61–62, 65–67
Aging, 132
Airway epithelial cells, 92
Amino acid sequences, 33, 41–43, 52
Amino-terminus, 42–43, 47, 53
Amphipathic nature, 44
Anabolism, 42
Angiotensin II, 12, 35, 108, 117, 125, 137, 146, 151, 162, 165, 168, 170, 261, 264
Angiotensin converting enzyme (ACE), 268
Angina, 214–15
Anoxia, 7–10, 20
Antiendothelin antibody, 171, 210
Arachidonic acid, 3–5, 111
Asthma, 81, 144
Atherosclerosis, 139–40, 231
Atractaspis engaddensis, 144
Atrial muscle, 146, 149
Atrial natriuretic peptide, 52, 149, 161, 166, 170
Autocrine action, 73, 91
Autoradiography, 82, 88–91, 183, 191

Baroreceptors, 262
Bay K 8644. *See* Dihydropyridine-based Ca^{2+} antagonist
Bepridil, 26
"Big" endothelin, 33, 162, 172, 181, 186, 191, 218, 232, 239, 249. *See also* Proendothelin
Bioassay, xi, 6, 31, 138, 243
Blood–brain barrier, 91, 204
Bradykinin, 129, 130
BRL 34915. *See* Cromakalim

8-bromo cGMP, 128–29
Bronchioles, 184

Caffeine-sensitive calcium store, 105
Calcium, 26, 68, 105, 107–8, 118
Calcium, cytosolic (intracellular), 27, 68, 83, 92, 139, 147, 163–64, 168
Calcium, extracellular, 23, 104, 119, 146, 163, 169
Calcium antagonist, 26, 59, 106
Calcium-free solution, 26
Calcitonin gene-related peptide (CGRP), 195–98
Calcium ionophore (A 23187), 3–7, 35
Calmodulin, 119
cAMP. *See* Cyclic adenosine monophosphate
Canine arteries, 3, 5–8, 11–12, 21
Canine veins, 3–4, 11–12
Carboxy-terminus, 47, 51, 53
Cardiac action potential, 145
Cardiac output, 159, 170, 224, 228
Cardiac tissue, 48–49, 52
Cardiogenic shock, 150, 153, 224–26
Catalase, 7
Cat arteries, 5
Catecholamine, 140, 226, 260, 262
Cell culture, 20, 25
Cellular growth, 137
Cellular processing, 36
Central nervous system (CNS), 179, 191–205
Cerebellum, 49
Cerebral vasospasm, 140
cGMP. *See* Cyclic guanosine monophosphate
Chemical denervation, 158, 238
Chick cardiac membranes, 58
Chromatography, high-performance liquid, 27, 44, 74
Chronotropic effects, 52, 144, 148

Circular dichroism studies, 44
Clara cell, 184–85
Cloning of receptor subtypes, 69
Competitive studies, 64
ω-Conotoxin, 151
Congestive heart failure, 221, 226–28
Conscious rat, 159, 162, 238
Contractile response, 23, 118–19
   anoxic, 12
   cyclooxygenase-dependent, 3–7, 125
   endothelium-dependent, 3–10, 19–20, 28
   hypoxic, 7–10, 20–21
   onset, 22, 103
   tetrodotoxin-sensitive, 5
   tonic component, 23, 103
Coronary blood flow, 148–49, 214
Coronary vasospasm, 141, 153
Coronary veins, 149, 151
Covalent cross-linking, 63
Creatinine phosphokinase (CPK), 228
Cromakalim, 102
Cross-reactivity, 74, 76
Cultured cells, no release of endothelin, 76
Cyclic adenosine monophosphate (cAMP), 249–51
Cyclic guanosine monophosphate (cGMP), 9, 126, 127, 246–51
Cycloheximide, 23, 25
Cyclooxygenase inhibitor, 3–5, 20, 22, 161, 166, 239, 249, 258
Cyclosporine-induced toxicity, 152, 171

Dazoxiben, 6
Deferoxamine, 7
Diacylglycerol (DAG), 28, 68, 103, 110, 113, 116
Dihydropyridine (DHP), 58–59, 69, 106
Dihydropyridine-based calcium antagonist, 28, 145, 148, 151, 163, 166, 192, 195, 199
Diltiazem, 12, 59, 106, 108
Dissuccinimidyl tartrate (DST), 61
Disulfide bonds, 42, 47, 53
DNA library, 72
DNA-synthesis, 137, 140
DOCA-salt hypertensive rats, 172

EDCF. *See* Endothelium-derived contracting factor
EDRF. *See* Endothelium-derived relaxing factor

Efferent arterioles, 163, 166
Eicosanoids, 241, 244
Electrophoretic mobility, 44, 61
Electrophysiological effects, 27
Endocrine action, 73, 81, 94
Endocrine system, effects on, 52
Endoperoxides, 7
Endotensin, 17, 20–21
Endothelial cell-conditioned media, 19, 22
Endothelial cell-derived constricting factor, 17, 21–22
   properties of, 21, 25
   see also Endothelium-derived contracting factor
Endothelial cells, cultured, xi, 10, 18, 21
Endothelial dysfunction, 140, 214, 260
Endothelin, action on vascular smooth muscle, 103–19
   basal release in cultured cells, 75
   binding to tissue membranes, 86
   biological activity, 41, 58, 68, 89, 96
   cellular binding, 81, 82
   cellular mechanisms of action, 118, 119
   chemical features, 41, 47
   constitutive release, 34, 35, 81
   contractile activity, 96, 118
   converting enzyme, 36–37, 232
   discovery of, xi–xii, 17–29
   genes of, xii, 37
   human, 32, 72
   hypoxic response, 10–14
   immunologically reactive, 74, 78
   in vitro studies, 10–13, 49–53
   like sequence, 36
   mature, 31, 36, 39
   molecular biology, 31–39
   mouse, 32
   mRNA in tissues, 76, 77, 97
   physiological roles, 67, 68, 81
   plasma levels, 80, 94, 97, 223, 230
   porcine, 31–32, 72
   production, inducers of, 38
   purification of, xi, 28
   radiolabeled ligands, 46
   rat, 32
   receptors, 43, 46, 58–69, 164
   specific high-affinity binding sites, 84–85, 88, 93, 144, 149–53, 163–64, 169–71, 189, 192, 203–4, 261
   spectroscopic properties, 43–44
   structure-activity relationships, 41–54
   synthesis, 72–98

INDEX

tissue concentrations, 79–80
tissue distribution, 79
tissue specificity of synthesis and binding, 72–98
vascular remodeling, 137, 172, 261
vasoconstrictor potency, 50, 52
Endothelin-1, 33–34, 49, 58, 63, 68–69, 72, 91, 104
Endothelin-2, 33, 49, 58, 63, 68, 72, 78, 104, 186–90, 194, 244
Endothelin-3, 33, 49, 58, 63, 68, 72, 91, 94, 159, 170, 186–92, 194, 239, 245, 252
Endothelin-β, 32, 72
Endothelin-1 mRNA, induction of, 35, 37
Endothelium, 202, 263
Endothelium-derived contracting factor, xi, 3–17, 158, 258, 263
  antagonists of, 24
  mechanism of action, 26
Endothelium-derived hyperpolarizing factor (EDHF), 261
Endothelium-derived relaxing factor, xi, 9, 52, 68–69, 125–26, 137, 161, 165, 221, 231–32, 239, 245–48, 253–54, 258–60
  interaction with endothelin, 14, 125–53
Endotoxin shock, 153, 171, 226
Epidermal growth factor (EGF), 139
Equilibrium binding study, 62
Ergonovin, 151
Esophageal musculature, 49
Essential hypertension, 140, 172, 209, 212–14, 229
5-N-ethylisopropylamiloride (EIPA), 116
Extracellular matrix, 137

Fibroblasts, 49, 137
Filtration, 163
Flunarizine, 10, 13
Free radicals, 153
Fura-2, 27, 82, 147

Gene expression, 31, 34, 73
Gene structure, 31, 33
Gene transcription, 73
Genomic library, 58
Glomerular capillary ultrafiltration coefficient, 167, 265
Glomerular endothelial cells, 164
Glomerular filtration rate (GFR), 165–67, 171, 221, 265
Goldblatt rats, 172

G-protein, 68–69, 111–12
Growth factor, 137
Growth-inhibiting substance, 137
Growth-promoting substance, 137
Guanine nucleotide regulatory proteins. *See* G-protein
Guanylate cyclase, soluble, 128

Half-life, 91
"Head up" tilt, 224
Heart membranes, 88–89
Hemodynamic effect, 158–63, 266
Hemostasis, 260
High polar region, 47
Homology, 41
[^{3}H]thymidine incorporation, 137–38
Human lung, 185–88
Human brain, 201, 203
Hybridization (in situ), 92–93, 179, 186–90, 194–97, 200, 231
Hydrogen peroxide, 7
Hydroquinone, 9
Hydronephrotic kidney, 166
Hydrophobicity, 42, 44
Hydroxyl radicals, 7
Hypertension, 133, 144, 153, 171–72, 261–62
Hypertensive response, 159–61
Hyperplasia, 140
Hypertrophy, 140
Hypothalmus, 199–200, 202, 204
Hypothermia, 153
Hypoxic facilitation, 7

Ileum, 49
Immunoaffinity procedure, 28
Immunocytochemistry, 80, 181–82, 185–94, 196–204, 231
Immunoprecipitation, 61
Immunoreactivity, 210–12, 215
Immunostaining, 185–91, 193–94, 196–98, 203
Inner medullary collecting duct cells, 166
Indomethacin, 6, 9. *See also* Cyclooxygenase inhibitor
Inositol triphosphate (IP$_3$), 68–69, 109–10, 139, 147, 151, 169, 204
Inotropic effect, 52, 144–48, 159
Interleukin-1, 35
Ionomycin, 114
Ischemia, 152, 231
Isoforms, 32, 88, 97, 103

Isolated heart, 25, 161, 166, 244
Isopeptides, 31, 58, 62
Isoproterenol, 146, 170

Kidney, 67, 68

Ligand–receptor interaction, 48, 85
Lipoxygenase, inhibitors of, 9
L-type calcium channel, 28, 106, 108, 151

Meclofenamate, 4
Membrane potential, 11
2-Mercaptoethanol (2-ME), 61
Mesangial cells, 137, 168–70
Methylene blue, 9
Microcirculation-hamster cheek pouch, 166, 253
Molecular dynamic studies, 45
Molecular weight of receptors, 23, 61
Monoclonal antibodies, 59–60, 154, 222
Motoneurons, 199
Myocardial infarction, 144, 154, 209, 215–17, 228–29
Myocardial ischemia, 149–50, 153, 228–29
Myocytes, 151, 161, 166
Myosin light chain kinase, 103, 118

Na^+–K^+ pump, 262
Na^+–H^+ exchange, 114, 116
Neuropeptide Y, 202
Nicardipine, 58–59, 106–8, 149, 166, 192
Nifedipine, 106, 108
Nimodipine, 12
Nitrendipine, 26
Nitric oxide, endothelium-derived, 12–13, 125, 161, 245–48, 253–54. *See also* EDRF
N^G-monomethyl-L-arginine (L-NMMA), 125, 127, 161, 239, 247
Nonendothelial cells, 22
Nonpolar region, 47, 53
Northern blot analysis, 31, 73, 78
Nuclear magnetic resonance (NMR), 44

Ouabain-sensitive oxygen consumption, 166
Oxygen, 9
Oxygen-derived free radicals, 6
Oxytocin, 92

Paracrine action, 73, 92–94
Patch-clamp technique, 106

Pharmacophores, 54
pH, intracellular changes of, 114–17
Phorbol, 12, 13
  dibutyrate, 114–15, 119
Phorbol esters, 28, 34–35, 103, 108, 111, 113
Phosphatidylcholine, 111, 116
Phosphatidylinositol, 27–28, 92
Phosphatidylinositol 4,5-biphosphate (PIP_2), 109, 116
Phosphodiesterase, 112
Phospholipase A_2, 9, 111, 161, 168
Phospholipase C, 28, 68, 109–10, 119, 139
Phospholipase D, 110
Phospholipid signaling mechanisms, 109. *See also* Inositol triphosphate
Plasma renin activity, 170, 226–27, 267
Platelet aggregation, 249–52, 260
PN 200–110 (calcium antagonist), 152
Polyacrylamide gel analysis, 114
Porcine arteries, 20, 22
Potential-operated channels, 106. *See also* Voltage-sensitive calcium channels
Prazosin, 151
Preproendothelin, 31, 34, 36, 39, 72, 164, 212
β-Preprotachykinin, 197–98
Pressor effects, 68, 162–63, 238, 264
Proendothelin, 33
Prolactin, 92
Prostacyclin, 3, 52, 67, 69, 125, 161, 221, 231, 241–42
Protein kinase C, 111, 113–14, 116, 119
Protein structure, 32
Protein synthesis, 138
Protooncogenes, 139, 169
Pulmonary artery, 20
Pulmonary hypertension, 221, 229–30

Rabbit arteries, 6
Radioimmunoassay, 36, 74, 181, 192, 210–12, 222, 243
Rat lung membranes, 65–67
Receptor, endoperoxide, 7
Receptor, muscarinic, 4
Receptor binding, 41, 46–49, 59
Receptor-operated channels, 106
Receptor, purinergic, 4
Receptor signal transduction, 108
Receptor subtypes, 41, 46, 58–69
  cloning, 69

Renal epithelial cell, 164
Renal failure, 144, 153, 171, 209, 217–18
Renal hemodynamics, 163
Renal plasma flow (RPF), 165–67
Renin, release of, 49, 52
Renin-angiotensin-aldosterone system, 170, 222, 264
Reperfusion, 152
Resistance arteries, 131
Respiratory tract, 179–91
Rhodopsin-like receptors, 69
Riboprobes, 179
RNA probes, 179, 199
RNA translation, 73

Sandwich-enzyme immunoassay (EIA), 209–10, 215
Sandwich preparation, 7–8, 13
Sarafatoxin, xii, 41–43, 49, 63, 106, 151–52
Saralasin, 25, 167
Sarcoplasmic reticulum, 105
Saturation binding curves, 87
Scatchard plot analysis, 59, 82–83, 95, 108
Sequence analysis, xii
Serotonin, 140
Shear stress, 35, 137, 172, 263
Shock, 221, 226, 228
SIN-1, 128, 134
[^{35}S]methionine incorporation, 138
Southern blot analysis, 32
Species specificity, 78
Spontaneously hypertensive rat, 3, 7
Staurosporine, 113
Streptozotocin-induced diabetes, 152
Stretch, 149, 172
Subarachnoid hemorrhage, 140
Substance P, 192–97, 199
Sucrose gradient sedimentation, 59, 60
Superoxide anions, 6–7
Superoxide dismutase, 6–7, 128

Tachyphylaxis, 23
Tail, carboxy-terminal, 42–43

Tetrodotoxin, 5, 151
TGF-β. *See* Transforming growth factor-β
Thrombin, 12, 127, 140, 214
Thromboxane A$_2$, 5, 7, 52, 67, 140, 161, 241–42
Thromboxane synthetase, 5, 6
Tissue plasminogen activator, 252
Total peripheral resistance index, 160
Tracheal smooth muscle, 49
Transforming growth factor-β, 34–35, 139, 214

Uterus, 49

Vascular permeability, 170
Vascular resistance, 223, 229, 268
Vascular smooth muscle growth, 137–41, 260, 267
  tone, 158, 171, 224, 259–60, 267
Vascular tissues, 48
Vasoactive intestinal constricting peptide (VIC), 32, 68, 72
Vasoactive peptide, 19
Vasoconstriction, endothelium-dependent. *See* Contraction, endothelium-dependent
Vasoconstrictor substance. *See* Endothelium-derived contracting factor
Vasopressin, 12, 35, 92, 140, 168, 199, 202, 226, 262, 265
Vasorelaxation, endothelium-dependent, xi
Vasospasm, 144, 172, 261
Venoms, 58
Ventricular muscle, 146, 149, 159
Verapamil, 23, 26, 59, 106, 149, 163, 166
VIC. *See* Vasoactive intestinal constricting peptide
Voltage-sensitive calcium channels, 26, 27, 58, 69
Volume expansion, 149, 170–71, 266

X-ray analysis, 44